SQUIRREL NATION

SQUIRREL NATION

Reds, Greys and the Meaning of Home

Peter Coates

REAKTION BOOKS

This one's for Mila, who'll soon be old enough to notice them

Published by Reaktion Books Ltd
Unit 32, Waterside
44–48 Wharf Road
London N1 7UX, UK
www.reaktionbooks.co.uk

First published 2023
Copyright © Peter Coates 2023

Published with support from the Marc Fitch Fund

All rights reserved

No part of this publication may be reproduced, stored in a retrieval system, or transmitted, in any form or by any means, electronic, mechanical, photocopying, recording or otherwise, without the prior permission of the publishers

Printed and bound in Great Britain by TJ Books Ltd, Padstow, Cornwall

A catalogue record for this book is available from the British Library

ISBN 978 1 78914 770 4

CONTENTS

PREFACE 7

1 A TALE OF TWO SQUIRRELS 20

2 RED BEFORE GREY 47

3 GREY AND RED 79

4 AMERICAN HUSTLE, c. 1919-39 109

5 WAGING WAR ON THE 'GREY PERIL', c. 1939-73 149

6 WANTED: RED AND ALIVE 176

7 LEARNING TO LIVE WITH (AND TO LOVE) THE GREY 224

REFERENCES 255
SELECT BIBLIOGRAPHY 317
ACKNOWLEDGEMENTS 320
PHOTO ACKNOWLEDGEMENTS 323
INDEX 324

Red squirrel at Formby, Merseyside.

PREFACE

Britain has two species of squirrel: the native Eurasian red (*Sciurus vulgaris leucourus*) and the introduced grey (*Sciurus carolinensis*). Since the latter arrived nearly 150 years ago, the fortunes of the two have differed wildly. The first squirrel is today seldom seen south of the Scottish Highlands, in a handful of sanctuaries in northern England and a few small islands off England's south coast. But the second is now near-ubiquitous, absent only from the Scottish Highlands. The Mammal Society's 'central estimates' of the number of remaining reds are 287,000 – only 38,900 of them in England – versus 2.7 million greys across the whole of the UK.[1]

The emotions that red and grey provoke vary widely, from admiration, affection, love, kinship and longing to feelings of anger, fear, loathing, contempt and even disgust. In 1962 *The Guardian*'s parliamentary correspondent, Norman Shrapnel, compared two debates in the House of Lords that occurred on the same January afternoon. The first lasted just eleven minutes. The other debate, which followed shortly afterwards, consumed four hours. The short discussion was about how best to control grey squirrels, and Shrapnel contrasted its impassioned quality with the lacklustre longer discussion, on NATO. The uninspiring debate on UK defence policy might have given visitors to the gallery the 'misleading impression that the House of Lords is more worth listening to on a subject like grey squirrels'.[2] Shrapnel's misgivings misread and misrepresented the genuine depth of interest peers of the realm – and British politicians more generally – took in squirrels, grey and red. The public at large is absorbed by squirrels, too.

A survey of squirrel coverage since 2000 in the *Daily Star*, a tabloid, shows reds in an entirely positive light. Meanwhile, greys are portrayed with overwhelming negativity. The red that emerges is a plucky underdog sorely in need of our assistance in its heroic fightback. A feature consisting of a compilation of fabricated text messages included one that encapsulates popular understandings of the imbalance of power between Britain's two squirrels: 'Cyril! Us greys r arder [are harder], smarter and more successful than u red wimps. Gilbert the Grey Squirrel.'[3] Royal help – extra feeding, erecting warning signs to reduce roadkill, culling greys on royal lands – always merits a mention in the *Daily Star*.[4] To-the-rescue stories – gene editing to produce all-male greys, or grey-hunting pine martens, among other examples – complement reports of dire straits.[5] Solicitousness for the endangered red includes provisions for planning permission: the former Liverpool FC captain Steven Gerrard was instructed to plant red-friendly conifers when building a mansion on a 2-acre plot near the National Trust's squirrel sanctuary at Formby, Merseyside, in 2016.[6]

A leading 1980s squirrel researcher (sciurologist) was 'never bitten by a grey squirrel, mostly because I always wear thick falconry gauntlets whilst handling them'. But, as she wore thinner gardening gloves when handling reds, she recalled that 'I have been bitten many times ... and have the scars to prove it.'[7] Tales of squirrels that bite abound in the *Daily Star* – though the biters are greys only. A girl of two whose face was scratched was the sixth victim of a series of biting incidents in Knutsford, Cheshire. 'Everyone's afraid to go out. The squirrel has caused chaos,' reported an adult who had been bitten on the leg. A squirrel entered a police station in Consett, Co. Durham, ran up a constable's leg and bit him on the hand, drawing blood. Swimmers fled a pool in Torquay after a squirrel jumped in and bit a man.[8] What takes the biscuit (peanut?), though, is a *Daily Star* story headlined 'Violent Squirrel Leaves Town "Afraid to Leave Their Homes" After Attacking 18 People' (28 December 2021). The residents of Buckley, northeast Wales, nicknamed the rampaging offender 'Stripe', a nod to the leader of the violent gremlins in the Christmas horror comedy *Gremlins*. Many other papers also picked up the story, most of them explaining that the RSPCA put down Stripe in compliance with legislation imposed in 2019 outlawing the release of captured greys (and any other member of a designated 'invasive' species, for that matter) back into the wild. 'I did feel incredibly harsh doing what I did, by putting it in a cage,' related the 65-year-old woman who had been feeding Stripe and then live-trapped him, 'but when people didn't feel safe in their garden, I had to do it. We were not safe. But it is all over now, thank God.'[9]

The RSPCA warns that squirrels 'may well bite if cornered which may result in a hospital visit or at least a tetanus jab'.[10] The ones that 'broke into' Medway Council's offices in Chatham, Kent, in 2016, which might have been carrying bacteria linked to Lyme disease, ate employees' packed lunches, set off alarms and gave the caretaker the runaround.[11] Other transgressive acts demonstrate swaggering impudence and buccaneering, such as 'Sid the Squirrel' invading Arsenal FC's pitch during a Champions League semi-final match, and (a different squirrel) pinching nuts from the visitor centre atop Snowdon, the highest mountain in England and Wales.[12] Then there are the greys that get into scrapes. The RSPCA rescued a 'greedy' squirrel stuck in a bird feeder, and prosecuted a window cleaner who drowned a grey that had been raiding the bird feeders in his garden (see Chapter Seven).[13]

Stories of outrageous individual behaviour are supplemented by tales of the species itself, which epitomizes the gnarly, wider challenge of invasive non-natives.[14] Consumption as a solution to the effrontery of squirrels coming over and swiping our nuts is another *Daily Star* staple: grey squirrel is and has been a novel filling for a Cornish pasty, a dish at London restaurants and meat sold by a Belsize Park butcher.[15] According to the manager of The Red Lion in Kent, 'Kentucky-fried squirrel', a staple of the pub's menu in 2019, tastes like a cross between rabbit and quail. He touts clean-living country grey squirrels as sources of eco-friendly animal protein.[16] Squirrel-eating people cannot be expected to do the job on their own, though. Restoring the lynx to the Lake District might serve as a useful biocontrol agent in one of the red's last English bastions.[17] The *Daily Star* also recently presented greys as a threat to governmental efforts to tackle climate change by planting woodlands, as they damage young trees.[18]

For the tabloid press, stories that name and shame individual squirrels, like Stripe and Sid, hold more appeal. Yet the individualized approach to squirrels (or other species) is not as flawed as it may initially seem, and certainly not on the slippery slope to anthropomorphism. Jaclyn Aliperti, a behavioural ecologist and science communicator, pointed out recently that studies of animal behaviour must recognize differences within species. Individual golden-mantled ground squirrels (observed in Colorado) share personality traits with humans. Audacity and aggressiveness can shape a squirrel's reproductivity and success, allowing an individual to access food and defend its territory. At the same time, shyness and caution (traits typically associated in Britain with native reds) can make life harder. But risk-taking may leave a squirrel prone to accidents or predation.[19] Look no further than Squirrel Nutkin's recklessness in Beatrix Potter's tale. There are not just species of squirrel, but individual squirrels, past and present.

Squirrel Nation joins a tradition of titles – Valerius Geist's *Buffalo Nation: History and Legend of the North American Bison* (1996), Seth Zuckerman and Edward Wolf's *Salmon Nation: People, Fish, and Our Common Home* (2003), Edmund Russell's *Greyhound Nation: A Coevolutionary History of England, 1200–1900* (2018) and E. Elena Songster's *Panda Nation: The Construction and Conservation of China's Modern Icon* (2018) – that foreground the connection between nature, in the form of animals, and the notion of the nation. I have consulted parliamentary discussions, newspapers at home and abroad, magazines, scientific publications, governmental records, documentary films, advertising and, of course, imaginative literature – for children and adults. I also draw on my own observations and experience.

I grew up with squirrels on my doorstep in Formby, where the pinewoods that host the National Trust's sanctuary were my favourite boyhood stomping ground. My landscape of childhood was simply red. The sign at the railway station at Freshfield, an area of Formby that includes the squirrel reserve, reminds alighting passengers that this stop is 'for National Trust Red Squirrel Reserve and Beach'.[20] I cannot recall when I saw my first grey; it might have been in Cambridge in the early 1980s. There are more greys than you can shake a stick at in the Victorian suburb of Bristol that has been my home for over thirty years. While researching and writing *Squirrel Nation*, I observed squirrels pretty much daily. I really got into the swing of things squirrelly while working from home during the first COVID-19 lockdown in England in the spring of 2020. A large horse chestnut tree dominates the view from the room that became my working-from-home study. A family lives in that tree, barely 10 metres away, and our garden is a hive of highly distracting, work-interrupting squirrel activity, especially since their kin have dreys in a clump of nearby elms. At the University of Bristol's Department of Historical Studies, the view from the window of my former top-floor office on Woodland Road is also dominated by a large tree: an expansive cedar of Lebanon that is home to squirrels as well. The branches brush against the eaves of the Victorian villa and I often used to hear the squirrels running around the attic space above me. (A maintenance man once tried to block their entry points with chicken wire.) A hazelnut left on the window ledge at the end of the day was invariably gone by the next morning. I often wonder if the squirrels at work and at home realized I was writing about them.

I have become an armchair sciurologist; the squirrel life I witnessed corroborated what I was reading. A book on British mammals of 1918 noted that 'one of the most conspicuous of its [a squirrel's] misdemeanours is its habit of attacking the early leaf-shoots of the horse chestnut . . . which in spring appear to be very palatable to it, and are ruthlessly torn from the tree and then thrown to the ground.' In an earlier book of 1907, a re-creation of a squirrel's life in Britain, I read that 'young horse-chestnut buds . . . make as good a breakfast as almost anything I know of.' The squirrel in both sources was red. But greys have the same habit. In *Sciurus: The Story of a Grey Squirrel* (1978), a book in Collins's factual 'animal lives' series for older children, Jan Taylor observed that in springtime *Sciurus* feasts on the buds of horse chestnut flowers and its tender leaf stems. Today's grey squirrel is just as fond of these buds and stems. And like their predecessors in the 1970s – and reds a century ago – they still leave a wasteful mess on the ground.

In the gift shop at the National Trust's Tyntesfield estate in northeast Somerset, 13 kilometres (8 mi.) from my home, you can buy a cuddly squirrel toy. It's red, of course. So, too, is the squirrel on the can of the National Trust's Kentish Red Ale ('dark and nutty', brewed by Westerham Brewery, Kent).[21] What if the National Trust was to sell a squirrel toy that was grey, and a 'grey ale'? How would members and visitors react? Perusing the paint shelf at my local hardware shop, looking for a gloss to repaint an exterior wooden door, I stumbled across a pot of 'squirrel grey' (and bought it, obviously). This sort of merchandise is not unprecedented. In 1920s Britain you could buy a soft toy ('Little Grey Brother') and stockings in a shade known as 'squirrel grey'.[22] Greys were not beyond the pale a century ago. In fact, as this book suggests, at no point since then have Britons considered them completely irredeemable.

The first chapter opens with three recent episodes that underscore Britons' close (even special) relationship with squirrels, red and grey. Exploring the place of squirrels, symbolically and reputationally but also physically within the landscape, leads me into a territory I call

National Trust
Kentish red ale.

Preface

Berger non-drip gloss paint in 'squirrel grey'.

the emotional ecology of home. Competition for Britain's first squirrel from its second squirrel has given rise to a distinctive form of squirrel-infused nationalism most strikingly demonstrated in how feelings about 'their' squirrels flavour British attitudes to Americans and things American.

Yet the native squirrel has not always engendered warm feelings. Before the first great British public debate about the grey's depredations in 1928–31, there were lesser debates about the damage done by an out-of-control common (or brown) squirrel. Chapter Two explores the red's history before it was joined by the grey, and then the period, into the early 1900s, when attitudes towards both squirrels remained fluid and the red's iconic status was in the early stages of formation. The 'mercilessly shot' squirrels 'destructive' to the fresh larch and fir shoots in a Gloucestershire arboretum were reds.[23] All the squirrels subject to inflammatory rhetoric and complaints in *Country Life* in the late 1800s and early 1900s were reds.

The first hints of a protective attitude were stirring, though. *Country Life*'s editor leaped to their defence: 'England is not rich in its wild mammalia, and to slay ruthlessly one of its most interesting is criminal.'[24] In early 1912, he had issued a fuller statement of support, in response to the Board of Agriculture's formal endorsement (in a leaflet) of extensive and intensive shooting. The editor claimed that many members of the public had taken to heart 'this most beautiful of all woodland animals'. Indeed, he insisted (if rather unpersuasively) that any efforts 'making towards its extermination' were likely to arouse 'a very loud protest throughout the country'.[25]

The grey squirrel in North America serves as backdrop to Chapter Three's examination of the period of introduction and internal transplantation between 1876 and 1929, the year of the last recorded internal relocation. The positive attributes many Britons assigned to the greys are a reminder that this species has not always been demonized. Charming, pretty, engaging, jaunty, cheeky, bold and acrobatic: these qualities made it a novelty in London and environs, whose green spaces became a congenial home in the early 1900s. One commentator, no friend of the grey, called it the 'most perfect of town dwellers'.[26] An emphasis on shared squirrelly qualities and a broad spectrum of opinion on the grey lasted through the 1920s, though calls tantamount to 'Yankee go home' grew louder, especially in country areas.

Before 1930, action against greys was taken mainly sub-governmentally, by private individuals such as farmers, foresters, gamekeepers, horticulturalists and other landowners or occupiers. Chapter Four investigates the hardening of attitudes to an animal increasingly dismissed as 'not really a squirrel at all but an American tree-rat'.[27] In 1917, naturalist Frank Finn reflected that the native squirrel 'has a way of disappearing unaccountably, sometimes in areas where the American invader has never appeared'.[28] I evaluate explanations for the grey's naturalization and vigorous spread offered at the time, emphasizing the difference between displacement and replacement, and weighing factors unrelated to or pre-dating greys. One of those factors, disease, highlights the growing clash of evidence between anecdotal reports and what the British Museum of Natural History's curators of mammals in the 1920s called 'a scientific point of view'.[29] Some rural residents believed that greys triumphed because males either castrated or killed, or both, their red counterparts, and crossbred and devoured young reds. Findings from the first national survey of the grey's distribution and impacts (1930–31) downplayed this form of direct assault, bolstering the mammal curators' stance.

In Chapter Five central government joins the fight against Britain's third worst farm pest and 'No. 1 forestry pest'.[30] The state incentivized, supported or led various efforts to combat grey populations via trapping, shooting and, later, poisoning. What victory would look like was unclear. Eradication? Reduction? Containment? Issuing free gun cartridges, paying bounties, eating them during wartime and dispensing poison did little to dent their numbers or to dislodge grey squirrels from a secure position among the fauna of England and Wales.

A government biologist's post-mortem on the 1950s bounty scheme flagged up that the Ministry of Agriculture and Fisheries (MAF) operated an equally ineffective – and 'not generally known' – bonus for killing reds between 1903 and 1917.[31] By 2021, the fact that a bounty on *reds* once existed (the post-mortem underlined the word 'red') was barely conceivable. A wildlife veterinary surgeon, Romain Pizzi, pointed out in his foreword to a paean to the species that reds 'now appear that very rare anomaly – a wild animal that no one in Britain maligns'. The ode to the red was by Polly Pullar, a Perthshire-based wildlife writer and rehabilitator of injured and orphaned young reds, who could watch reds from her kitchen window while washing the dishes. Reared on stories such as *Squirrel Nutkin*, Pullar echoed Pizzi's sentiment: 'there are probably few British native wild mammals . . . that are so universally adored,' stressing that 'I have yet to find anyone who says anything negative about the red squirrel.'[32]

Chapter Six looks at the red's valorization and sanctification in tandem with the grey's criminalization and demonization. 'Of the immigrant grey squirrel, that aggressive coloniser, the vilest stories of murder and pillage will eagerly be consumed,' a journalist noted of rural audiences in 1962. 'But the red,' he continued, 'being both beautiful and persecuted, has acquired the aura of a saint.'[33] Leading up to Christmas 1963, the red squirrel and the robin were the most popular motifs for greeting cards produced by naturalist organizations and wildlife groups.[34] Prominent landmarks in the elevation, post-1945, of the 'home' squirrel to quasi-sacred status were the creation of Tufty

Fluffytail and the establishment of a road safety group called the Tufty Club – to which the six-year-old Pullar belonged – as well as the red's adoption for an anti-litter campaign. In 1987 two Zoological Society of London (ZSL) researchers reckoned that, in England, reds were 'almost universally welcomed'.[35]

Part of this veneration process was the sprouting of a thicket of protective organizations in the 1990s and early 2000s. If these various units, of what we might be tempted to call the red army, were to choose a supreme commander, then I suspect it would be (the former) Prince Charles. For the king, when still heir to the throne, 'nearly raises the ghost of Churchill when he talks about saving the pointy-eared, admittedly handsome redcoat cousin of the gray', reflected an American writer.[36] Applying the latest advances in squirrel science (equipped, since the 1960s, with radio-location telemetry), defence groups try diverse tools and tactics to level the playing field, from food supplementation (reds) to reproductive control (greys).

By the late 1950s, the grey's status as the 'most successful and most hated' of animal introductions to Britain was cemented.[37] Their bad reputation, it seemed, could sink no lower. This hostility provides a textbook example, if unappreciated, of sociologist Stanley Cohen's notion of 'moral panic', wherein a 'condition, episode, person or group ... emerges to become defined as a threat to societal values and interests; its nature ... presented in a stylized and stereotypical fashion by the mass media'.[38] Grey squirrels can easily be added to Cohen's original list (1972) of juvenile delinquents and deviants ('folk devils'), which includes Mods and Rockers, Teddy Boys, football hooligans and skinheads. Greys can also be bracketed with more recent (non-specifically youthful) examples of 'folk devils' that inspire moral panic, which Cohen identified thirty years later, in a third edition (2002) of his book: refugees and asylum seekers that – in examples of what Cohen calls 'tabloid rhetoric' – 'flood into' and 'over-run' Britain as a 'human tide', 'spongers', 'scroungers' and 'cheats' that 'milk' a country offering a hospitable environment ('soft touch' Britain).[39] For a moral panic to take root, Cohen identifies three ingredients: a *suitable enemy*, a 'soft target, easily denounced, with little power' and a *suitable victim* ('someone with whom you can identify').[40] The grey provides the enemy and the red supplies the victim. Greys are also easily denounced; whether they lack power is debatable.

In 1963 the American writer Joseph Langland published a poem about squirrel-hunting farm boys that included the refrain: 'Squirrels, beware, beware; Red squirrel, run! The fixed ideas are coming to hunt you down.'[41] Change the colour to grey, and the setting from American oaks to English oaks: the notion of fixed ideas coming to get you then works nicely as a description of the prevailing British mentality. Yet, since the 1960s, however much its presence is bemoaned, and no matter how overblown the reiterative rhetoric of fear and panic, Britons have become increasingly accustomed – and resigned – to the grey's permanent membership of Britain's fauna. In the early 1980s, a squirrel scientist concluded that the 'robust' grey 'is to be admired for its success'.[42] Also involved, though, is warm embracement. At the twentieth century's close, not all Britons shared foresters' and timber merchants' conviction that the grey was a neighbour from hell.[43] American journalist William Montalbano realized this too, quoting the observation of ZSL's Peter Cotgreave that 'for many people the gray is just as nice [as the red].'[44]

The final chapter explores growing evidence of reputational rehabilitation and positive emotional engagement with Britain's second squirrel, particularly among younger Britons in urban/suburban settings who may well never have seen a red in the flesh. Peter Lurz, who had worked on red squirrel conservation for twenty years, disclosed in 2014 that 'I do mourn their disappearance from my old garden in Cumbria and would sorely miss them if they disappeared completely.' Alongside this statement, Lurz brought up another form of disappearance: the loss, among youth, of a sense of what makes a red squirrel special. A sighting 'can even be met with bemusement by members of the younger generation in the south, who only see red squirrels while on holiday'.[45] (Lurz had the Lake District and Scotland in mind.)

The grey, by contrast, has become almost omnipresent. In 1979 a specimen was spotted at circa 900 metres (2,950 ft), near the top of Snowdon.[46] A 1989 advertisement for lager underlined and reinforced that more and more people are not just making their peace with greys, but are actively pro-grey. Another sign of shifting sentiments is growing resistance to defamation and culling and the allegation of 'squirrel racism' from the animal welfare and rights sector. As early as 1942, an official at MAF warned in an internal memo on 'infestations' that 'any [radio] broadcast inciting the public to the slaughter of the grey squirrel would at once raise a terrific volume of protest from animal lovers, misguided though they be.'[47]

On patrol with a division of grey-cullers in the Lake District in 2017, *The Guardian*'s environment correspondent, who, like most Britons, was habituated to greys, reacted to the first red he saw as if it was an exotic species ('so dainty and alertly pretty, with fine tufts of hair above its ears as extravagant as the eyebrows of Denis Healey').[48] A recent nationwide survey of 3,758 people over the age of eighteen – factoring in the variables of age, gender, presence of and type of squirrels in areas of residence, as well as level of squirrel knowledge (conservation and control efforts) – indicated not only that reds are valued more highly than greys, but that tree damage committed by reds is more tolerated. Yet it also revealed that many Britons like to see greys in their gardens and parks as well as in the countryside and that 59 per cent of respondents (though largely unaware of problems associated with greys) disapprove of lethal control methods and think both species merit conservation.[49] 'The charge sheet against the grey squirrel', emphasizes a prominent British sciurologist, 'is based on hundreds of peer-reviewed research papers. There really is no defence for it.'[50] But is the grey more maligned than malign?

Time's passage blurs awareness of a species' origin. How many know (or care) that the rabbit – the first creature Potter enshrined, in *The Tale of Peter Rabbit* (1902) – was introduced by the Romans or the Normans? Is the grey's fitness in present-day Britain more important than the technicality of its place of origin a century and a half ago? The 150th anniversary of 1876, the year that, by general consensus, the grey squirrel was first introduced to these shores, is approaching. So is there a more suitable animal icon for twenty-first-century Britain than an endangered native species such as the red squirrel? Environmental journalist Patrick Barkham had posed a similar question in 2013, introducing a BBC *Wildlife* reader poll to choose 'Our

National Icon'. 'What better symbol of modern Britain than an exotic newcomer?', he asked. The brown hare or the horse chestnut might be suitable. But 'one of the most successful introductions, the grey squirrel, was probably too controversial a candidate.'[51] A decade later, is it finally time to learn to stop worrying and love a species that should perhaps be called the British grey squirrel?

I

A TALE OF TWO SQUIRRELS

What do Coleen Rooney (née McLoughlin), Paul McCartney and Nick Clegg have in common? The answer, obviously, is: squirrels. Three recent episodes featuring the celebrity, the rock star and the former politician reveal the claw-hold of Britain's two squirrel species on British life and how they have punched well above their modest weight and size.

Episode One: The Ring in the Sanctuary

Media interest in Britain's dwindling reds rivals engrossment with so-called WAGS, a term for footballers' wives and partners popularized during coverage of the England team at the 2006 World Cup. Two years earlier, red squirrels and WAGS had combined as if heaven-sent. A 'tabloid sleuth' reported that eighteen-year-old Everton and England star Wayne Rooney had paid for sex in December 2002.[1] Before the story broke, Rooney came clean to his fiancée, Coleen McLoughlin (also eighteen). She then reportedly stormed out of their mansion on Victoria Road, in Formby, Merseyside, headed for the nearby pinewoods.[2] On arrival, she apparently flung her engagement ring into the trees.[3]

Those woodlands, owned by the National Trust, represented one of the red squirrel's last English bastions. 'A friend' of the couple told a reporter that 'the squirrel sanctuary is full of trees and bushes. It will be a nightmare to find the ring.'[4] National Trust spokeswoman Debbie Peers, fearing that an onslaught of treasure hunters armed with metal detectors would upset the reserve's three hundred reds, warned that 'anyone ... seen disturbing the squirrels will be asked to leave'.[5] At the time, and in her semi-autobiography *Welcome to My World* (2007), Coleen Rooney denied chucking her ring into the 'squirrel park'. But her alleged deed's consequences for the squirrels were widely reported, nationally and internationally.[6] A reporter for *The Sun* joined the ring-hunting throng, all in vain. 'It has probably gone already,' commented squirrel reserve warden Louise Mitchell a month later; squirrels stash shiny objects in their nests (dreys).[7]

Episode Two: Come On You Reds

Formby's reds – a reporter dubbed them the 'hidden victims' of the ring 'saga' – were precisely those for whom a high-profile fundraising activity was organized in 2000, four years before the ring incident.[8] Everton manager Walter Smith was the only figure from the footballing world among the luminaries that supported the fundraiser, known as 'Scribble a Squirrel'. But more than two hundred personalities rallied to the cause from the worlds of sport (for example, Geoff Boycott), music (Phil Collins, Elton John, Barry Manilow, Paul McCartney), comedy (Rory Bremner), fashion (Heather Mills), television and radio (Joanna Lumley, Norman Wisdom, Bruce Forsyth), stage and screen (Sean Bean, Judi Dench, Michael Caine, Richard E. Grant, Rolf Harris) and various others, including Richard Branson, the duchess of York and Alan Titchmarsh. They all drew a squirrel, and their drawings were auctioned online during that year's Red Squirrel Week to benefit a campaign for red squirrel conservation organized by the Northumberland Wildlife Trust, Red Alert North East and the Calvert Trust.[9] The Trust provides activity holidays for people with disabilities and their families, in Northumberland's Kielder Forest, whose dense stands of non-native conifers, like Formby's, provide another, even more important red redoubt: nearly two-thirds of England's reds call Kielder home. At the auction, McCartney's scribble, 'Pink Squirrel', commanded the highest price (£4,120). And despite its colour, McCartney's squirrel was reckoned to be the closest match to the protagonist in Potter's *The Tale of Squirrel Nutkin* (1903).[10]

McCartney later devised another squirrel, named after the peninsula separating the Mersey and Dee estuaries. The 'cheeky Scouse nut-muncher' – unambiguously grey – debuted in a video accompanying McCartney's children's song 'Tropic Island Hum' (2004). Wirral the Squirrel became the central character in McCartney's first children's book, *High in the Clouds* (2005). Displaced when an evil developer lays waste to his woodland home, the teenage Wirral flees a hunter by grabbing a ride in a balloon that takes him and a 'plucky' red squirrel, Wilhelmina, to a secret island sanctuary. On groovy Animalia, where all creatures live in harmony and freedom, Wirral falls for Wilhelmina, a symbolic act of red–grey coexistence.

Episode Three: Silly Squirrel Rules

Five years after *High in the Clouds* was published, squirrels cropped up unexpectedly as part of the Conservative–Liberal Democrat coalition government's political and constitutional reform initiative. One of Deputy Prime Minister Nick Clegg's first duties was to oversee an unprecedented exercise inviting the public to nominate pointless and illiberal laws for scrapping.[11] On BBC1's *Breakfast* programme on 1 July 2010, Clegg spoke of 'lots and lots of old stuff on the statute books that we should get rid of for starters'. Then he gave an example. 'I've just discovered . . . would you believe it, that there's still an old law . . . that says it's an offence if you don't report a grey squirrel in your own back garden.'[12] Clegg had targeted a clause in the Grey Squirrels (Prohibition of Importation and Keeping) Order of 1937. Authorized by section 5(2) of the 1932 Destructive Imported Animals Act (which applied exclusively to muskrats, another import from North America), the clause imposed a fine of £5 (equivalent to around £250 in 2017) for failure to 'give notice to the appropriate department' – namely, the pest control branch of MAF, or, in Scotland, to the Scottish Secretary – if aware of a grey on your property.[13]

Clegg's singling out of the £5 fine alarmed red squirrel supporters. Was he complacent about the grey threat to native squirrels? Moreover, would Clegg's example of silliness offend fellow Liberal Democrat cabinet member Danny Alexander? Now Chief Secretary to the

Treasury, Alexander was a former head of communications at the Cairngorms National Park Authority, which oversaw Caledonian pine forests representing one of the last places in Britain where reds still flourished on a large scale. (A press release with Alexander's signature declared 'Your Squirrels Need You! The Quest Is On to Save the Highland Red Squirrel – And You Can Help!'[14]) Alexander also represented a constituency in northeast Scotland where reds still ruled the roost.

Alexander was redheaded, too. In her speech at the Labour Party's Scottish Conference in October 2010, deputy leader Harriet Harman included an ill-considered jibe: 'Many of us in the Labour Party are conservationists and we all love the red squirrel. But there is one ginger rodent that we never want to see again in the Highlands – Danny Alexander.'[15] Accusing Harman of 'outrageously discriminatory remarks', Clegg made a statement consistent with an interpretation of his stance on the £5 fine as pro-grey or, at least, suggestive of greater tolerance: 'I think any form of discrimination against rodents or ginger-headed folk is wrong.'[16] The *Express on Sunday*, though, felt Harman should do more than just apologize. She should volunteer 'for [Alexander's] old bosses at the Cairngorms National Park, where the fight to save the red squirrel goes on. It would do her good to spend a weekend in the mud and rain baiting grey squirrel traps.'[17] Reacting on Twitter to his defamation, Alexander pronounced that the red squirrel 'deserves to survive, unlike Labour'. The Conservative deputy leader in the Scottish Parliament, Murdo Fraser, joined the assault on Harman's remark by reiterating that the red was 'held in very high regard by the Scottish public'.[18]

Squirrels and the Emotional Ecology of Home

Four years later, in 2014, the imminent repeal of the offending provision in the 1937 Order gave red-approving and grey-disapproving journalists another opportunity to publicize the plight of England's surviving reds (much-loved means much-missed). The silly squirrel law's repeal provided the occasion for the government to formally concede defeat in its long war against Britain's other squirrel. As the

Solicitor-General for England and Wales Oliver Heald informed a Commons Public Bill Committee:

> The order requires occupiers to report the presence of grey squirrels on their land to facilitate the eradication of that species. However, it is no longer considered feasible to eradicate grey squirrels, so the requirement to report their presence on one's land is no longer useful or observed ... people will no longer be required to report sightings of grey squirrels on their land.[19]

A spokesperson for Defra (Department for Environment, Food and Rural Affairs, which in 2002 replaced MAFF (Ministry of Agriculture, Fisheries and Food, which itself had superseded MAF in April 1955)) downplayed the repeal's significance. There was no evidence that the fine had ever been imposed. Besides, repeal hardly left greys above the law. The Order's other provisions were superseded by the Wildlife and Countryside Act of 1981, under which the grey is a pest killable, humanely, by anyone with a firearms licence, without seasonal restrictions or quotas, though killing is not a statutory requirement.

Though never enforced, not everyone considered the 1937 provision preposterous. Defra sought to reassure the public that the government had not abandoned the red and remained committed to culling greys in certain areas under certain circumstances ('culling' being a euphemism more publicly palatable than 'killing'[20]). Nonetheless, red advocates thought the state had surrendered. Conservative MP Andrew Bridgen, a member of the Public Bill Committee on the scrapping of the fine, proclaimed 'British trees for British squirrels.' And Janet Wickens, director of the Red Squirrel Survival Trust, explained its position: 'while I can see the sense of getting rid of [the clause], we said we would prefer it to stay on the books. We didn't feel comfortable supporting this move because it's one step closer to accepting an invasive non-native species and giving it the right to live here.' It was 'not OK', in her opinion, for verminous greys that carried a disease lethal to reds – the squirrelpox virus (SQPV)

– to be 'part of our wildlife'.[21] Looking ahead, Wickens believed scrapping the law would make it harder to reintroduce reds to their former homes, as it absolved landowners of the requirement to tackle greys on their properties.[22]

Particularly upset were northeast England's red campaigners and champions among politicians. Chi Onwurah, Labour MP for Newcastle Central, wanted Heald 'to recognise the fact that red squirrels are still present in Northumberland', a county with more reds than any other in England. She felt that the government's defeatist attitude ignored, even disrespected, the efforts of a host of local volunteers, organizations and businesses (including butchers selling grey squirrel meat) to secure the red's future by culling greys.[23] A cause that could unite Newcastle's first black British MP with peers on vast rural estates packed unusual power.

For the politicians that signed up, the red cause was not just bipartisan but apolitical. Being against red squirrels was like being against motherhood or apple crumble. Rooting for reds was undeniably bipartisan, but there was no national consensus. The red's champions admitted as much by being perplexed by the grey's popularity. In 2006 the Conservative MP Peter Atkinson – whose Hexham, Northumberland, constituency, he claimed, housed 80 per cent of England's red squirrel population – confessed he was no closer to understanding why so many were so fond of greys than he had been a decade earlier. After he advocated killing greys with a rat poison (warfarin) in the House of Commons in 1996, about five hundred letters deluged his office. Each one accused him of being a 'cruel and evil man' hell-bent on exterminating greys.[24]

Britons enjoy a special relationship with dogs.[25] But the bond with squirrels is special too. Like dogs, squirrels share – sometimes invited, sometimes uninvited – the intimate, everyday spaces of our homes and daily environments. A home includes not just domesticated, privately owned dogs, cats, budgerigars and squirrels. The animals in our homes and the process of pet-making also encompass so-called wild species, collectively owned. By 1900 the invited experience of indoor cohabitation with squirrels was largely over; keeping reds as caged pets went out of fashion. Reports of greys' uninvited presence inside the home began to appear in the 1950s. If greys gain access to a loft or attic, they can nest in the rafters. Even if they do not set up home in your home, they can chew through the PVC-coated insulation on electrical cables – a serious fire risk – and hone their incisors on lead telephone cables.[26] 'Sharing their home ... with a family of squirrels' was an apt description of this novel living arrangement.[27] By 1964 the pest experts at MAFF were fielding complaints from householders on London's 'fringe' (also about squirrels' use of fibreglass wool to build dreys indoors).[28] In a House of Lords debate on forestry in 1981, Lord Hale complained that the ubiquity of greys in his London garden suburb extended to his home, shared for years with greys that threatened to 'bore into virtually anything'.[29]

Home is not confined to a dwelling's interior. Home extends to the garden.[30] Home also stretches beyond the immediate domestic space of individuals, families and other small units to encompass a local area. Ecological meanings of 'home' reinforce this notion of 'home environment'. Ernst Haeckel's coinage of the term 'ecology' (*Ökologie* in German) in 1866 is often traced to *oikos* (ancient Greek for 'house' or 'home').[31] *Oikos* is also closely associated with the concept of family, immediate and extended. A century after Haeckel inserted 'ecology' into the lexicon of science, the Apollo astronauts' paradigm-shifting photographs of our planet – *Earthrise* (1968) and *Blue Marble* (or *Whole Earth*) (1972) – led the Earth to enter collective Western consciousness as a planetary home, an abode shared not just with people of other nations but with myriad other species.

Yet, as the Chinese environmental historian Xia Mingfang reminds us, we inhabit many small universes 'within the big universe'.[32] Somewhere between the smallest and the biggest – and arguably more meaningful and manageable than the grandiose planetary scale – is another understanding of home. This is home as the nation we live in, as in the home nations of England, Scotland, Wales and Northern Ireland that make up the United Kingdom. Home as nation – as in homeland – in which private and public spheres converge, is central to this exploration of squirrels in and of Britain.

A synonym for nation is country, as in 'fellow countrywoman'. And, like home, country is multilayered. *The* country means a rural area. No matter that Britain (especially England) has been one of the world's most urbanized nations since the mass industrialization of western Europe and North America in the nineteenth century, with 83 per cent of its population classified as urban in 2021. For many who live in the country(side), rural Britain *is* Britain and rural England *is* England. Wartime posters encouraged identification with the countryside as the real Britain (by default, the real England). The most recognizable is Frank Newbould's *Your Britain: Fight for it Now* (1942), a composite image for the Army Bureau of Current Affairs of southeast England's South Downs: rolling pastureland, sheep, sheepdog, shepherd and beckoning farm nestled among deciduous trees.[33]

The demands of homeland security extend beyond protection of the strictly human world. Just as borders and sovereign national territories require defence against human invaders, the repulsion of non-human incursions is something about which some people feel strongly. In the Lords in 1919, during a debate on establishing the Forestry Commission (in which the grey squirrel was barely mentioned), Lord Phillimore spoke up for the native squirrel in the face of substantial anti-red sentiment. Though the owner of a conifer plantation in Oxfordshire, he characterized the much-maligned red as 'one of the most beautiful objects in English natural life'.[34] And so, as this feeling spread and the grey expanded its domain, Britain's squirrel history became closely linked to themes of patriotism, nationalism, belonging, citizenship, territoriality and defence of the realm.

How ironic (and humiliating), a journalist observed in 1946, that a country so effective in repelling human invasions over the centuries enjoyed such paltry success against representatives of 'lower creation' in the shape of grey squirrels.[35] As Britain's leading sciurologist of the 1950s and '60s, Monica Shorten, observed, the grey found a 'fat and vacant living waiting for it once the only native squirrel had been struck low by disease'.[36] Forty years later, David Stapleford, who had bred more than six hundred reds for reintroduction purposes in an enclosure in his Norfolk back garden, reckoned that the grey ('oversexed, overfed and over here') confronted Britain with its 'most successful invasion since the Norman Conquest'.[37] Post-1945 references to the grey's 'occupation' of British territory smack of military takeovers, as in Roman Britain or (more recently) German-occupied Europe.[38] Regaining control of occupied territory was a tall order, even three-quarters of a century ago. But action could still be mounted at sub-national level to prevent further loss of territory. From the grey's perspective, of course, invasion is homemaking.

Nature and nation are as intimately connected as nation and home. Australia, Canada, New Zealand, South Africa and the United States – the former white settler colonies (so-called) – are the nations primarily associated with eco-nationalistic investment in certain native forms of nature as an integral (pre-cultural, apolitical) component of national identity.[39] Yet, as this book suggests, an 'old' country like Britain also fashioned its sense of self from non-human indigeneity.

Squirrel Nation is about the squirrel's place, not least, within 'Deep England'. This mythic pastoral landscape has a heavy, normatively southern bias – often to the total exclusion of northern Britain – typified by the Home Counties enveloping London that Newbould's propaganda evoked.[40] By place, I mean not only the squirrel's position within the imagined landscapes of Englishness and Britishness. I am equally interested in the squirrel's position within actual environments, urban, suburban and rural.

A handful of cultural geographers have examined English animal landscapes, specifically those constituted through the hunting of otter and wildfowl, and placed the bittern and the coypu, the latter another invasive non-native from the Americas, within the Norfolk Broads.[41] Other analysts of landscape, Englishness and Deep England have not really considered the role of animals, imagined or material – perhaps because Arcadia is mainly conceived as pastoral or watery, the abode of sheep or otter. But an English landscape can be arboreal too. The regional case studies in Paul Readman's *Storied Ground* (2018) include Northumberland, the New Forest and the Lake District. More treed than the English norm, these are all prime squirrelscapes. Yet Readman's interest in the animality of place and of Englishness between the late 1700s and early 1900s is restricted to a mention of the Herdwick Sheep Association (1899) and Potter's interest in this distinctive Lakeland breed.[42]

Animal historian Hilda Kean noted two decades ago that the red squirrel became a 'symbol of an idyllic rural Britain' in the 1930s.[43] In 1944 a reporter dubbed it the 'home' squirrel.[44] Since then, it has become an almost mythical creature in a bucolic homeland of leafy Deep England. Kean also observed that the red became 'consolidated as a motif of England's heritage' in the early 1980s, assuming a hallowed position in the pantheon of Englishness alongside red phone boxes, cricket bats and warm beer (that is, served at room temperature).[45] Within the recent atmosphere of resurgent and insurgent English nationalism, this poster animal is clung to even more fiercely than warm beer and embraced even more tightly than red phone boxes. If anything is considered deservedly next in line to take back control of their country, my impression is that it is the red squirrel. And with the prospect of even closer ties to the United States as post-Brexit Britain distances itself from its nearest neighbours, geographically speaking, a book that approaches feelings of Englishness and Britishness through animal history may be particularly timely.

Animals, observed Henry David Thoreau, 'are all beasts of burden, in a sense, made to carry some portion of our thoughts'.[46] Foregrounding what animals represent to us is a familiar approach to animal history. Middle-class Victorian Britons, for example, cast their pet dogs as devoted friends; strays they associated with undesirable human elements.[47] Meanwhile, across the Atlantic, nineteenth-century Euro-Americans constructed the wolf as an outlaw, later as Nazi and 'Jap', and, more recently, as 'lone wolf' terrorist.[48] The dogs and wolves themselves, of course, have little say over how we ascribe meaning to them, negatively or positively, unflatteringly or flatteringly, regardless of the niceties of behavioural realities and advances in scientific understandings.

Animal history is more than a branch of cultural history, however. Instead of dwelling on animals as projections of ourselves, some animal historians prioritize the more proactive role of the material (corporeal) creature, highlighting otherness and difference, trying to make sense of them more on their own terms. Total animal history demands both perspectives. Looking at human culture, values and society through an animal lens does not have to overshadow the animal that acts more autonomously and, in animal and environmental historian Susan Nance's words, inhabits 'parallel realities shaped by their own priorities, instincts, and experiences'.[49] Grey squirrels excel at making our world theirs. But their world is also ours. Instead of simply reflecting British ideas of belonging, home and nationality, squirrels have helped to shape them. They are co-constituters.

The squirrelscapes that squirrels create as furry, scampering entities are rich places in which to explore the emotional ecology of home.[50] Squirrels arouse strong emotions. 'From time to time,' fulminated Lord Middleton in 1981, 'I become consumed with fury and exasperation at the sight of young sycamores and beeches stripped of their bark and killed by grey squirrels.' (Not all Britons have been preoccupied by the grey's threat to its native counterpart.) So 'consumed with fury', in fact, that Middleton sometimes felt the 'appropriate penalty for the introducer – his name escapes me for the moment, but if it was the grandfather of one of my noble friends I apologise – would have been hanging, followed by drawing and then quartering!'[51]

'Emotions *do things*,' gender studies scholar Sara Ahmed emphasizes, adding that 'they align individuals with communities ... through the very intensity of their attachments.'[52] Powerful feelings supply ties that bind individuals into groups that fight for reds and against greys. Those who love reds might be seen as squirrel nationalists. Those who love greys could be called squirrel internationalists (or squirrel eco-cosmopolitans). Ingrid Tague, a cultural historian of animal–human relations, approaches emotional attachment to animals largely in terms of affection, companionship, sympathy and grief.[53] Yet dislike and its extreme form, hate, are kinds of emotional attachment too. Britain is a nation of squirrel haters as well as of squirrel lovers. The grey squirrel featured among the fourteen mammalian species included on the International Union for the Conservation of Nature's list of *100 of the World's Worst Invasive Alien Species: A Selection from the Global Invasive Species Database* (2008). Love for reds and hatred for greys can be different sides of the same coin, shaped, at least in part, by the binaries of native and non-native, eligible and ineligible, good and bad. Greys are among the *world's worst*.

To love reds, you do not have to hate greys. Wildlife managers, especially, are pro-red rather than anti-grey, more likely to approach grey 'control' dispassionately. But many who trumpet 'our' squirrels as superior have a low opinion of greys, though 'ours', they concede, are definitely not better in competition with greys. Within the 'ecology of nationhood', the 'naturalisation of nationality' and 'naturing the nation' (terms that anthropologists Jean Comaroff and John Comaroff use), this devotion to reds, whose higher value is taken as a given, can slide into eco-nationalism (that is, eco-ethnocentrism), even eco-jingoism (eco-nativism and eco-chauvinism).[54] Whether there were any repercussions for the grey in the climate of heightened English nationalism during the run-up to the Brexit referendum of 2016 is largely a moot question. It is hard to see how animosity could have been ramped up any higher in certain quarters. More germane, for me, is that the grey squirrel originally came from the United States, not Lithuania, Poland or Romania.

Squirrel Nationalism and Squirrel-Flavoured Anti-Americanism

On the first page of his seminal study of the grey's spread across Britain, biologist Arthur Douglas (A. D.) Middleton, Britain's first sciurologist, reported, 'I know of more than one patriotic Englishman who has been embittered against the whole American nation on account of the presence of their squirrels in his garden.'[55] Among the various American commentators who quoted Middleton's observation was *New Yorker* staff writer Eugene Kinkead. In a set of reflections in 1975 on the eastern grey, he noted they sometimes had to be chased out of the Metropolitan Museum of Art, located on Central Park's fringes. Next, Kinkead alluded to an 'inanimate' squirrel ever-present at the Museum: 'Squirrel in a hazelnut tree', a postcard in the gift shop, showed a detail from the sixth tapestry in the seven-part *The Hunt of the Unicorn* (*c.* 1500). In it, one can see a chunky, brownish-red squirrel in the bottom left-hand corner, nibbling on a hazelnut. He explained that this squirrel was a native European red, adding that, in Britain, 'our gray squirrel is a rampant and generally unwelcome immigrant.' After quoting 'a British writer on squirrels' – Middleton's above-quoted statement – Kinkead announced that disease unrelated to greys (something Middleton had pinpointed) was actually responsible for the red's parlous condition, but 'British squirrel people' blame the 'newcomers'.[56]

Shorten (who inherited Middleton's mantle of leading British sciurologist) likewise suspected that Britain's 'other squirrel' was 'further condemned by the British on account of its alien nature'.[57] As well as being the 'other squirrel', as in additional squirrel, the grey was also 'the other' in the sense of not the right one. Improperness was not confined to individual greys or groups; the species itself was wrong. 'Otherizing' a species strips it of the credentials the right one possesses. As Shorten explained, facing an ever-lengthening list of crimes and misdemeanours, the grey perpetrator was portrayed as a 'tree rat' that 'killed off' Nutkin's real-life version. No other analogy could have packed anything close to the clout of conflation with the reviled rat. 'Tree rat' de-squirrellized the grey.[58] Defined relationally and morally, a red squirrel is a squirrel that is not a grey squirrel; and a grey is a squirrel that is not a red. In short, a grey is unworthy of squirrel-ship – a sub-squirrel.

Disdain for Americans and their country was not, of course, a necessary concomitant of disapproval of an animal that even zoologists dubbed the 'American immigrant'.[59] A U.S. reporter assured American readers in 2012 that red squirrel champion Wickens 'insists that she doesn't hate all Americans. Just the squirrels'.[60] Nonetheless, the righteous, proprietorial indignation over the unruly American trespasser that Middleton reported (*their* squirrels in *his* garden) is frequently encountered in stigmatizing manifestations of squirrel nationalism. Former country diarist for *The Guardian* William Condry, fond of condescending digs at Americans and America, included holders of anti-American sentiments among grey-hostile groups in the early 1970s. 'If you are a forester, a gardener, a farmer, a bird-lover, or simply a hater of all things American,' he observed, 'you will probably say there are far too many grey squirrels and that their number ought to be reduced drastically.'[61] Forty years later, a London property manager particularly enraged by the damage they inflicted when infiltrating roof voids quipped that 'like many American imports the grey squirrel has outstayed its welcome.'[62]

The American origins of the two stuffed greys in the British Wildlife display at the Bristol Museum and Gallery are bluntly underscored. 'Howdy, I'm originally from the USA', reads the caption for the specimen mounted on the ground amid autumnal leaves. American sources often mistakenly credited Americans resident in Britain with the first introductions.[63] On the other hand, Americans in Britain often felt defensive ('don't blame me for them').[64] Yet not even the most virulently anti-American of Britons have tried to pin responsibility on Americans. The European Squirrel Initiative's chairperson, Miles Barne, told *USA Today* that benighted British landowners thought that what turned out to be 'great big bully boys' that 'terrorized' the 'nervous' native reds 'would be a nice addition to their forests'.[65]

The narrative of conquest that cast the red as dispossessed and the grey as dispossessor received fictional expression in British novelist Michael Tod's *The Silver Tide* (1994). 'I got the idea', Tod explained, 'from thinking about what the Europeans did to the American Indians, believing that they had no culture so the invaders could do what they liked to them. The grey squirrels [in Britain] did the same thing to the red squirrels.'[66] An American specialist in furry fandom, a subculture fascinated by anthropomorphic animal characters in genres such as science fiction and cartoons, agreed that the novel's 'arrogant' colonizers brought to mind 'the stereotype of the American settlers who considered the native Americans a "decrepit bunch of savages" who did not deserve the land that they lived on'.[67]

Tod's depiction of greys as silver-coloured was not unprecedented. Within a decade of *Nutkin*, Potter switched from red to grey. One of the greys in *The Tale of Timmy Tiptoes* (1911), set in North America, is Silvertail. There was no suggestion in *Timmy Tiptoes* that Timmy and Silvertail were dubious newcomers from across the Atlantic, let alone part of a deluging wave. But Tod's 'silver tide' signifies an unstoppable force that creates a nation within a nation, a home away from home, that the greys call 'New America'. The action is set in 1961 in the southern part of New America – specifically an area once known as Dorset (Tod's birthplace) – from which the 'peace-loving' red 'community' has been evicted.[68] Gabbro, one of the pioneering pair of male greys, sits on a fencepost 'tearing the limbs off a blackbird fledgling with his sharp yellow teeth'. Meanwhile, the other pioneer, Marble, licks the fledgling's blood from Gabbro's lips, 'proud to be an Explorer, Missionary and Disturber of the Peace'. Their home base is an actual place, Woburn Park, the Bedfordshire estate that was the most consequential of the first places in Britain to host greys. The greys' leader, the Great Lord Silver, despatches Gabbro and Marble to seek out fresh territory to relieve overpopulation pressure at Woburn. The grey troops' instructions are to 'get rid of all the Brown Jobs, kill them if need be. We need more Leaping-room.'[69]

The greys introduce themselves as 'Squirrels of the Silver Kind' who hail from the 'Great Lands far away over the water'.[70] Unable to bring himself to call them squirrels, the red known as Oak the Cautious refers to 'grey creatures'. The newcomer–native encounter turns on the distinction between ownership and guardianship. Possession of place is alien to the more egalitarian and communitarian reds.[71] The indigenes' last stronghold is 'Ourland', an island refuge off New America's south coast.[72] 'Ourland', Tod explains in his epilogue, corresponds to Brownsea Island, Dorset, a red squirrel sanctuary currently with a population of approximately two hundred.

Unwanted intrusions from 'far away over the water' that Britons have compared to 'squirrels of the silver kind' include attainment tests for children, linguistic 'imports', hamburgers, human resources departments (replacing personnel departments), Starbucks coffee chain ('the Yankee grey squirrel of the High Street'), venture capital strategies and cupcakes ('bold and brash, much bigger and brighter . . . trying to edge out our own little fairy cakes').[73]

Seven of the eight direct introductions Middleton logged were from the United States.[74] That the British press rarely mentioned that the grey's native ground included areas of southeast Canada is therefore unsurprising. (Exceptionally, a country diarist for *The Guardian*, W. D. Campbell, noted, on 4 April 1973, that 'this undesirable alien' might be a 'possibly Commonwealth' immigrant.) Government officials monitoring grey activities occasionally adopted a geographical approach, as in references to the North American grey squirrel or the eastern grey squirrel of North America. Most descriptions, though, were of the American grey squirrel.[75] Given this almost reflexive association of greys with the United States, studies of British attitudes to the United States and experiences of American influences are remarkably squirrel-less. Patrick Deer, a scholar of British war literature, has remarked that 'global U.S. culture marks almost every aspect of everyday life' in Britain. This observation could be usefully expanded to include grey squirrels as revealing evidence of a pervasive American presence.[76] John Lyons's *America in the British Imagination: 1945 to the Present* (2013) includes a chapter on 'Culture Wars: American Global Supremacy and British Nationalism, 1990–2001'. It would be easy to add a chapter on 'Squirrel Wars: American Grey Squirrel Supremacy and British Eco-Nationalism'. The depth of British emotional investment in reds, the conviction that Britain's treetops rightly belong to them, cannot be fully grasped unless situated within a larger context of American invasion, cultural, commercial, political and military – however ostensibly friendly or peaceful. Grey squirrels cast a whole new light on notions of the ugly American (that is, uncouth, greedy, unmannerly, loudmouthed, unstoppable, super-sized American), the unloved American and the American Peril.[77]

'While I welcome visits from our U.S. human cousins, this little export needs to go,' remarked journalist Janice Atkinson-Small.[78] Other British commentators did not distinguish so clearly between human and non-human. Conflation proved hard to resist. British newspaperman Peter Pringle spent long stints as a U.S. correspondent for British papers. In 1996 he recounted a recent visit to the UK. Staying with a childhood friend in an Oxfordshire village, gunfire woke him one morning. Peering out, he saw eleven dead greys on the lawn, shot by his host to protect his walnuts and hazelnuts. 'Whenever I come over here,' Pringle exclaimed, 'I never fail to draw fire about the awfulness of something American. Never mind that Britain continues to adopt American innovations (spin doctors come to mind).' When he complained about the racket, his host fired back: 'It's all your fault. Grey squirrels come from America. And they're such bullies they've forced our much nicer indigenous squirrels from their natural habitat. You rarely see a red squirrel any more.'[79]

Like Pringle's Americanophobic host, greys got under the skin of British popular historian Paul Johnson, though he was a renowned Americanophile. (In 2006 President George W. Bush awarded him the Presidential Medal of Freedom.) Johnson observed the devious 'operations' of a pair in Hyde Park in 2001 and pondered the paradox of difference in attitude towards rats (as you would expect) and greys (oddly positive, despite doing their thing 'in the open, in broad daylight, and not with the underground furtiveness proper to rats').[80] A year later, six months after 9/11, with the first Bush administration's 'war on terror' in full swing, Johnson returned to greys' misdeeds. The crime scene was his own London home, where he related his ongoing battle against a particularly incorrigible and crafty individual (Randy/ Ran), who was trying to get at the nuts in his bird feeder. 'Garden terrorist' Ran was 'vigorous, determined and persistent, immensely greedy and quite ruthless in satisfying his appetite'.[81]

Johnson's dislike for Ran was matched by a profound respect for his adversary's intelligence. What he failed to grasp, though, was that American squirrels' hegemon had given 'over here' sentiments a new lease of life following the closure of U.S. bases in Britain after the Soviet Union's demise in 1991. Six years later, touring Britain to promote his tome *A History of the American People*, he encountered minimal anti-Americanism. Johnson was delighted to report that 'overpaid, over-sexed and over here' was no more than a 'historical curiosity', incomprehensible to Britons under the age of forty (he was nearly seventy).[82]

Johnson's demotion of British anti-Americanism to a relic of the past somehow overlooked the squirrels that disrupted his verdant domestic tranquillity. But William Montalbano, the *Los Angeles Times*'s London bureau chief, had figured out what was really going on. Earlier in the year of Johnson's book tour, presenting the British success of the 'classic American East Coast peanut cadger' as a democratic challenge to exclusivity, Montalbano reflected:

> Overfed, oversexed and over here. Half a century ago, such good-natured grousing was aimed at American GIs who came to liberate a continent. In these high tech times it is squirrels

that rouse the English angst. In scarcely more than a century, grey squirrels, imported from America, have toppled the British red squirrels from the perch of treetop privilege they have enjoyed since the ice ages.[83]

At the same time, the grey squirrel stood for more than the pinnacle of American ascendancy. As Johnson's reference to backyard terrorist suggests, as well as serving as a broad conduit to channel pejorative feelings about Americans and American life, the grey has become an all-purpose reference point for an assortment of threats, undesirables, deplorables and all things regrettable. 'The bird and animal kingdoms have their own terrorists in the form of the grey squirrel,' complained a letter-writer to a paper in northeast England. 'Is there anything this "pushy" American asylum seeker may be put off by?'[84]

Earlier (pre-9/11) analogies riffing off pushiness and toughness – and identifying a direct casualty – that were not specifically American included 'more boisterous' football's eclipse of cricket with its early start to the season; capital gains tax ('once alien . . . now as firmly established as . . . fir trees and grey squirrels'); the displacement of the smart, stylish dresser by the 'unwashed hippie look' at Eton; PLO 'assassins' operating from refugee camps in Lebanon ('it is like the grey squirrel; if he gets about he will always assassinate the local inhabitants'); British lager louts running amok in Ibiza (threatening to 'drive out the higher income, more respectable tourists'); Lake District incomers from 'pushier' southern England occupying the 'old stone houses' of the 'indigenous humans of Westmorland'; predatory, American and Australian-owned global media empires that 'drive weaker competitors into extinction'; and GM maize jeopardizing adjacent organic and traditionally grown crops (equated with reds).[85]

Greys also provide an off-the-peg example of the problem of too much. 'A glut of anything', a journalist contended, 'can be a problem – butter, babies, grey squirrels.'[86] Greys' replacement of reds also provided fodder for non-judgmental comparisons regarding evolutionary and technological advancements. After living side-by-side in Europe for around 10,000 years, for instance, Neanderthals were supplanted by *Homo sapiens*, perhaps 'outperformed for food, as the red squirrel is outperformed by the grey'. Another non-censorious analogy was flat screen TV's supersession of the cathode ray tube.[87]

Squirrel Reputations

The regular influx of non-native invasive species since 1876, when the first greys to establish a naturalized population were introduced, has not diluted their reputation as an exemplar of the disadvantages of the global reshuffling of species. Japanese knotweed might top the list of Britain's invasives in monetary cost. In terms of emotional toll, squirrels outrank knotweed. Interviewed at the Sixth Non-Native Species Stakeholder Forum, in 2009, Mark Hill, lead author of English Nature's *Audit of Non-Native Species in England* (2005), disclosed that 'top of my list would be the grey squirrel. These animals are not especially new but they're horrible things.'[88] Familiarity (it was identified in 2015 as 'probably' Britain's 'best-known example of an established invasive species'[89]) and the passage of time does not automatically bring wider acceptance.

As for the red squirrel, though not officially Britain's national animal, it might just as well be. Defined by environmental historian Sandra Swart as 'any creature that over time has come to be politically identified with a nation-state', many countries have enshrined a national animal.[90] The beaver is Canada's national animal; the bald eagle is the symbol of the United States. The springbok is South Africa's national animal, the kangaroo is Australia's, the kiwi bird is New Zealand's and the brown bear Russia's. Though the UK lacks a designated species, within England, the lion and the bulldog are widely recognized national animals. Despite the lion's non-native status, England's royal coat of arms, adopted in the late twelfth century by Richard I ('The Lionheart'), displays three golden lions. Three blue lions feature on the team crest of the England men's and women's football teams; the latter's nickname is 'the Lionesses'. The bulldog is a more informal symbol, but also primarily associated with England. John Bull, a distinctly English figure appearing in the eighteenth century, was often depicted with a bulldog. Stocky, undaunted and far from pretty, the breed embodied Churchill's steadfastness during the Second World War. More recently, far-right groups have appropriated the breed: the magazine *Bulldog* (published between 1978 and 1985) was subtitled the *Paper of the Young National Front*.

The hedgehog – which Potter also immortalized, in *The Tale of Mrs Tiggy-Winkle* (1905) – recently joined the bulldog and the lion as premium national animals. Attracting 42.2 per cent of 9,108 votes cast, the hedgehog topped the 2013 BBC *Wildlife* magazine poll to identify the 'best natural emblem for the British nation'.[91] The red squirrel was on the shortlist of ten, alongside the badger, otter, water vole, robin, swift and ladybird. But the red's even more precarious position than the hedgehog – between 1 and 1.5 million hedgehogs remain – did not guarantee success. Championed by the Red Squirrel Survival Trust, whose patron, Prince Charles, hailed it as 'a national mascot', the red came fourth, with 8.01 per cent of the vote.[92]

Despite coming fourth in 2013, the red's trajectory, reputationally, was upward. When BBC *Wildlife* had invited its readers to select their favourite animal in 2000, the dolphin emphatically won. The red squirrel came seventh.[93] When more than 2,000 readers of BBC *Wildlife* voted for Britain's ten favourite mammals in 2008, the red ranked fifth, behind the fox, badger, hedgehog and – the winner – the otter.[94] By 2013, it had climbed above fox and otter. Regardless of polling, based on the amount of press coverage received, the red squirrel is indisputably the number one British animal and the most heavily nationalized of species. It is difficult to imagine a jokey journalistic description of the hedgehog comparable to this: 'Red squirrels are British. Red squirrels have a stiff upper lip. Red squirrels cry at the national anthem. Red squirrels have a fundamental understanding of decent British values.'[95] The emotional temperature of the conversation over the hedgehog's future is lower, not least because a non-native congener does not make its life harder.

Squirrelscapes and the Squirrel's Place in Britain

Loves and hates, like other powerful emotions, often remain private. As a contribution to the environmental history of emotions – and what has been called the more-than-human historical geography of emotions[96] – *Squirrel Nation* can only explore expressions of emotion articulated in public, committed to the record by voting in a magazine poll or sending a letter to the editor of a newspaper. Animal historian Chris Pearson (with dogs in mind) pinpoints the importance of an individual's 'emotional, physical, and mental encounters'.[97] In this book, I pay close attention to a series of direct individual encounters: a red or grey spied through a window, leaping from branch to branch in a garden; a grey hopping onto the bench in a local park, expecting and duly being given titbits; a red spotted hurtling down a pine tree in a sanctuary; a grey live-trapped in a garden or woodland, then bashed over the head or shot in the back of the head. These are examples of 'emotional practices', 'the habits, rituals, and everyday pastimes'.[98] The emotive language we use to express our feelings is another emotional practice. Accessing how squirrels react emotionally to us, though, is beyond my powers, not just as a historian but as a mere human. To paraphrase the renowned statement about bats: we can never know what it is like for a squirrel to be a squirrel.[99]

These gardens, parks and woods, these places of squirrel encounter, are squirrelscapes. The place of the squirrel in Britain more abstractly – we might call it the 'squirrel question' – has been a contested matter for longer than the grey has called Britain home. How have different elements of British society included and excluded squirrels, spatially and attitudinally? How far have notions of what and what does not belong shifted since 1876? The squirrel's place in British history is fraught with paradox. The red has not always been an icon of Britishness – and of Englishness, in particular – nor an uncontested emblem of exalted native nature. Nor has the grey always been a generally, let alone universally, accepted emblem of undesirable, dislikeable non-nativeness.

In 1909 a journalist reckoned that if you were a resident of Highgate, a village transitioning into a north London suburb, then there was little to choose between the exploits of the 'English brown squirrel' and those of the grey, freshly arrived in town. Looking beyond Highgate, though, and switching from a garden-owner's to a forester's perspective, the writer worried about 'the substitution of a larger, stronger, more destructive animal for the undoubtedly destructive animal we already possess'.[100] Greys might turn out to be worse, but reds were bad enough. Others considered reds worse. Natural history writer Frank Finn conceded that 'our own kind' was more destructive. Reds were also far less sociable. In an urban environment where the squirrels were grey, their greater sociability was an 'unmixed benefit' for children and adults.[101] Between the extremes of the unimpeachable red and the irredeemable grey resided a diversity of feelings and positions.

During a Lords debate on setting up the Forestry Commission in 1919, the Earl of Crawford warned colleagues not to be seduced by 'appeals to sentiment and mercy' for an animal (the red squirrel) that 'smashes up the seed pod in a reckless manner' and 'will dig up acorns ... planted to seed a young oak wood'.[102] Highlighting the views of William Ling Taylor, the Commission's former director-general, the Fauna Preservation Society pointed out in 1951 that trees fit for purpose as telegraph poles were hard to find during the Second World War because reds had girdled many suitable specimens.[103] How an animal once widely regarded as a common pest was converted into a national treasure is another example of a fairly recently invented tradition. There was no pre-grey harmony between Britons and reds. Nor was it a matter of its straightforward victimization by a bigger and badder American cousin.

Attitudes to reds depended on where you lived, how squirrels behaved and whether it was 1826, 1876, 1926, 1976 or 2016. Long before 1876, most rural folk considered the 'common' squirrel – as Britain's native squirrel was usually referred to, pre-grey – a fast-breeding vermin species no different from rats, mice and rabbits. The 'squirrels destroying young pheasants' and the 'squirrels eating young pheasants' in 1890 and 1912, respectively, were red.[104] Gamekeepers, foresters and farmers denounced the red as robber, poacher and tree wrecker, and 'persecuted' it (their term) every which way, in some places to the verge of extinction.

Historians have studied 'great' massacres of cats in 1730s Paris, rats in Hanoi in 1902 and donkeys in Bophuthatswana in 1983.[105] The great squirrel massacre in the Scottish Highlands should be added as a slow violence sort of massacre. In Scotland, now home to 80 per cent of the UK's reds, deforestation and pest control effectively eliminated them by 1800. English and Scandinavian stock revived Scotland's red populations within the context of reforestation. Then, in the early 1900s, they were again slaughtered en masse.[106] Yet in areas where reds were losing out to the 'alien enemy' grey, such as the Home Counties, they were a lamented loss.[107]

In Scotland, among those with timber interests, hostility lingered into the early 1980s. The Duke of Atholl's Perthshire seat, Blair Castle, was surrounded by extensive coniferous woodlands. Though greys were absent from his lands in the early 1960s, he requested statistical documentation of their damage to trees (elsewhere) in 1962. Any subsequent interest he had in squirrels was restricted to reds, which were plentiful at Blair Castle. Noting reds' fondness for seed cones, grabbed before foresters could gather them, in 1981 he supported limited culling when they interfered with propagation of rare firs.[108]

Moreover, those cherished reds whose remaining homes are now sanctuaries – such as the reserve near the Rooneys' residence in 2004 – are not as native as they may seem. Like many of Britain's indigenous squirrels, Formby's may be descended from continental European stock imported to compensate for native populations' widespread demise.[109] Few of today's British reds do not share DNA with their counterparts across the Channel and the North Sea.[110] Afforestation with non-native conifers is good for Britain's reds because lots of their ancestors evolved in Scandinavia's and Germany's coniferous forests, whereas restoration of native broadleaved woodland for biodiversity's sake is good for greys, approximating their North American homeland.

Many of the red's proponents regard it as the truest of true Brit animals. But as the British Isles and Ireland were once physically connected to the European mainland, they share various species – among them not just the squirrel, but the bear, beaver, otter, wild boar, wolf and weasel (though many of these species are long extinct over here).[111] Britain's squirrel is a subspecies of the Eurasian squirrel species (*Sciurus vulgaris*, Linnaeus, 1758). One of the 'bushy-tailed facts' about 'Brit squirrels' cited recently in a UK tabloid is that 'they're not just a British species – they are found across Europe' (the reporter

A Tale of Two Squirrels

Richard Kearton, *Squirrel*, hand-coloured photogravure, negative
c. 1895–1905; print 1905.

also claimed that the species is Denmark's national animal).[112] Moreover, the British Isles represent the westernmost fringes of a range that stretches longitudinally from Ireland to Kamchatka and latitudinally from Lapland to Italy's toe and heel (probably the biggest range of any squirrel species). If Britain's subspecies were to die out, this would be a local extinction at the species level. Populations are 'stable' elsewhere, apart from northern Italy, the only other part of Europe, aside from the island of Ireland, where greys have also been imported. Across much of its range, the Eurasian squirrel remains truly common. In a report of 2006 for the animal protection law firm, Advocates for Animals, wildlife biologist Stephen Harris and others concluded that 'even if red squirrels do eventually disappear from Britain, they could easily be reintroduced again [provided effective and publicly acceptable methods of reducing grey populations are devisable]. Globally, red squirrel populations are not threatened and the conservation effort in Britain is of little importance.'[113] Harris is no stranger to reds, recalling them from Thetford Forest, Norfolk, in the 1960s and '70s. His position, though, can be summed up as 'the red has lost – so accept the grey.'[114]

Other Britons, raising the British Isles subspecies to the importance of a species, feel that preservation efforts are worth the candle. Picture this scenario – 'The Sad Tale of the Late Squirrel Nutkin' – conjured by John Dean for the *Northern Echo*, a newspaper in northeast England, on 5 December 2000. The date is New Year's Eve, 2099. The location is a pub in northeast England. The occasion is the inaugural meeting of the national Lost Animals Society, formed in the wake of the piteous announcement that England's last red has died in a Northumbrian forest.

As early as 1912, a member of the ZSL speculated that interspecies squirrel competition would eventually eliminate reds. And they took a markedly different position from the Advocates for Animals report on the relationship between species and subspecies. Since the native red was a 'race or sub-species' confined to the British Isles, Britain's squirrel race was 'irrecoverable if once lost'.[115] Then, in 1920, 'W.S.B.' reported 'the opinion of many eminent zoologists' that the grey's 'overbearing attitude towards our own brown squirrel . . . may ultimately result in [its] extermination'. And what 'extermination in this country' meant was 'utter extinction, as the race is restricted to the British Isles'.[116]

Since the 1930s, the grey's place in Britain has been hotly debated. No matter that its status (like the rabbit's) has long been that of 'domiciled alien'.[117] It is a simple matter, for some, of reds being in place

A Tale of Two Squirrels

and of greys being out of place. More than a question of geography and of zoology, this is also about right and wrong. Meanwhile, attitudes to greys have become more complex. In some places, inclusion is now largely accepted, not least because it is so difficult to exclude, expel or enclose. In the June 2011 'Wildlife to Work' survey, the grey topped the list of mammal sightings (thirty species in total); almost half (141) of the more than three hundred readers of BBC *Wildlife* magazine who responded spotted them on their 'daily journeys'. The hedgehog trailed way behind the rabbit and fox, with only 25 respondents reporting sightings.[118]

That the grey had replaced the red as the truly common squirrel was underscored – as was the sharp divide between town and country – by the brouhaha over an episode of BBC Radio 4's *Gardeners' Question Time* in June 2010, during which panellists dispensed advice to the programme's 2 million listeners on how to eliminate greys. The panellists received mail denouncing what Andrew Tyler, director of Animal Aid, called their 'hateful and bigoted' suggestions. The chairperson of the show, Eric Robson, commented that 'people in towns are funny creatures and think how sweet grey squirrels are. In fact, they're tree rats to the countryman.'[119] 'Do you think squirrels belong in domestic gardens?' was the question posed in an online poll by *The Guardian* on 17 June 2010, following the *Gardeners' Question Time* row. In total, 62 per cent of respondents answered affirmatively ('Yes, they're cute'); 38 per cent disagreed ('No, they're a menace'). The poll attracted 61 comments, one commentator wondering how many 'yes' votes were from 'townies' whose gardens raised only flowers.[120]

If greys have become quasi-public pets in many gardens and parks, their presence elsewhere remains discordant and illegitimate. In the countryside, especially where or near where reds cling on, some feel they should be excluded. The demand on resources may be high but the homeland must be made more secure for the right kind of squirrel. 'I don't like doing this,' conservation biologist Craig Shuttleworth explained as he was about to bludgeon a skull, 'but they don't belong here.'[121]

Britain's squirrelscapes are moral landscapes. The assignation of species to categories of good and bad, useful and useless, harmless and harmful, attractive and ugly, based on their place within human cultures, is called the sociozoologic scale.[122] Sociozoologic ratings are themselves influenced by phylogenetic scales, the biological ranking of organisms from higher to lower on the scale of evolutionary development. These two scales, the sociozoologic and the phylogenetic, determine moral considerability, our ethical obligations to a species.[123] Considerability pivots on relatability. Self-described 'anti-disciplinary' scholar Eva Hayward posits the key question: how easy – or hard – is it to 'map our bodies onto' theirs?[124] Crabs, insects and invertebrates generally rank low in affection and esteem.[125] And so relatively few oppose the culling, or attempted eradication, of invasive non-native crustaceans such as the Chinese mitten crab. Designation as invasive species, pest or vermin undercuts a species' moral standing – in short, its right to life. Pest or invasive status, environmental humanities scholar Thom van Dooren contends, makes deaths 'unacceptable' from a strictly ethical standpoint seem 'somehow acceptable'.[126]

A blanket definition of all members of a species as 'categorically killable' (to quote science and technology studies scholar Donna Haraway) can be complemented by a blanket definition of all members of a species the killable species adversely impacts as categorically non-killable.[127] Until the Wildlife and Countryside Act (1981) bestowed full protection on red squirrels (extended to Northern Ireland in 1985), culling them was legal in the UK. Culling greys so reds can live is a striking example of 'killing for conservation', a practice for which Swart's term 'beastly nationalism' seems appropriate.[128] Killing for conservation runs into particular problems securing public support when the targeted species is mammalian – and considered cute to boot. Cuteness, often linked to furriness – the hedgehog perhaps the exception that proves the rule – enhances charisma and interferes with expendability.

2

RED BEFORE GREY

WHEN Britain had only one squirrel – and after the new squirrel arrived but before it challenged the first's exclusive status – the native was generally known as the common squirrel; common as in plentiful.[1] There was no need to specify colour. If mentioned, it was usually brown. As the second squirrel spread, and the common squirrel was less commonly found, Britain's first squirrel became red. And yet, as the newly red squirrel receded and the second squirrel became more visible and increasingly perceived as problematic, Britons largely forgot 'our' squirrel's history.

Before 1900, noted a forestry official, 'British foresters were fortunate in that they had only one species of squirrel with which to contend.'[2] Before the grey 'established itself as a British resident' (to quote another forester), Britons chastised their original squirrel for the same misdemeanours and crimes.[3] 'As regards habits, the grey squirrel, like our native brown species, has many offences laid to its charge,' observed naturalist Hugh Boyd Watt in 1918. And in 1930, A. D. Middleton observed that 'there is every reason to believe that the grey squirrel would become quite as formidable a pest as the native species.'[4] Thirty years earlier, the problems greys caused in their homeland, such as nibbling off leaf buds, were the point of reference for Scots who dismissed Britain's common squirrel as an irredeemable, 'sadly destructive agent'.[5] The 'squirrel colonists' in question hailed from England and mainland Europe, misguidedly imported to boost declining stocks.

Nor were the reds that colonized Scotland's conifer plantations in the latter 1800s the only British populations targeted. While English

Above, two common squirrels sitting on a branch of a tree; middle, a house mouse; below, a rat. Lithograph by Jemima Blackburn, from Adam White, *The Instructive Picture Book*, 3rd edn (1859).

landowners were planting the first greys on their estates, 'persecution' of reds was the norm in England too. Jane Ellen Panton, a writer on rural themes, came across a gamekeeper's tree in the borderlands of Kent and Sussex on which were strung up the corpses of jays, stoats and 'even that harmless marauder, the red-coated squirrel'.[6] *The Field* frequently reported that reds, seasonally, were carnivorous. No bird's nest was safe.[7] Children's stories also noted flesh-eating. 'Thoughtless' schoolboys were often accused of stealing eggs and fledglings. Others identified reds as the real culprit.[8] When 'Mr Squirrel' tells a boy that he enjoys eggs, the lad – adopting the moral high ground, having outgrown the pastime of egg collecting – reacts with shocked disbelief: 'very wicked of you'.[9] 'Mr Squirrel' reassures the boy that he never eats *hen's* eggs. And in the semi-factual *Scud: The Life Story of a Squirrel* (1907), when his brother, Rusty, raids a hedge-sparrow nest, Scud issues a disclaimer: 'eating eggs is a thing which is considered by well-bred squirrels to be thoroughly bad form.'[10] Yet Scud readily confesses to earning the enmity of the forester, who 'could not have had any friendly feelings for us, as we bit the tips off his young larches'.[11] When the financial toll greys exacted during the Second World War preoccupied MAF, the Ministry acknowledged that the red itself was once pestiferous (if 'nothing like so bad').[12]

'Our' squirrel's acquisition of its cherished status was neither quick nor smooth. Britain's pre-grey squirrel history may also surprise in other respects. Reds undertook their own colonizing venture into territory north of the Grampians, where, before a release on Lady Lovat's Beaufort estate, on the Beauly Firth, west of Inverness, in 1844, evidence of their presence is inconclusive.[13] Moreover, not long before the first greys were added to the English countryside, reds were still being reintroduced to areas of central and southern Scotland where they had been shot out, stripped of habitat or succumbed to disease. (Reds were also planted south of the firths of Clyde and Forth, where, according to the Scottish biologist John Alexander Harvie-Brown, evidence of indigeneity is sparse.)[14] The elimination of tree cover in England in the early modern era also shoved reds towards the edge (in Ireland, reds were extinct by 1800, perhaps earlier). Introductions from continental Europe to save the species gave today's English reds as complex a genetic make-up as Scotland's.[15]

The Fall and Rise of Scotland's Reds

In the 1890s, Harvie-Brown summarized, in three words, the red's history in Scotland before the first release of greys in circa 1890: 'decline, resuscitation, and increase'.[16] Centuries of deforestation (for fuel and shipbuilding, to oust wolves and to clear land for cultivation and sheep pasture) brought substantial decline, accentuated by shooting to protect new afforestations. Numbers were so severely slashed by 1800 that some believed the species had become extinct in many parts of Scotland.[17] By the early 1800s, the squirrel was an unfamiliar creature in counties such as Dumbartonshire, where, some reckoned, they had been extinct since 1791.[18] As such, 'the apparition of a squirrel [in 1830, probably a reintroduction] made as great a sensation among the men of Dumbartonshire, as if a flight of flying foxes had descended in the Kentish hop gardens.'[19]

The injection of 'fresh red squirrel blood' from southern England (and, to a lesser extent, from Cumbria) at various locations between the early 1770s and late 1860s resuscitated Scottish populations.[20] The first release (c. 1772), the Duchess of Buccleuch's initiative, was at the menagerie in Dalkeith Park, near Edinburgh.[21] Some were sourced from Norway or Sweden, including the 1793 reintroduction at the Duke of Atholl's estates near Dunkeld, Perthshire, where they thrived among the larch plantations.[22] Eight or nine reds that crew members brought back from Riga, a Baltic port, were landed in Arbroath, Forfarshire, in 1817, ending up in the local plantation (Guymel).[23]

Augmented by those that readily escaped into the woods around Dalkeith menagerie, populations burgeoned. New generations spread west, north and south in step with the maturation of fir, larch, spruce and Scots pine planted extensively in the late 1700s. As the plantations on novelist Walter Scott's Abbotsford estate in the Scottish Borders shot up, his gamekeeper and forester, John Swanston, observed that squirrels became thicker on the ground.[24] Biologists characterized the reintroduced squirrel as an intrepid 'wanderer' that even swam across rivers and lochs.[25] By the 1850s, advance parties had crossed the Don into Aberdeenshire. By 1919, it was clear that Scotland's squirrels were strongly influenced, genetically, by stock from England and Scandinavia. The Scottish naturalist James Ritchie called them a 'fresh race'.[26] A Scot apparently unfamiliar with Ritchie's and Harvie-Brown's squirrel histories, but who considered himself knowledgeable, was Lord Burton. During a Lords debate in 1971 on an amendment to permit the poisoning of greys in England and Wales, Burton gave a mini-lecture on Scotland's squirrel history that mixed accuracies with inaccuracies. Burton – who supposedly inspired the Highland laird, Hector MacDonald, in the BBC series *Monarch of the Glen* (2000–2005) – owned a 16,200-hectare (40,000 ac) estate in Inverness-shire. Pine martens and buzzards, he claimed, had killed off the original red population north of the Tay; no mention of persecution. And his observation that the current Highland population was descended from Scandinavian reintroductions was only part-accurate; no mention of stock from southern England.[27]

The English Red's Heritage

Other parts of Britain also experienced a noticeable nineteenth-century upsurge in reds. 'This country did not suffer from squirrels until fifteen or twenty years ago,' remarked George Grey of Milfield, Northumberland, a village near the Scottish border, in 1889. 'I have heard old men say', he continued, 'that they remembered when a squirrel here would have been looked on as a rare animal.'[28] The likeliest source of growth was trans-border migration. Northeast England's population was reinvigorated by way of specimens of southern English and Scandinavian ancestry arriving from Scotland. This complicated genetic demography was replicated further south. Specimens brought over from Scandinavia in the late 1700s and early 1800s reinforced dwindling populations and restocked squirrel-less woodlands.[29]

More than forty subspecies of the Eurasian red squirrel have been described, based on variations in morphology (among them skull size and shape, brain case, palatine bone and teeth) and coat colour. Some subspecies were identified entirely on the criterion of colour, a dubious

Two Eurasian red squirrels beyond the UK (probably Ukraine).

practice as subspecies often undergo two annual colour phases. As such, certain reputed subspecies have turned out to be synonyms for other subspecies. In 1971 the Polish zoologist Jerzy Sidorowicz whittled the number of subspecies down to eighteen, dispensing with, among others, Korea's *Sciurus v. coreae* and *S. v. croaticus* (which was endemic to the lands covered by the former Yugoslavia).[30]

The subspecies Sidorowicz identified are still widely accepted. The list includes *S. v. lis* (Japan's central archipelago); *S. v. rupestris* (Sakhalin island, far eastern Siberia); *S. v. manchuricus* (Hokkaido, Japan's northernmost island); western Siberia's *S. v. argenteus*; *S. v. infuscatus* (Spain and Portugal); *S. v. fuscoater* (Central Europe, the most widely distributed subspecies in Europe); *S. v. balcanicus* (the former Yugoslavia, Bulgaria and Greece); *S. v. italicus* (Italian peninsula, except for northern Italy); and Norway and Sweden's *S. v. vulgaris* – though northern Norway and Sweden, also Finland and Arctic Russia, have their own subspecies, *S. v. varius*.[31] The twelve western European subspecies include the British Isles' and Ireland's *S. v.*

leucourus, which Robert Kerr first named in 1792.[32] 'Since Britain became an island,' explained a UK government-made film in 1959, 'a distinct race of red squirrels' developed.[33] In the mid-1960s, the Mammal Society of the British Isles reiterated that the 'British race ... [was] not found elsewhere.'[34]

However unreliable as a measure of a subspecies, colour, nonetheless, is a striking marker of difference. The seasonal variations that mainland Europe's subspecies undergo are all darker than their British counterpart's colour polymorphism. Kerr, for instance, dubbed the Scandinavian variety *S. v. rufus*.[35] Ear tufts and tail that bleach to pale blonde, even cream, during summer, were the definitive traits of *S. v. leucourus* that Kerr originally identified.[36] Continental reds do not bleach. Until zoologist Oldfield Thomas cleared things up in 1896, though, certain naturalists wondered whether light-tailed squirrels were a particular subset of the subspecies, an irregularity rather than the subspecies' signature feature.[37]

The Britishness of the British Squirrel

Did introductions render *S. v. leucourus* less distinctively British? In the early 1930s, 'continental stock' – probably specimens of *S. v. fuscoater* – were released among the Austrian (black) pines between Birkdale and Formby, in coastal southwest Lancashire (now Sefton, in Merseyside).[38] In the early 1960s, field work by Liverpool Museum noted that the local reds lacked the 'typical pelage of the British race *leucourus* Kerr'. The red squirrel's 'little known moult-sequence', commented C. Simms, 'gave rise to numerous erroneous reports of "Grey Squirrels"'.[39] This release in the 1930s might explain the Formby population's broad colour spectrum, from light to dark red.[40] Reds with bright red coats and tails, recorded near Huddersfield, Yorkshire, were also reckoned to be descendants of *S. v. fuscoater*.[41]

An early twentieth-century infusion from abroad seems to have hauled the red population of Epping Forest, Essex, back from the brink, allowing 'our own beloved' squirrel to survive there beyond the 1930s – surprisingly late for an area close to London.[42] At a time, circa 1910, when disease was felling reds elsewhere in both Essex and other southern counties, local landowner C. E. Green bought a batch of 'Continental Red Squirrels' at London's Leadenhall Market and released them in Epping. By 1916, F. J. Stubbs noted black-coated and dark-tailed squirrels (black being a distinctive continental colour phase of the Eurasian red) near Theydon, west Essex.[43] Their continental origins may have rendered them immune to disease.[44] Stubbs believed they were *S. v. fuscoater*; others thought they were *S. v. vulgaris* (endemic to Norway and Sweden). The skin of a squirrel taken from Epping Forest in January 1936, the British Museum of Natural History confirmed, was indeed *S. v. vulgaris*.[45]

Squirrel scientists have recently confirmed the overlap between British and continental squirrels. After all, British squirrels have evolved separately over the relatively short period of 8,000 years since the last ice age; the earliest fossil record in Britain, from Binnel Point, Isle of Wight, has been radiocarbon-dated back approximately 4,500 years.[46] Moreover, that geographical apartness between the British Isles and mainland Europe was overridden when specimens of *S. v. vulgaris* were imported to Perthshire in 1793. The 'British race' might have existed once, but based on analysis of 214 skulls and skins, squirrel researchers V. P. W. Lowe and A. S. Gardiner reckoned that, by the 1700s, it 'probably was already local if not rare' in England.[47] By the early 1980s, the only populations that still matched the pelage associated with late eighteenth-century descriptions of *S. v. leucourus* were those of western Cumbria, which has no record of introductions.[48] The area's native sons, such as the chairperson of the Forestry Commission, Lord Clark of Windermere, reported in 2004 that 'many scientists' believed Cumbria's squirrels were 'the only indigenous English red squirrels extant'.[49]

Yet the observations of a network of local residents in the 1970s had begun casting doubt on the thoroughly British status of squirrels even in Beatrix Potter and Squirrel Nutkin's patch of the Lake District: few of their tails bleached completely in summer.[50] Around 1980, it seems, Cumbria's reds also started to lose their genetic and morphological distinctiveness (the British subspecies' skulls were previously characterized as significantly smaller). Phylogenetic analysis in 1999 that compared 207 specimens from twelve UK locations, two German sites and a Belgian locale indicated no evolutionary divergence between Britain and the rest of Europe: 'UK samples showed no obviously greater phylogenetic affinity with each other than with mainland European sequences.' This finding thus 'refutes the current classification of the UK red squirrel as a distinct subspecies, *S. v. leucorus*'.[51]

Genetic exchange took place through anthropogenic introductions and 'natural' means. What facilitated 'natural genetic exchange' in northern England and southern Scotland, a 2003 study concluded, was conifer afforestation. Plantings of Sitka spruce, Norway spruce and lodgepole pine expanded Kielder Forest's size substantially during the 1950s and '60s. The enlargement of Kielder (the first trees were planted in the 1920s), located mostly in Northumberland but with about one-fifth spread into Cumbria, knitted together forest fragments. Defragmentation (especially as the Sitka and Norway spruce matured) promoted squirrel connectivity.[52] Separate populations merged after 1980, though the direction of gene flow was predominantly from north (Northumberland and southern Scotland) to west (Cumbria).

Analysis of skulls and pelts collected by museums in Newcastle (Northumberland) and Carlisle (Cumbria) between 1918 and 2000 revealed a 'startling' genetic transformation during the 1980s. Moreover, by the early 2000s, gene introgression affected Cumbrian populations up to 100 kilometres (62 mi.) southwest of Kielder.[53] Further efforts, in 2004, to clarify the genetic role of 'artificial translocation' (deliberate restockings) were based on DNA analysis of circa 180 museum pelts collected between 1861 and 2002, mostly in northern England but also from Italy, the Netherlands, Spain and Sweden. The data indicated that most of Britain's surviving red populations contained continental, particularly Scandinavian ancestry. That this Nordic heritage was most evident since 1960 and strongest in northeast England suggested that squirrels with powerful Scandinavian genes were better fitted to flourish among the non-native conifers dominating Kielder than British reds with no or low Scandinavian ancestry. 'All of the evidence collected to date', the researchers concluded, 'suggests that of the extant northern English populations [in 2001], only the western region [Cumbria] prior to 1980 is likely to have contained *S. v. leucourus.*' And so, 'there seems little doubt that the majority of the extant populations of British

S. vulgaris are of continental European subspecies ancestry.' Even the eight specimens from an extinct population in Dorset that the British Museum acquired in 1894–5, before any recorded introductions to southern England, contained HI, the signature Swedish squirrel haplotype (a unique combination of DNA sequences).[54]

The Ups and Downs of Britain's First Squirrel

Press coverage of the study's DNA investigations highlighted the finding that Cumbrian reds were perhaps the 'only surviving descendants of Britain's original red squirrels'. 'Our research', team leader Peter Lurz observed, 'shows Cumbrian squirrels are very special.'[55] At the time, though, this special population barely amounted to a thousand. As numbers evaporated, it was easy for press, politicians and public to assume life was rosy for reds before the greys arrived. Yet numbers in England and Wales were already falling in 1876 for various reasons: capture for food and as pets; 'persecution'; 'sport' hunting; and bouts of epidemic disease. Maurice Burton, Deputy Keeper of Zoology at the Natural History Museum from 1949 to 1958, pointed out in 1951 that if the grey was indeed impacting negatively on the red, then it was simply carrying on the work humans had already begun.[56]

Food

Drawing on a footnote in the 1837 edition of Gilbert White's *The Natural History and Antiquities of Selborne*, Maurice Burton referred to large sales of common squirrels in London in the 1830s. The source, one William Herbert, was astonished to hear,

> from a man who kept a bird and cage shop in London, that not less than twenty thousand squirrels are annually sold there for the *menus plaisirs* [small or simple pleasures] of cockneys, part of which come from France, but the greater number are brought in by labourers to Newgate and Leadenhall markets, where any morning during the season four or five hundred might be bought.

The informant related further that he sold roughly seven hundred a year, though the domestic population was cyclical, meaning that every seven or so years there were practically none available for sale.[57]

Whether or not daily numbers for sale were as high as four hundred to five hundred, many were destined for the plate. Recognizing the palatability of North American squirrels, a British commentator conceded that the 'common English squirrel is not bad in the nutting season . . . of excellent flavour' (though 'there is little of him').[58] 'To the excellence of the [red] squirrel as an article of food I can myself testify,' observed Harvie-Brown:

> The flesh is pinky-white, like young rabbit, and sweet. The epicure has only once to taste them, and if he possesses a squirrel-haunted wood, he won't grudge the squirrels a fair share of his filberts, hazel-nuts, or cherrie-stones. They can afterwards be cooked in as many ways as a rabbit, and are wholesome and excellent food.[59]

A British gourmand, who likened the native squirrel's flavour to that of a young cat ('well cooked') and noted that the Eurasian squirrel was a popular dish in Sweden and Norway, observed in 1889 that Britain's 'lower classes' occasionally ate it.[60] And the *New York Times*, intrigued by MAF's recent effort, in 1943, to persuade meat-deprived Britons to eat grey squirrel as a wartime food-conservation measure, underlined that squirrel pie was once a 'favorite country dish' across the pond.[61]

Pet

Yet for many Londoners who were not 'cockneys', *menus plaisirs* meant keeping squirrels for amusement. Since the Middle Ages, the squirrel had been a common companion animal across Europe for women from the ruling classes and among the well-heeled.[62] Art historian and museum curator James Rorimer argues that the squirrel in the hazel tree in the sixth of the Unicorn Tapestries suggests the tapestries were created to mark the marriage of Anne of Brittany to Louis XII in 1499. He based his case on a separate tapestry depicting Anne with

Charles VIII of France and a pet squirrel, as well as another tapestry portrait of a squirrel and Anne.[63] Regardless of how often Anne was depicted with squirrels, the squirrel's inclusion in these tapestries aligns closely with elite pet-keeping practice (caged squirrels were frequently encountered in nunneries, too[64]).

The squirrel as pet was accommodated in a hutch or cage, usually indoors, and when taken out and about, might ride on its owner's shoulder. Pet squirrels, observes the historian of medieval pets, Kathleen Walker-Meikle, were routinely 'described and depicted as being fitted with a collar [sometimes belled] and chain, usually finely crafted in silver'. Holbein the Younger's *A Lady with a Squirrel and a Starling* (c. 1526–8) suggests the practice persisted into the sixteenth century. John Lyly's play *Endymion* (1588) contains the following exchange:

> SIR TOPHAS: What is that the gentlewoman carrieth in a chain?
> EPITON: Why, it is squirrel.
> SIR TOPHAS: A squirrel? O gods, what things are made for money![65]

In the early 1600s, Edward Topsell noted that squirrels, if captured young, 'grow exceedingly tame and familiar . . . for they run up to men's shoulders, and they will oftentimes sit upon their hands, [and] creep into their pockets for Nuts'. If they did not nibble woollen garments, they would be 'sweet-sportful beasts' and make 'very pleasant playfellows in a house'. (Moreover, if eaten young, their flesh was sweet and wholesome.)[66]

Not all caged squirrels were pets. In the late 1700s, a squirrel in a revolving wire cage served as signage for one line of business. According to one of Harvie-Brown's informants, 'cages with climbing squirrels and bells to them were formerly the indispensable appendages of the outside of a tinman's shop, and were, in fact, the only *live* sign. One, we believe, still [1826] hangs out on Holborn.'[67] A late nineteenth-century visitor to London Zoo's squirrel house interpreted this kind of sign as testament to the animal's universal popularity, attributable to 'our English' squirrel's status as 'the prettiest of British

In the New Forest, a pair of reds engage in a 'sort of game of hide-and-seek, running among the branches, making doubles and turns': 'Squirrels at Play', in *Chatterbox* (1903).

quadrupeds'.[68] This bizarre tinsmith's practice faded out in the early 1800s. But by the 1830s, notwithstanding the clothes-nibbling which was off-putting to Topsell, keeping squirrels was a democratized pet-keeping practice on both sides of the Atlantic.[69]

Squirrels' popularity as pets expanded with growing affluence.[70] Making squirrel cages was a thriving cottage industry in London.[71] 'What little boy or girl does not love a squirrel?', enquired a children's annual.[72] An adult writing for juvenile owners recommended a roomy cage. One measuring at least 4 feet long and 4 feet high, and 3 feet wide (1.2 × 1.2 × 1 metres), would accommodate three to four.[73] The cage usually came with a revolving wheel. The owner of a squirrel called Filbert described his 'gambols' in a 'round-about' cage as 'diverting as those of a kitten'.[74] The kitten analogy also commended itself to John Ruskin, who appraised the squirrel as 'innocent in all his ways, harmless in his food, playful as a kitten, but without cruelty'.[75]

Nonetheless, Ruskin's feelings were mixed. The frantic, purposeless circular movements of a caged squirrel provided the critic of modernity with a metaphor for his denunciation of pell-mell industrial production and all-consuming mechanization. Referring to the equestran ride through Hyde Park along Rotten Row, Ruskin observed that 'the tendency of the entire national energy is therefore to approximate more and more the state of a squirrel in a cage … Hyde Park, in the season, is the great rotatory form of the vast squirrel-cage; round and round it go the idle company, in their reversed streams.'[76] For Ruskin, all Londoners were effectively squirrels on a treadmill, ever in motion but getting nowhere.[77]

Ruskin's main concern was for squirrel-like people, not squirrels. The pet trade, reckoned an objector to the caging of any creature, was the main threat squirrels faced in the mid-1880s.[78] Many squirrels sold as pets in London – often by hawkers who extracted the upper teeth so they could not bite – were snatched as kits from dreys in Windsor Park.[79] They were 'destroyed out of' Richmond Park, another hunting ground for pets, 'many years' before 1897, because, allegedly, 'the 'Arries of London … break down branches and injure the trees in attempts to capture' the 'charming little graceful fauns of the forest'.[80]

The snatched kits – up to a dozen – were then placed 'under the charge of a cat that has been deprived of her kittens'.[81] (Stories of baby squirrels raised by a devoted feline foster mother were a firm Victorian favourite.) 'No sooner are these little creatures exposed for sale, by boys in smock-frocks ("just come from the woods!") than they find ready purchasers,' William Kidd explained.[82] Kidd was the owner of Scaramouch, the best-known pet squirrel in nineteenth-century London. Kidd recommended squirrels because they were 'all life and vigour' and full of 'fun' that 'knows no bounds', continually inventing 'new tricks'. Dismissive of Ruskinian-style disapproval, Kidd explained that 'these little fellows delight in flying about like lightning, and always mope if not so indulged'. 'That squirrels go mad from frolicking in rotary cages, is an old wife's fable,' he continued.[83]

At the same time, Scaramouch enjoyed plenty of liberty. He rode on Carlo the spaniel's back in pursuit of the cat and hoarded sugar from the breakfast table ('he was constantly offending; I was as constantly forgiving').[84] Kidd's first 'English squirrel', Skuggy, was also uncaged to breakfast with his human family.[85] After feasting on sugar, milk, bread and butter, egg and marmalade, he 'would put his paws, one on each side of our face, and lick our chin all over with his rough tongue'.[86] Accounts of the madcap escapades of uncaged squirrel pets were wildly popular.[87]

By 1900, squirrels were firmly established as pets, commonplace as cats, dogs, parrots, magpies, ravens and guinea pigs. That the young Scud, a fictional red of the Edwardian era, is captured to become a child's pet is telling.[88] Potter kept two reds in London in 1903, though her experience was rocky. As she explained to a former neighbour, because those she bought and kept in a cage 'weren't a pair', they 'fought so frightfully that I had to get rid of the handsomer – and the most savage one', which she called Nutkin.[89] Unfortunately, the looks of the other male, Twinkleberry, who had a more pleasant disposition, were marred: 'half of one ear has been bitten off.' Twinkleberry lived in a house Potter made from a soapbox hung inside a large bird cage. 'I don't think either of them are as nice as my little rabbit,' she told a neighbour's children, 'but I daresay I shall make a book about them before next Christmas.'[90]

Pursuit and Slaughter

How heavy a toll was taken by the removal of kits from the wild for the pet trade is difficult to quantify. Numbers might have been high, but red populations were rebounding in the late 1800s. Non-native conifers extensively planted in the late 1700s and early 1800s matured between 1860 and 1890, furnishing congenial conditions. More squirrels meant more complaints, which helps explain why so many early

The British red squirrel was 'very easily tamed, lives well in captivity, and makes an amusing and interesting pet'. The illustration accompanies these lines from a poem:

'I wonder, would he let me stroke
　　His fur, so soft and brown?
Come, Willie, you and Edward try
　　To lift him gently down.

But off he scampers with a bound,
　　And antics not a few,
And hidden by the leafy screen
　　Is quickly lost to view.'

– D. B., 'The Squirrel', in *Chatterbox* (1885).

twentieth-century homes included stuffed ones. The caption under a display case containing a pair of reds at the Horniman Museum, southeast London, explains that they were 'easily available (killed in large numbers as a forestry pest) and could be afforded by modest households'. Taxidermists reinvented a pest of the woods as an ornament of the home.

Animal Aid activists have recently reminded us that before greys were targeted, reds were hounded just as hard.[91] Britain's reds were not a game species like eastern greys in North America, pursued for sport and/or the pot according to hunting regulations.[92] They were fair game without limits on numbers and huntable year-round, though especially in winter, when unprotected by foliage. Whereas Victorian and Edwardian campaigners for humane treatment targeted the pet trade, a nature writer identified reds' main 'enemies' as 'the country boys'.[93] In *Scud*, 'mischievous boys' throwing stones or brandishing 'squailers' (short sticks with tips weighted by heavy lumps of lead) threaten woodland peace.[94] Other accounts collared a certain kind of Londoner. Squirrel shooting along the Thames was apparently a popular pastime of 'cockneys' in the 1880s.[95] On the urban fringe, William John Stokoe later reported, 'the "sporting instinct" of ignorant people has led them to kill or mutilate the Squirrel with sticks and stones.'[96]

Those primarily responsible for squirrel 'massacre', though, were gamekeepers and foresters.[97] The 'War declared against our race' in *Scud* was waged by the local keeper because Scud's brother snatched a young chicken from a coop, dined on 'pheasant food in the coppice' and committed a third (suspected) 'crime' of snaffling pheasant chicks. Buckinghamshire's beechwoods, *Scud*'s author claims, are practically squirrel-less because of the keeper's mission to make them safe for pheasant chicks.[98] Scud remonstrates that such transgressions are exceedingly rare. Harvie-Brown also wrote this off as an occasional habit of unrepresentative individuals.[99]

Harvie-Brown noted that songbird advocates also assigned a 'bad character' to squirrels, whose 'misdeeds are in every one's mouth'.[100] In Birnam, Perthshire, in 1892, in the garden of Heath Park, Potter observed a squirrel 'in the laburnum under the window'. Fortunately, 'as it was mobbed by about thirty sparrows and some chaffinches . . .

it did not get a spring at them.'[101] By contrast, wildlife artist George Edward Lodge saw no firm evidence of this 'bad habit of robbing small birds of their eggs and young'. Their 'villainy' was largely confined to nibbling top shoots off trees and de-barking firs – a price worth paying for their graceful and beautiful presence.[102]

Foresters and plantation owners largely repudiated Lodge's views. An unorthodox school of thought believed squirrels were good for trees, because they never managed to find many of the acorns and other nuts and seeds cached over a wide territory. For some, the forgetful squirrel's inadvertent role as forest regenerator justified its membership of the woodland community. 'Many of those oaks which are called spontaneous are planted by the squirrel,' a journalist reminded readers. As such, 'this little animal has performed an essential service to the British navy.'[103]

The prevailing view, though, was that killing substantial numbers of squirrels was an indispensable plank of forest management. And so – in language transferred to the grey within decades – a correspondent warned that 'instead of being admired for his extreme grace and agility, he will be doomed as a pestiferous ... nuisance.'[104] In the New Forest, managers meticulously recorded numbers killed. Between 1880 and 1927, the Crown Forest and its successor, the Forestry Commission, shot 21,352 reds.[105] The first bag was 167. Numbers peaked in 1889 at 2,281. After 1893, the annual figure never exceeded 1,000. That the annual kill had plummeted to 35 by 1927 had nothing to do with greys, which did not arrive in the New Forest until the 1930s. The modest tally reflected the return to low levels in the demographic cycle: 'the most violent partisan of the Red Squirrel must acquit the Grey of any blame for that,' noted naturalist Richard Fitter.[106] The New Forest's (red) squirrel population in 1927, when culling ceased, was broadly comparable to numbers in the late 1870s. Before 1880, numbers were simply insufficient to inflict damage – and therefore to require shooting.[107]

Local males slew unrecorded numbers on top of the officially sanctioned kills. 'He is regularly hunted by mobs of lads armed with sticks,' lamented *The Field*. A 'regular feature of village life', formerly, is how the *New York Times* characterized the Sunday morning hunts that provided a filling for pies.[108] In the New Forest's flagship Boxing Day hunt in the early 1900s, men and lads hunted in large groups from dawn to dusk. Armed with catapults, stones and squailers (or scoggers), an outfit could bring in up to eighty or ninety squirrels. Squirrel hunting was a 'very ancient sport among the lower orders', observed the New Forest's deputy surveyor, Gerald Lascelles.[109]

In Scotland, the reintroduction-fuelled resurgence also brought an upsurge in human assault. When one of the northward-migrating 'new-comers' appeared on his lawn at Altyre, near Forres, Morayshire, in the autumn of 1855, Alexander Gordon-Cumming (3rd Baronet of Altyre) was unamused. With 'the instinct of a keen forester', his granddaughter Constance Frederica (a travel writer and painter) recounted, he 'very quickly despatched this poor little precursor of the destructive army which so quickly followed'. Then, having exhausted their winter stores of nuts, they 'invaded', in early spring, the Altyre Woods – the 'rapidly multiplying host of immigrants' feasted on sap-rich young larch.[110]

In midsummer, the sappy phloem (the vascular tissue that transports soluble organic compounds made during photosynthesis to other parts of the plant) is thickest and highest in sugar. To access phloem, squirrels, red and grey, strip the bark of saplings and maturing (ten- to forty-year-old) broadleaves, especially beech, oak and sycamore. Barking (stripping) provides a route for fungal and insect penetration, bringing decay. If barked all the way round (girdled), a tree can perish.[111] If the portion of a tree above the ring-barked zone – often the top quarter – dies, strong winds can snap it off.[112]

Gordon-Cumming and his son, according to Constance Frederica, warred incessantly against the 'beautiful but mischievous little creatures'. Peak shooting season was autumn, when the 'nibbling armies' quit the conifers for the hardwoods to gather acorns, beechmast and other nuts, discarding substantial amounts while searching for the perfect morsel, leaving the forest floor 'strewn with their rejected fragments'. Incentivized by a bounty payment, the annual kill rate on the Altyre estate (which engaged a marksman 'solely devoted to squirrel slaughter') during the 1870s averaged 1,000, dropping to around six hundred in the early 1880s.[113] On the Beaufort

Castle estate, further west, annual bags ranged from 1,055 in 1879 to 728 in 1883.[114]

In the 1840s, there were reportedly no squirrels in Moray, Banff and Nairnshire. By the 1870s, however, they were 'as common as rats and mice'. Scarcely a wood in Nairn or Moray was 'free from their ravages'.[115] At Aboyne Castle, near Huntly, Aberdeenshire, woodlands were 'beyond recovery'.[116] And a correspondent of Harvie-Brown reported from Dalkeith Park in 1878 that 'you will see the nimble little rogue leaping from twig to twig and biting off the buds by the hundred – I may safely say hundreds daily, as the snow below is *thickly strewn* with the *debris*.'[117] On the Stobhall estate, near Perth, reds were culled for sixpence per tail over nearly two decades.[118] In Roxburghshire, in around 1870, the 'great pest' was 'hunted down without mercy'. Between 1862 and 1878 on the Cawdor estates (Nairnshire), where large-scale afforestation began in the late 1700s, 14,123 reds were killed at a yearly rate of approximately 1,100 for a total reward sum of £213 13s. 2d. (3d. per head), a sum that converts into roughly £14,000 (and just under a pound a head) in today's money.[119] Barking was so severe that some estates had 'altogether stopped planting' by the late 1870s.[120] Since the early 1800s, Harvie-Brown observed,

> innumerable complaints have been made, and letters written, and means taken for their destruction. Unanimously, my correspondents condemn the squirrel as one of the most destructive animals which frequent our forests. Scarcely one has a good word to say for it in this respect, and it would, I imagine, be very difficult to undertake, with any chance of success, a case in defence of it. So abundant, indeed, is the proof given of its destructiveness, and, I may add, so patent is the destruction done, to any one accustomed to travel in the woods, that it might hardly he considered worth while to give these proofs in detail.[121]

Scottish authorities on wildlife frowned on the red's exploits in Scotland in much the same way that British biologists more generally regarded the activities of introduced non-native species in Britain (and beyond) as ill-advised and regrettable. They bracketed the reintroduced and colonizing common squirrel in Scotland alongside the European rabbit, fox and weasel in Australia, the 'English' sparrow and European starling in North America, and the North American rainbow trout, brook trout and grey squirrel in Britain.[122] James Ritchie's authoritative study of animal life in Scotland, published in 1920, included the 'common red' next to rooks, choughs, rats, moles, sparrows, rabbits and hares in a section on vermin and pests.

The reference point for concerns about the grey's impact was bad experience with reintroduced reds ('too successful', complained Mark Anderson, professor of forestry at Edinburgh University in the 1950s[123]). Within 25 years, Ritchie noted, the grey squirrel had 'taken possession of a strip of country' in west Scotland 32 kilometres (20 mi.) long and 24 kilometres (15 mi.) wide. This threatened to unleash a 'plague as grievous as that which has rewarded the well-meant efforts of the enthusiasts who set the Common Red Squirrel free in our woods'.[124] And just as the absence of 'natural enemies' over here was identified as a reason for the grey's flourishing, a dearth of predators was also seen to work in the red's favour. In the 1930s, around Loch Alvie in the Cairngorms, where the red still 'reigns supreme', Thomas Coward, a country diarist for the *Manchester Guardian*, attributed its abundance to gamekeepers' elimination of pine martens and wild cats.[125]

An association representing subscription-paying estates with extensive timber interests emerged to spearhead culling. They set up the Highland Squirrel Club in 1903 – the year of *Squirrel Nutkin*'s publication – in response to depredations in eastern Ross-shire, parts of Sutherland and Inverness-shire north of the Caledonian Canal, where, prior to the mid-1840s, there were no squirrels. The Club's bounty began at 3*d.* per tail, rising to 6*d.* in 1940 (equivalent to 98 pence, in 2017's money).[126] The first year's kill rate was 4,727, rising to 7,199 in 1909. Figures for 1935 and 1936 respectively were 1,780 and 2,254. The figure for 1937 was 753 and the downward trend continued (only 331 in 1938) until the Club was wound up in 1946 (the figures for its final three years were 69, 53 and 93 squirrels, respectively). The kill rate over its lifetime was 102,900 and it expended a total of £1,504 across its 43 years of operation.[127] In the 1930s, debaters of the grey squirrel 'problem' armed themselves with these kill rates; the grey's proponents reminded their opponents that the increasingly embattled and fondly regarded red had not always benefited from such affection and sympathy.[128]

British opinion on reds in the late 1800s and early 1900s was split. In the first camp were those ('ladies' were often specified) who thought native squirrels were 'ornaments' of 'our woods and glades'. In the second were those who judged them to be not simply vermin, but 'of all vermin the most verminous'.[129] A prominent defender of the Eurasian red in the 1890s was William James Stillman, an American painter and war correspondent. In retirement in England, he published *Billy and Hans* (1897), a non-fiction account of two squirrels from Germany's Black Forest. Stillman had kept them as pets in Rome while serving as the U.S. consul to Italy during the American Civil War.[130] Stillman deplored the persecution, in England and Black Forest alike, of the 'frolicsome spirit of the woods'. They occasionally ate bark and shoots, he conceded, when other foods were unavailable. The good condition of his own woods at his home of Deepdene, in Surrey, underlined their 'trivial' impact and that the magnitude of persecution was unwarranted (pine beetles were responsible for most of the damage).[131] For Stillman – who kept an American red squirrel as a boyhood pet – there was 'nothing in the English landscape so beautiful as the common squirrel'.[132]

In 1899, *Country Life* compared efforts to exonerate reds of the 'capital offence of gnawing bark off saplings' to historians' attempts to reinvent Henry VIII as a 'good husband'.[133] In a long letter to the editor published in the magazine in 1912, two photographs supplied by the writer underscored reds' injuriousness. The first depicted grotesquely bulbous callouses grown over some of the bark left on the mostly bare trunk of a Lawson's cypress in a north Cornish plantation. Curled strips of bark remained, dangling. The second photograph showed that the upper part of the main stem was dead. A third photograph revealed damage of a different sort inflicted by an 'American grey squirrel'. In the further-flung reaches of Kew Gardens, the new squirrels' teeth had extensively scraped the edges of identification labels, made of lead a twelfth of an inch thick, on oak and beech trees.

The letter writer added, though, that the 'brown ones' also engaged in this teeth-sharpening exercise, if on a smaller scale.[134]

In 1954 Britain's leading squirrel researcher, Monica Shorten of the Bureau of Animal Population (BAP) at the University of Oxford, pondered the ironies of British squirrel history. Around 1900, when its numbers remained 'superabundant', the red, though 'more timid', had been just as criminalized as its grey counterpart was now. Still entirely capable of inflicting the sort of damage now almost exclusively associated with greys, the red was considered pretty innocuous in the 1950s as numbers were comparatively low.[135] In the mid-1950s, only thirteen of 538 forests in England (two), Wales (four) and Scotland (seven) reported damage by reds, with conifers alone affected.[136] And so, Shorten observed, we tended to forget – or forgive – the red's 'misdoings'. Fading numbers, reflected a reviewer of her book, *Squirrels* (1954), meant it 'can retain our instinctive goodwill for a native species'.[137]

Malady

Shorten agreed that there were too many greys. Nonetheless, like Middleton, her predecessor at BAP, she attributed the red's decline since around 1900 principally to deforestation, viral disease and parasitic infections. This was clearly news to some reviewers of her book, one of whom learnt from it that the 'imported grey squirrel is probably *not* responsible for the partial extinction of the charming native red squirrel' – a conclusion, remarked another, that many readers would find 'particularly surprising'.[138]

Wildlife populations are affected by more than predation, human and non-human, and impacts such as habitat loss. Many small mammals – particularly rodents, the order of gnawers to which squirrels belong – experience what Middleton called pronounced 'periodic fluctuations'.[139] A leading cause of cyclical variations among reds was outbreaks of latent disease, perhaps triggered by overpopulation, which in turn caused food shortages that left many squirrels susceptible to sickness.[140] 'Virulent epidemics', Middleton explained, sometimes caused precipitous falls to the verge of extinction within months; other populations underwent a slower reduction over a span of five to twenty years.[141]

The specific ills that afflicted squirrels across Britain between the 1890s and 1920s were variously identified, and usually inconclusively. From Hampshire to Nairnshire, 'lay observers' such as gamekeepers and amateur naturalists often suspected mange, a wasting disease. Enteritis, distemper, myxomatosis and parainfluenza were mentioned too.[142] Post-mortems also indicated coccidiosis, a form of enteritis. Coccidian parasites attack the gut or liver, bringing on acute diarrhoea. Infection is usually lethal; otherwise, multiple reinfections kill.[143]

The wave of epizootic disease hitting reds in the early 1920s was viral. Parapoxvirus kills within two weeks. Though now well established in parts of southern England, the grey's role in transmission is unclear. The Eurasian squirrel is a species especially prone to parapoxvirus, but greys in Hertfordshire and Kent succumbed to it in sizeable numbers in 1923–4. Then, in the autumn, winter and spring of 1930–31, greys were hit from Cheshire to Yorkshire and, further south, from Oxfordshire to Kent.[144] But when MAFF's veterinary scientists detected parapoxvirus in a healthy grey specimen in the 1990s, they confirmed the new squirrel's transmission capacity, which suggested that they might have brought the virus over with them. Still, scientists figured that only animals under stress (from food shortages, for example) were likely to capitulate. This was precisely the 'choosy' red's situation.[145] Moreover, whereas coccidiosis struck greys too, parapoxvirus was 'benign' in them.[146] Yet it was difficult to implicate greys when they were absent from areas, such as Norfolk, where parapoxvirus was rife in the 1920s.[147] Whatever the vector of transmission, these epidemics reversed a period of red expansion that encompassed broadleaf and mixed woodlands as well as the conifers increasingly regarded as the native squirrel's 'natural habitat'.[148]

From Common/Brown Squirrel to Red Squirrel

During this late nineteenth-century growth in numbers, aside from sporadic grumblings about 'persecution' and the casualties of the pet trade, substantial concern about the red's future was lacking. In 1877, the year after a pair of greys was released in Cheshire, the social reformer and National Trust founding mother Octavia Hill spoke on the National Health Society's behalf about London's open space provision. She did not mention squirrels, but a report on her talk observed of Kensington Gardens that 'even now, in spite of the encroachments of modern builders, the squirrel can still be seen skipping and hopping from tree to tree.'[149]

The following year, however, another observer contradicted this soothing reassurance: 'the squirrel, we are afraid, is no longer to be seen bounding from bough to bough in Kensington Gardens.'[150] Beyond London, though, little anxiety was registered during the 1880s. Concern was restricted to expanding urban zones. In the Bristol area, for instance, the residential development that had sprouted on the Somerset side of the Avon Gorge, at Leigh Woods, since the Clifton Suspension Bridge opened in 1864, was identified as invidious to squirrels.[151] Yet away from cities, the status quo seemed intact. 'So plentiful' was the 'common squirrel', 'in nearly all our woods and forests', according to naturalist-parson J. G. Wood and his son, Theodore, that it 'sometimes works considerable mischief'. (Even urbanites, they added, were widely familiar with what they termed *Sciurus Europaeus*, if, regrettably, as 'unfortunate specimens exhibited for sale by ... itinerant hawkers'.)[152] 'To thrill at ... the clatter of a squirrel among the boughs' was a timeless feature of a rural childhood for Sidney Colvin, keeper of prints and drawings at the British Museum, who had grown up on a large Suffolk estate.[153] And the pinewoods and heaths near the Surrey village of Witley were breezily characterized as the squirrel's 'immemorial territory' by painter Alfred Herbert Palmer.[154]

Before and after the grey's arrival – into the early 1900s – those primordial squirrels enjoying their ancestral rights were often characterized as 'common' (even 'ordinary').[155] Common was the translation of the Latin *vulgaris*, denoting the usual, most common form of something, as in *Sciurus vulgaris*. Even when the grey had become *the* squirrel in Regent's Park by 1912, reds seen in other London parks were still dubbed 'the common kind'.[156] Some commentators focused on colour rather than commonness. But they did not describe reds as red. Russet, rusted and reddish-brown were hues sometimes chosen.[157] Brown, however, was the main designation (as in the 'brown flash of a squirrel' in Kew Gardens).[158]

Mostly, though, squirrels were simply squirrels.[159] A poem about London's natural treasures did not need to specify the colour of the squirrels leaping 'gracefully' in Kensington Gardens, Richmond Hill and Kew; nor did a report on the squirrel's delightful antics, whether enjoying their 'forest freedom' or in the home as a pet.[160] Similarly, the squirrels in an account of Windsor Great Park were just squirrels.[161] Likewise, the squirrels in a piece of nature writing set firmly in English gardens, orchards and wood were never directly identified by colour. Only twice was their colour hinted at: they were 'indistinguishable among the red leaves'; an autumnal oak's foliage was 'squirrel-tinted'.[162] The only indication of colour in the squirrel chapter of *Tommy Smith's Animals* (1899) resides in a description of movement: 'a red streak down the trunk of a beech tree ... and up the trunk of a pine tree'.[163]

And when the colour red was specified, the purpose was not necessarily to assert the unique or independent status among squirrels of the English or British variety. A discussion of squirrel provenance prompted by a visit to London Zoo's squirrel house in 1894 highlighted

Nine different specimens of squirrels (family Sciuridae) including the flying squirrel, shown in their arboreal habitat, etching by James Stewart, from Oliver Goldsmith, *A History of the Earth and Animated Nature* (1862).

the 'English red' squirrel's cosmopolitan family connections. Among the house's residents were the tree squirrel, flying squirrel, ground squirrel, Trinidad squirrel, Chinese squirrel, Siberian grey squirrel (a subspecies of *S. vulgaris*), Hudson's Bay squirrel (American red squirrel) and the North American grey squirrel. After an exposition of the wondrous colour variation within *S. vulgaris*, from place to place and from season to season, the correspondent explained that the 'English red' was found throughout Eurasia's northern reaches.[164]

Descriptions of squirrels in England as English squirrels sometimes appeared in popular natural histories, as in 'ordinary English squirrels' and 'common English squirrels'.[165] Scientists also attached nationality to squirrels in Britain, taxonomically and patriotically. In 1899 a Methodist minister-cum-biologist emphasized separateness, referring to a 'race of squirrels now distinctly English'.[166] By contrast, M.A.C. Hinton, deputy curator of mammals at the British Museum of Natural History, and Britain's leading rodent expert, emphasized connectivity: the 'British or, Light-tailed Squirrel' was 'nearly-allied' to the 'continental species'.[167]

Squirrel Nutkin and the Nutkin Effect

Identifications of core character traits supplemented distinctions and commonalities between subspecies and characterizations of colour. In the late 1800s and early 1900s, squirrels in children's stories on both sides of the Atlantic were commonly described as frisky – an adjective typically applied to a horse likely to prove a handful for an inexperienced rider, or a gambolling lamb full of the joys of spring, a boisterous puppy, or a squirrel. *Frisky the Squirrel* (1869), by British author Charlotte Elizabeth Bowen, is about a native red ('and being very frolicsome, "Frisky" became its name'). On clear winter days, Scud reports in *Scud* (1907), 'we were as frisky as ever.' When *Frisky the Squirrel* was published stateside in 1873 (with a subsequent edition in 1889), the squirrel in the illustrations metamorphosed from red to grey (Bowen's text is colour-blind). Likewise, in *The Tale of Frisky Squirrel* (1915), an American story, Frisky ('a lively little chap . . . very bold, too') is an eastern grey.[168]

Cover of Charlotte Elizabeth Bowen, *Frisky the Squirrel* (1889).

A British squirrel also headlines *Frisky Tales: True Nature Stories* (1928). Found on the ground after falling from a drey, Frisky is taken home to a Suffolk village and raised as a beloved family pet (though liberated when she becomes a bit too frisky). Frisky is 'a red one' by default, the author explains, 'for the nasty foreign grey ones, which are killing off our own British born, had not yet reached our part of the world.'[169] Frisky, as this story suggests, was a popular name for a pet squirrel. When Potter was thirty years old, during one of her numerous summer visits to the Lake District since age sixteen, she visited Mrs Frisky, the newly acquired pet of Miss Molly, a girl in Near Sawrey. Apprehended in a cage trap a few days earlier and brought home, Molly had released Mrs Frisky from confinement and she was now up in the loft with a brood of kits.[170]

'Frisky' has since acquired racier undertones, as in Frisky Friday. This trend also applies to the adjective 'saucy', another favourite British and American storybook descriptor of squirrels. The main meanings of 'saucy' have shifted since 1919, when Allan Wright's *The Story of the*

Saucy Squirrel was published in London (and in the United States in 1926). In the 1930s, 'saucy' acquired connotations of the ribald, rude, titillating, lewd and suggestive. Prime examples are innuendo-laden British seaside postcards and the pin-up photos of swimsuited actresses such as Betty Grable that U.S. servicemen carried on tour in the Second World War.

In the nineteenth century, however, in American English and British English, in addition to 'frisky', 'saucy' variously meant jaunty, rakish, sporty, raffish, impertinent, disrespectful, forward, stylish, sprightly, brassy, perky, cheeky and mischievous – as in a British squirrel that was 'busy', 'gay', 'flippant, pert, and full of play'. Britain's red was also 'saucy' in its uninhibited pursuit of ripe filberts (a type of hazelnut).[171] And a seventeen-year-old English girl was 'saucy as a squirrel and as agile as one'.[172] In New England, Boston Common's greys were famously saucy. An American red squirrel elsewhere in Massachusetts, an American poet observed, created a 'saucy din'.[173]

The quintessential saucy – and frisky – squirrel is Potter's Nutkin. *The Tale of Squirrel Nutkin*, which Potter also illustrated, is set in the Lake District. While on holiday with her family at Lingholm, an estate whose grounds extend to the western shores of Derwentwater, in September 1901, she wrote to a child of her former governess that 'there are such numbers of squirrels in the woods here.'[174] In 1905, she bought Hill Top Farm at Near Sawrey, about 40 kilometres (25 mi.) to the south. Derwentwater and one of its islets, St Herbert's Island, supplied the setting for Nutkin's madcap misadventures. Potter's description of Nutkin as a 'little red squirrel' was not a reaction to a larger squirrel's presence; there were no greys at large in London, where she wrote Nutkin, until the 1910s – and certainly none in the Lakes. Squirrels were routinely described as little, regardless of colour.

Potter did not describe Nutkin as saucy or frisky. But Nutkin is frisky to a fault in his dealings with Old Brown, the owl (almost certainly a tawny owl) who lives in a hollow oak and regulates access to the nutty bounty (acorns and hazelnuts) of Owl Island (St Herbert's).[175] Red squirrel rehabilitator Polly Pullar describes Nutkin as 'a little devil overflowing with mischief and playfulness'. That is too charitable.[176] Nutkin is downright impertinent and disrespectful to Old

Brown, fooling and prancing around and singing riddles, ever more manically. Meanwhile, the other squirrels industriously gather winter stores, crossing to Owl Island multiple times on rafts, using sticks as oars and tails as sails (a feat Potter encountered in an American account of squirrels voyaging down a river).

A professor of medicine has argued (whimsically) that Nutkin, who displays 'boundless energy and extreme motor, vocal and cognitive restlessness', behaviour that becomes 'increasingly erratic', may have had Tourette's syndrome. Tourette's is characterized by 'uncontrollable gesticulations and verbal outbursts' and a 'fascination with rhymes, riddles and word play' ('Hum-a-bum! Buzz! Buzz! Hum-a-bum buzz!').[177] Undiagnosed case of Tourette's or not, Nutkin finally leaps crazily at the owl's head – almost his last act. Old Brown's patience finally snaps and the irritating, irreverent and work-shy Nutkin ends up in his waistcoat pocket.

Nutkin puts up a fight, though, and escapes becoming supper, but only by leaving the best part of his bushy tail in the owl's clutches.[178] His associates behave impeccably in deference to the island's owner (poaching nuts is unthinkable). Nutkin, on the other hand, commits the 'nursery arch-sins of rudeness and disobedience'.[179] This behaviour hardly sets him up as a role model for middle-class children, or for the propertyless working classes. The conventional reading of Squirrel Nutkin is as a morality tale: a warning that disobedient children will be punished. Edwardian parents were doubtless uncomfortable with Nutkin's flagrant flouting of authority in the shape of Old Brown. And for them, the take-home lesson was surely that Nutkin's tail loss underscores the consequences of rebellion.

That a subversive squirrel was precisely Potter's intention cannot be ruled out, however. Signs of working-class muscle-flexing in the late 1880s may have inspired her, or been anxiety-inducing.[180] Perhaps Potter wanted children to identify, at least a bit, with the not just naughty but downright rebellious Nutkin, rather than with his dull and obsequious, anything but frisky and saucy fellow squirrels.[181] Nutkin's bad boy status is surely integral to his tale's enduring appeal for children. He certainly embodies the ambiguity of British attitudes to red squirrels at the turn of the century: delightful in appearance, graceful in motion and charmingly frisky but capable of excess. If 'not so mischievously destructive, who would destroy animals of such grace and beauty?', inquired a contributor to *The Field*.[182] Potter's message, that squirrels could be endearingly but maddeningly naughty, would not have surprised readers who kept them as pets.[183] Loveable nuisances, for sure, but there are more sinister hints in Potter's tale. The offering of a fresh-laid hen's egg to Old Brown on the last of the six days of island nut-gathering implies thievery. Besides, the fat mice and moles with which the squirrels previously propitiate the owl suggest that even the best behaved of reds are carnivorous.

The opening line of Shorten's *Squirrels* nods to Nutkin's impact: 'We all met squirrels in our nursery picture books.' *Squirrel Nutkin*'s sales were high and Potter's young readers thrilled.[184] Published in August 1903, by that Christmas Potter's father found a toy squirrel 'already being sold as Nutkin'.[185] Contemporary reviews, however, were not that favourable. Most critics, British and American, ranked

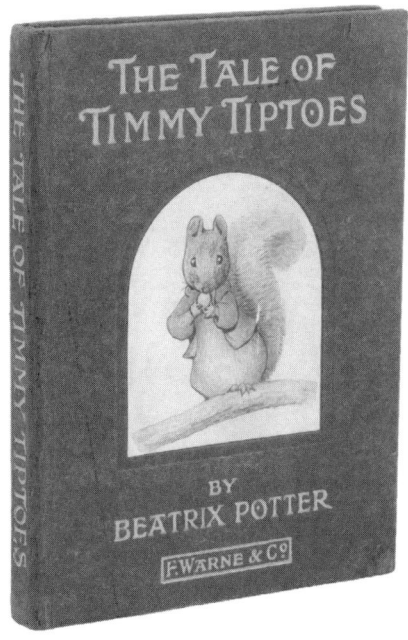

Beatrix Potter's two squirrels: covers of *The Tale of Squirrel Nutkin* (1903) and *The Tale of Timmy Tiptoes* (1911).

it below its immediate predecessor, *Peter Rabbit* (1902).[186] Over time, though, Nutkin's reputation rivalled that of the equally naughty rabbit. Potter, who died in 1943, did not live to see Nutkin become red squirrel preservation's most potent recruitment device. But his transmogrification into a poster child of endangered native nature came at a price. Gone is the transgression, disobedience and impudence that endeared him to generations of children. Many of Nutkin's defining characteristics – audacity, insolence and disrespectfulness – were precisely those that many Britons have associated with greys since the 1920s. The role of squirrel bad boy was assumed by the real-life version of Potter's lesser-known squirrel, which Shorten called her 'other squirrel': Timmy Tiptoes.

3
GREY AND RED

The year 1905, when Potter acquired a Lake District home, was also when London Zoo acquired its first greys from Woburn.[1] Nothing suggests that the grey's swift spread into the city beyond nudged Potter to add it to her cast of characters. Ingratiating herself with her expanding American readership and keeping her publisher happy by maintaining the brisk flow of books appear to be the main reasons for writing *The Tale of Timmy Tiptoes* (1911).[2]

As punishment for stealing their nuts (a false accusation), the other greys incarcerate Timmy in a hollow tree, stuffing him through a woodpecker's hole. He survives by eating the nuts that he and his wife Goody stashed inside, but he overindulges. Too chubby to extricate himself, he must wait until a lusty wind snaps off the treetop. The American identity of the 'little fat comfortable grey squirrel' is not mentioned. Nor are those of Chippy Hackee and Mrs Hackee, a pair of eastern chipmunks (a species of ground squirrel) and an unnamed American black bear. Nonetheless, the setting is familiar – the woods near Potter's Lakeland home at Hill Top. And it is hard to picture storybook squirrels in the United States drinking tea and huddling under an umbrella.

As Potter had never previously cast animals that were not British species, her second squirrel book was a massive departure. She drew Nutkin directly from life: 'Twinkleberry' and 'Nutkin', her London pets, and the Lake District's wild squirrels.[3] She also studied dead reds. A gamekeeper reportedly shot one for her, which she boiled down to a skeleton to ensure maximum accuracy.[4] The location of

Squirrel Nutkin is also unmistakably precise, the high quality of its visual depiction reflecting Potter's deep familiarity with the details of place. But Potter had never visited North America. Her illustrations in *Timmy Tiptoes* were based on a black bear at London Zoo, a pet chipmunk borrowed from a cousin, a taxidermy text by Rowland Ward and a reference work at the Museum of Natural History.[5] That Timmy and Goody wear clothes, unlike Nutkin and his chums, may betray Potter's lack of confidence in her ability to portray them persuasively.[6] The setting is even less convincing. Nonetheless, sales were respectable and the book was favourably reviewed.[7] *Timmy Tiptoes* has not aged well, though. The inferior authenticity of animals drawn at second hand, and the dissonance between creatures and backdrop, makes it one of her least inspired and lowest-rated fables.[8]

More important for Britain's squirrel history is that Timmy Tiptoes is no bad egg. In 2006 the Conservative MP Bill Wiggin, Shadow Minister for Defra – substantiating his knowledge by explaining that 'I have very young children' – demonstrated an impressive command of the details of Potter's squirrels. Whereas Wiggin's colleagues – not to mention journalists – overlooked Nutkin's behavioural flaws, he thought it was 'probably worth mentioning . . . that Squirrel Nutkin . . . is not a particularly good squirrel compared with Timmy Tiptoes, the grey squirrel, who, although greedy, is a better behaved squirrel'.[9]

Greys at Home

Had Potter crossed the North Atlantic as part of her research for *Timmy Tiptoes*, she would have learnt that tensions between reds and greys were not unique to England. If she saw greys at large during spells in London, she did not record an opinion. 'I always tell nice Americans to send other nice Americans along,' Potter wrote to an American woman who was about to visit her.[10] It is a pity that we do not know whether she considered the grey squirrel a 'nice' American animal.

An American living in Britain in the 1990s, writing for young Americans accustomed to considerable squirrel diversity at home, explained that, in the UK, 'we have only two kinds of squirrels.'[11] In addition to the eastern grey (as Americans often called it) and the American red (*Tamiasciurus hudsonicus*; sometimes referred to as chickaree), North America has another three arboreal (tree-dwelling) species: the fox squirrel (*Sciurus niger*), Abert's squirrel (*S. aberti*) and the western grey squirrel (*S. griseus*).[12] The western grey is larger than its eastern counterpart, but restricted to parts of California, Oregon and Washington.[13]

The eastern grey largely inhabits hardwoods below the coniferous forest belt, ranging from southern Canada down to Florida and from the eastern seaboard across to Minnesota, Kansas and Texas. Conifers, east and west of the Mississippi, were typically home to the red, also known as the northern pine squirrel.[14] The eastern grey is second to the red as the most widely distributed North American tree squirrel, but nowhere enjoys exclusive occupation. When British settlers arrived in Virginia, the Carolinas and New England, they encountered grey, red and fox squirrels (not to mention Virginia's 'flying grey'[15]) living alongside each other. The settlers described greys with reference to the squirrel they already knew, as 'much of the Nature of the English, only differing in Colour' and 'almost like our English squirrels'.[16]

The equating of unfamiliar grey with familiar red downplayed the difference in size. The grey is also bigger than the American red, which is unrelated to its Eurasian counterpart. In North America, however, their roles are effectively reversed. Ruminating on the British government's efforts to encourage consumption of greys during the Second World War – to kill the two birds of food scarcity and burgeoning grey numbers with one stone – an American reporter contrasted power relationships between squirrel species in Britain and North America. In the UK, the grey 'overcomes' the red. In the United States, it 'takes a beating' from the red, a 'fiercer' and 'unscrupulous fighter'.[17] Shaking off the arch-territorial red where they overlapped was a prime motivation for the eastern grey's migrations.[18] An American naturalist, basing his observations on activities in the trees around his home in Connecticut, informed British readers that reds, though barely half the size, 'will whip the greys in a running fight every time'.[19] A prominent early twentieth-century American wildlife conservationist, William T. Hornaday, insisted that the 'quarrelsome' red, which 'makes up in courage' 'what it lacks in size' and 'grace', was the only kind of squirrel justifiably hunted and undeserving of protection.[20] Given the red's reputation as expeller of 'the more desired gray squirrel from its territory', an American wildlife biologist in 1929 thought it more useful 'to record the exceptions to this behavior, rather than the rule'.[21] Sciurologist A. D. Middleton drew on these observations to remind Britons that the red was 'invariably the more pugnacious of the two' in its homeland.[22]

Fictional accounts reflected this power dynamic. The protagonist of Ernest Thompson Seton's children's story *Bannertail: The Story of a Gray Squirrel* (1922) is orphaned as a kit when a farm boy kills his mother. Adopted by a farm cat, he lives in a barn. When his home burns down, young Bannertail makes a new life outdoors. Looking for pine cones, he is 'set on' by Redsquirrel (the 'Red-headed one'), 'scold of the woods'. Bannertail retreats, but is pursued into his native hickory territory, where he must defend his nesting tree against Redsquirrel, who conducts his 'pestiferous' business 'with the energy and fury that so often go with red hair'.[23]

The British nature writer William Beach Thomas, in 1945, drew on a recent *New York Times* exposition of the attributes of greys and reds in North America that both united them and set them apart. 'What is more graceful than a full-tailed grey squirrel making its swift way among the upper branches?', asked the American reporter, who celebrated 'each movement' as 'a ripple of sleek agility'. Though 'equally accomplished' in that department, the reporter denounced the red squirrel as 'a scoundrel and a quarrelsome neighbor to all around him . . . a nest-robber and a nuisance in the woodland, a bullying little pirate who makes life miserable even for those of his own kind'. Any British naturalist, Thomas concluded, would see the roles of red and grey in Britain as 'exactly reversed'.[24]

If American reds menaced eastern greys, the latter were regarded as more damaging to human interests. Reds pilfered apples, pears and grains, but greys were perceived as a more incorrigible orchard and crop pest.[25] An Act passed in 1728 required:

Every master of a family [in two Virginia counties] to kill or cause to be kill'd six squirrels or six crows for every tithable person in his family, under the penalty of three pounds of tobacco for every one he is deficient in, to be applied for lessening the County levy. And this to continue for three years, by which time the people are in hopes to free themselves from the injuries they receive from these enemies to their crops.[26]

In 1749, at a reward of 3*d.* (or 6 cents) a tail (accounts vary), the colony of Pennsylvania paid out £8,000 on 640,000 specimens in the colony's agricultural districts.[27] In the late 1700s, a 'squirrel dinner' was considered more effective than a bounty in the Appalachian backcountry. Greys were staple items in the larder of frontier homesteads; reds were too small and considered less flavoursome.[28] But squirrel was not actually served at a squirrel dinner, a meal after a squirrel-killing challenge between two groups of contestants. The aim was to bag more dead squirrels than the opposition during a time-limited period; the losing side bought dinner.[29] These contests sometimes arraigned two hundred men on each side. In Madison County, Kentucky, a two-day hunt in the 1790s yielded 5,589 skins.[30]

Over the past two centuries, though, American farmers and foresters have rarely considered greys a serious pest.[31] But they are labelled a nuisance in areas of North America to which they have been introduced – and there have been at least 29 translocations within North America, from Hawaii to Manitoba.[32] A species becomes non-native, strictly speaking, when transplanted beyond its historic range. West of the Rockies, the eastern grey was as non-native as in the British Isles, Australia (introduced to Melbourne, *c.* 1880s, and, subsequently, to Adelaide, in 1917), South Africa (brought in *c.* 1900), Ireland (transplanted in 1911 from Woburn) and Italy (released in 1948).[33]

In 1909, eastern greys were released in British Columbia at Stanley Park, Vancouver, and later into the province's interior.[34] In the 1920s settlers from Missouri brought the 'perky, pesky' and 'sassy' eastern grey to Miles City, Montana, with subsequent introductions to other eastern Montanan cities.[35] By the late 1990s, exploits familiar to

Britons – raiding bird's nests and bird feeders, eating flowers, digging up bulbs; and, indoors, chewing up insulation and wiring – generated similar grumblings among residents of Vancouver and cities in Montana. Some also reported that it was detrimental to the native pine squirrel.[36] So far, in Montana, the grey remains largely confined to the mature oak and hickory hardwoods along city streets and in the gardens and parks of urban areas. But further west, in Washington state, the grey is spreading beyond urban parks and country clubs, the sites of introduction, into the remaining forested habitat of the less adaptable western grey.[37]

Coming over Here

Midwesterners who migrated to Montana imported the eastern grey to recreate home in a strange environment. Settlers around the colonized, neo-European world shared this need for reassuring familiarity. Animals from home were integral to colonization. The introduction of greys to a colonizing country like Britain requires a different explanation. Curiosity and a penchant for the novel and the exotic were the key drivers.

Before 1876, greys were occasionally found in Britain, however briefly. Benjamin Franklin brought over what was perhaps the first. When living in London in the early 1770s as representative for Pennsylvania, he asked his wife, Deborah, who remained at home, to procure some greys as gifts for a daughter of English friends, the Shipleys, whose Hampshire home (Twyford) became his rural retreat.[38] A squirrel named Mungo was presented to sixteen-year-old Georgiana, who, though 'remarkably fond of all Squirrels', preferred him to the familiar 'European Squirrels' because he was 'more Gentle and Goodhumored and full as lively'.[39]

Deborah Franklin had reservations about how tameable the wild specimens she obtained would be, as well as their longevity as pets.[40] Her fears were quickly realized. Mungo only lasted a few months. The 'fine large grey Squirrel you sent . . . is dead,' Franklin wrote. Having 'got out of' his cage and away from the house, a dog killed him 5 kilometres (3 mi.) from Twyford.[41] Mungo became a minor celebrity after Franklin penned an elegy, apparently intended as the squirrel's tombstone engraving.[42] This elegy included a precursor of what invasion biologists later called the enemy release hypothesis, an explanation for the success of a non-native species in a new environment that emphasizes the absence of the population-limiting pathogens, parasites and predators with which it had co-evolved. Had Mungo remained safely in captivity, he would have avoided such an 'unfortunate End'.

> Alas! poor *Mungo*!
> Happy wert thou, hadst thou known
> Thy own Felicity!
> Remote from the fierce Bald-Eagle,
> Tyrant of thy native Woods,
> Thou hadst nought to fear from his piercing Talons.
> Nor from the murdering Gun
> Of the thoughtless Sportsman.

Among the chief enemies from which Mungo's species had been released by coming to Britain were raptors such as the bald eagle. Franklin had not anticipated a canine adversary. Nonetheless, Mungo's fate provided a teachable moment for those, 'Whether Subjects, Sons, Squirrels or Daughters', who 'blindly wish more Liberty'. The blessings of 'apparent *Restraint*', which offered 'real *protection*', arose again in connection with Mungo's successor.[43]

Deborah despatched a replacement 'Squerel' for the bereft Georgiana.[44] Beebee fared better than his predecessor.[45] Moreover, as Georgiana wrote to Franklin eighteen months before the Declaration of Independence, Beebee was enjoying 'as much liberty as even a North American can desire' (her father, an Anglican bishop, was passionately 'pro-Colonial').[46] As much freedom as it was possible to grant a pet squirrel without incurring the risks of being *at* liberty, and vulnerable to predation, is presumably what she meant. In the event, Georgiana's second 'American squirrel' lived to at least six years old, and though looking worn by that age – blind, too – Beebee remained spirited.[47]

The next greys that entered Britain alive – rather than as a muff, boa tippet or the lining or trimming on a woman's winter pelisse – were zoo-bound. In 1833 Liverpool's Zoological Gardens received a 'present' (the source is unspecified) of an 'American squirrel'.[48] (Museums, meanwhile, acquired dead ones.[49]) Other grey sightings were hard to verify, such as an 1820s report from Wales of a breeding population of squirrels lacking the red's signature ear tufts. 'I have been informed', the correspondent observed, 'that the grey squirrel monopolizes the woods [Cwm Llwynog, Montgomeryshire] and that the common red kind are seldom seen near them, which appears reasonable enough, for the size and strength of the grey animal renders him more than a match for the other.'[50] Other grey presences logged before 1876 were equally sketchy.

In 1929 a landowner in Kent reported that her elderly gardener, a former local wood-reeve, was convinced that there had been a large population of 'British grey squirrel' in Eggarton (Egerton) Wood when he was a boy in the 1860s. In his view, 'they had always been there.'[51] Moreover, a respondent to Middleton's request for information, in 1929, about the grey's presence and impact recalled seeing a pair of the 'American breed' in a well-wooded back garden in 'Burton-crescent', Windsor, circa 1876–7, when he was a boy: 'How they came there I never heard.'[52] These so-called greys in Wales, Kent and Berkshire (the gardener in Kent reckoned they were roughly the same size) were most likely a seasonal colour variant on the native squirrel. In winter, a naturalist explained, grey hairs dominated reds' fur.[53]

'How came this attractive American stranger to find a footing and make a home in our midst?', enquired a British naturalist in 1923.[54] The grey has been so heavily vilified since the 1920s it is worth recalling the positive attributes – among them jaunty charm, endearing boldness and thrilling acrobatics – that underpinned its transplantation and initial popularity. The first 'turning out' resulting in a widely agreed-on presence 'in the wild' – there may have been earlier, but inconsequential introductions – was in 1876, at Henbury Park, near Macclesfield, Cheshire.[55] The instigator was Thomas Unett Brocklehurst, a member of Macclesfield's leading silk manufacturing and banking family, who brought back two pairs of eastern greys from New York.[56]

Of the five (very similar) subspecies of eastern grey, Brocklehurst's almost certainly belonged to *Sciurus carolinensis pennsylvanicus* (*S. c. leucotis*), which ranges from the Alleghenies of Pennsylvania northward to southern New Brunswick and southern Ontario. Noting that New Yorkers dubbed this subspecies the 'Central Park squirrel', an American reporter later speculated that Central Park might have been the source of Brocklehurst's quartet.[57] This is unlikely, as New York City was virtually squirrel-less by the mid-1850s, and the construction of the park itself was not finished until the mid-1870s. Repopulation of the completed park with eastern greys began in 1877–8, with releases from the Central Park Menagerie into the Ramble, a thickly wooded area of nearly 16 hectares (40 ac).[58]

According to R. E. Knowles, a Cheshire-based ornithologist and natural historian whose family owned coal mines and brickworks, Brocklehurst initially displayed his squirrels in a cage on the outside of his house.[59] Within a year, he released them into his grounds.[60] Perhaps the thrill of showing off his exotic pets wore off. Whatever the reason, 1876 is widely considered the most portentous date in Britain's squirrel history. Many popular accounts give the impression that for Britain's reds, the year 1876 was the equivalent of 1492 for the Americas' indigenous peoples. Reacting to a *New York Times* article about Britain's 'squirrel wars', published the day before Columbus Day in 2007, Eric Mendelsohn, a Canadian academic, reflected on the grey's role over there: it 'reminded me too uncomfortably of how we Europeans colonized America'.[61]

The American businessman George Shepherd Page, temporarily resident in Richmond, southwest London, instigated the next verifiable introduction, circa 1889–90. Another American who had moved from New York recorded that he sorely 'missed these [gentle] companions ... when we came to make our home in London'.[62] Page, a keen outdoorsman with an estate in New Jersey, probably shared this hankering for a recognizable creature. Or perhaps he wanted to give Britain a living gift from America: a British newspaper referred to the 'benevolence of a wealthy gentleman from New York'.[63] Later, the British sciurologist Monica Shorten offered a different scenario. Late nineteenth-century concern about the grey's future in its homeland, triggered by deforestation and overhunting, she speculated, may have led the grey's American champions 'to seek a new refuge for it in Britain' as a sort of insurance policy.[64] Whatever Page's motives, he 'turned out' five specimens directly imported from New Jersey in Bushy Park, west London, the second largest of the capital's royal parks.[65]

But transplantations – almost invariably of *S. c. pennsylvanicus* – were mainly British initiatives.[66] Brocklehurst wanted to prettify and enliven his new seat (he had acquired Henbury in 1875[67]) with a novelty animal. This desire was shared by other estate owners who visited North America, whether industrialists or landed gentry. A trip across the Atlantic or a North American contact was not essential, though. Among the panoply of exotics for purchase in 1887 from Britain's leading animal dealership, Cross's Menagerie, in Liverpool, were cockatoos, cardinals, a lion, a jaguar, black panther, Russian bears, lions, tiger cubs, mongooses, juvenile Burmese elephants and, in a single cage, thirty grey squirrels.[68]

Many were headed for zoos and travelling menageries. Others were obtained as pets. The 'American grey squirrel' was a recommended addition to the standard list, possessing the 'charm of novelty' and 'prettier' than 'our own red species'.[69] A natural history writer (who also suggested the mongoose and prairie dog to the jaded pet owner) differed on the question of looks ('scarcely as pretty as our red squirrel') but described it as a 'hardy species, which becomes very tame'.[70] Some pet greys escaped or were set free. Liberty-craving squirrels, and careless or exasperated owners who decided its 'quick, resentful temper' rendered it an unsuitable pet after all, doubtless played their part in the grey's naturalization.

More instrumental, though, were the squirrel activities of a particular British aristocrat.[71] Page's released cohort failed to establish a self-sustaining colony in Bushy Park. But he had greys to spare. When Page returned to the United States, the Duke of Bedford acquired them.[72] Bedford's seat at Woburn (80 kilometres (50 mi.) northwest of central London) is often erroneously designated the grey's 'original English home'.[73] Yet Woburn was indubitably a more consequential introduction site than Henbury. The batch of ten that

Bedford turned out in 1890 provided the progeny for additional releases from 1892 to 1913 at sites in Oxfordshire, Denbighshire, Yorkshire, Buckinghamshire and Cheshire, as well as in Birmingham, London and Ireland, which formed the nuclei for naturalized colonies.[74] Scotland's first colony, on the other hand, originated with a pair from Ontario released at Finnart on Loch Long, Dumbartonshire, circa 1890.[75]

The grey squirrel was just one of many exotics that landowners brought over to their estates. Yet Bedford's enthusiasm for acclimatization – Woburn was Britain's premier playground for experimentation – charmed his visitors more than those living nearby. Few of the duke's large, showier beasts from overseas could cope without assistance, let alone establish themselves in the wild. Even among Woburn's non-native rodents, the grey's success was the exception. Whereas North American chipmunks and flying squirrels flopped, greys flourished to the point of acquiring pest status locally. The proprietor of a local hostelry 'shot seven that were eating his peas'; Bedford was reportedly incandescent.[76]

Whether this lethal response to a raid on vegetables was the first documented adverse reaction to a grey's activities is difficult to say. But it was a harbinger of the complaints that began as a trickle but thickened into a steady stream by 1914. The superintendent of London Zoo, Reginald Innes Pocock, observed in 1906 that they had already 'overrun' Woburn and 'driven away' its resident 'red British squirrel'.[77] At Woburn, they 'increased so rapidly that it became desirable to reduce their numbers, and it is stated that about 1,000 were killed during a recent winter, and 300 in one week'.[78] By 1917 the grey had occupied 'all the neighbouring woods' and 'ousted the smaller native squirrel over a considerable extent of country'.[79] In Cheshire, Henbury's thriving population was also a source of annoyance. Before his death in 1918, William Walter Brocklehurst (who inherited Henbury from his father, Thomas, on his death in 1886) apparently 'gave orders to kill them'.[80]

Fictive greys, however, were firmly confined to North America. *Timmy Tiptoes*'s setting was the norm for children's stories featuring greys.[81] And only once does Scud, the eponymous squirrel in *Scud*, identify with a particular species. In his youth, living as a house pet, he meets a fellow pet, a flying squirrel from Mexico. 'We English squirrels', Scud cogitates, might boast of our capacity to leap considerable distances. But this feat pales beside the jumping prowess of 'these American cousins of ours'.[82] Apropos of which, 'The Grey Terror' is a chapter title in *Scud*. The terror in question, though, is an enormous, grey-furred wild cat. Other non-human foes enumerated are domestic cats, dogs, polecats, rats and weasels. Grey squirrels are conspicuously absent.[83]

Regardless of how storybook squirrels self-identified, there was still little reason to worry about 'our' squirrels in 1918. The British Museum's mammologist Martin Hinton was satisfied that this 'slenderly built rodent' of 'peculiarly graceful and elegant appearance' remained 'common in all wooded localities of Great Britain, except only those in which its numbers are kept in check by persecution'. With the pine marten largely gone and aside from a hard winter, 'man' now stood out as its only real enemy. In *Scud*, too, 'man' is always the 'worst enemy'.[84] And so the status of the 'British Squirrel' in England required 'no special comment' in Hinton's 1918 account, except to note that, across much of northern England, it was reportedly 'of comparatively modern reintroduction'.[85]

At the same time, grey clouds were starting to gather. When Hinton's account was first published in 1910, it carried a single entry for 'Squirrel' under 'Land Mammals – Order Rodentia'. In the 1918 edition, there were two: 'British squirrel' and 'American grey squirrel'. Moreover, Hinton's 'British squirrel' entry included a caveat: should the grey secure 'a good footing here', which looked increasingly likely, it would constitute a 'most formidable rival for our native species'.[86] Hugh Boyd Watt struck the same equivocal tone a few years earlier: in general, the new squirrel's 'distribution seems to be circumscribed as not to justify any apprehensions – at present anyway'.[87]

By 1917 the squirrel from the other side of the Atlantic had begun to dislodge 'man' as the red's primary adversary. Any subsequent references to 'grey terror' were synonymous with American squirrels. Letters to the editor were already captioned: 'The Grey Squirrel: More Evidence for the Prosecution'.[88] This creeping sense of unease was expressed in a shift in nomenclature. The squirrel that faced the 'big foreigners' was increasingly identified by nationality and colour, metamorphosing from the common squirrel into the English red (or brown) squirrel.[89] The grey squirrel put the red into the red squirrel. A 1921 book about wild life in arboreal environments identified the 'native Red Squirrel' as 'very seldom seen', apart from in the 'gloomy surroundings' of a fir forest, which relatively few British animals frequented. It was becoming rare because the 'American Grey Squirrel' was 'far too numerous'.[90] And the qualities that British commentators assigned to the grey – unruliness, brashness, restless ambition and unseemly confidence – were identified as calculated to deliver success in inter-squirrel competition, particularly in proliferating urban environments.

Taking the City by Storm: An American Animal in London

According to geographer Peter Atkins, we classify animals in the city in four ways: as useful (for transportation or food); enjoyable (songbirds in gardens, for instance); desirable (as companion animals) and transgressive (unwanted, out of place, vermin and pests, like rats and pigeons).[91] Squirrels (not mentioned) highlight the limitations of attempts to impose order by fitting into at least three of Atkins's categories: they provide pleasure and are pet material (private and quasi-public) but also cause annoyance and anger. Greys and reds – though especially greys – also slot into the useful category as they are (sometimes) eaten.

In *Billy and Hans* (1897), William James Stillman observed that the grey's acclimatization since the late 1840s in the parks of eastern seaboard cities from Boston to Baltimore represented an unqualified improvement in the quality of city life. Turning his attention to the other side of the Atlantic, he recommended Britain's native red for this role of a trusting and sociable species. Stillman's faith in the red's urban credentials ('nothing prevents this charming sight from being common in the English parks') was puzzling. Elsewhere in *Billy and Hans*, he observed that the 'European squirrel' was 'by nature one of the timidest of animals'.[92] Frank Finn, former editor of *The Zoologist*, was less ambivalent about the red's suitability for city life. In the countryside, where they practised the same destructive habits, there was little to choose between red and grey. By 1917, however, the grey had proven more adaptable (in other words, more flexible behaviourally) than the 'shyer' red, the 'native beast', to 'life in the vicinity of man'.[93]

Eastern greys were transplanted within their native American range, environmental and animal historian Etienne Benson explains, to 'beautify and enliven' the cityscape for the gratification of swelling numbers of city folk.[94] Greys were brought to London for the same reasons and became equally cherished members of the urban community. In turn, London's open spaces, public and private, large and small, offered the same opportunities and support for greys as America's urban frontiers. Between 1890 (the Bushy Park release) and 1916, at least five plantings occurred within 32 kilometres (20 mi.) of St Paul's.[95] Greys quickly became Cockney.

Approximately ninety specimens arrived at London Zoo from Woburn in three batches between 1905 and 1907.[96] These were neither the zoo's first greys nor its first non-native squirrels.[97] But previous greys' existence was precarious. Seeking food on the ground, the 'little grey' risked being pounced on by 'some jungle-cat or jackal'.[98] There is no indication that pre-1905 populations survived. The arrivals from Woburn were initially housed in a moated open-air enclosure (measuring 40 × 50 m/ 44 × 55 yd) on an island in Three Island Pond. When sufficiently tame, they were released into the wider zoological gardens. By the summer of 1906, a few had already escaped into adjacent Regent's Park by leaping onto and over the boundary railings. ZSL then released a dozen to see if the 'common American grey squirrel is capable of establishing himself in Regent's Park'. Pocock identified this as an ideal site to emulate the squirrel success of Central Park, where, according to a British reporter, they were 'as tame as London pigeons'. For there were no reds to harass or fruit trees to damage in Regent's Park.[99]

A visitor in the spring of 1909 estimated that twenty 'North American' greys were at large at the zoo (their tracks in the snow 'a good deal deeper and heavier than an English brown squirrel's' daintier feet).[100] At the zoo and in Regent's Park, they swiftly won favour, particularly among children. (Being diurnal is a precondition of squirrel appeal.) Photographer J. C. Warburg observed that the grey squirrel, unlike the elephant, did not engender a mixture of fear and wonder, children having 'no dread of the furry little squirrels sitting up to beg or chasing one another with agile leaps'. In 1918 children and wounded soldiers alike let them 'walk up their backs and shoulders' and snaffle nuts 'out of their hands'.[101] Within months they spread to park-fronting villas' gardens, where they also received (at least initially) a warm reception from residents welcoming squirrels' urban reappearance.[102]

Regent Park's grey 'colony' was already so 'strong' by the autumn of 1908 that it was 'impossible' to traverse the park 'without making the acquaintance of a bright-eyed little fellow in a fluffy silver and brown suit'.[103] Life for London's new squirrels was probably safer here than in the zoological gardens. At the zoo, George, the great mandrill (West African baboon), apparently developed a taste for the tame 'North American squirrel', which he lured into his cage with food scraps.[104] A reporter encouraged visitors to Regent's Park to 'make friends with the little strangers by feeding them'.[105] Familiar individuals were regularly fed, photographed and named. This matched the practice of parkgoers in American cities, where the living for freshly arrived squirrels was also made easier by visitors offering them peanuts.[106]

Denunciations of begging by squirrels in the United States and complaints that peanut (monkey nut) providers encouraged indolence, pauperization and a culture of dependency failed to deter most park visitors.[107] In London, concerns about the begging squirrel as symbol of moral decay were milder; and monkey nut vendors were not ostracized as they sometimes were in American parks. A reporter noted, though, that the Regent's Park squirrel 'begs for dainties in brazen defiance of the vagrancy laws'. If squirrels flouted laws criminalizing 'wandering abroad and begging', so did parkgoers, who found 'London's new pets' more fun to feed than pigeons, seagulls and ducks, and who could buy peanuts at the Refreshment Pavilion.[108] A photograph of children dispensing peanuts along Regent's Park's vertically bisecting route the Broad Walk, was captioned 'The Squirrel

Girl feeding a squirrel in Central Park, New York, 1904.

Likes to Pick Out the Largest Monkey-Nut in the Bag'. The accompanying piece celebrated this area of the park's conversion into a 'squirrel's playground'.[109]

The grey's benefactors (usually female, according to Warburg) waited on benches with a 'lapful of nuts'. And to help them through the winter months, observed Leslie G. Mainland (author of various popular histories of London Zoo), sympathizers left flowerpots filled with filberts, 'Spanish' nuts (another hazelnut variety, also known as the Barcelona nut) and 'monkey nuts'.[110] In summer, when filberts and 'Spanish' nuts were out of season, sweet treats enlivened the London park squirrel's peanut diet. One of Warburg's photographs – all of them were taken in Kensington Gardens – depicted a male grey crouched on a park chair, nibbling a morsel of biscuit offered by a nursemaid strolling with a pram.[111] Squirrels were apparently also partial to chocolate almonds.[112]

The only parkgoers squirrels were wary of, Warburg reckoned, were stone-throwing boys and dogs (especially Irish terriers). Yet the 'artful dodger' always seemed to elude stones and teeth, hopping from tree to tree via their upper branches, which were 'what the Elevated Railway is to the New Yorker'.[113] The only other threat was the 'London cat'.[114]

Regent's Park's well-fed and relatively secure greys were typically characterized as American or North American. Unusually among Britons, the English author Thomas Anstey Guthrie cast them as 'acclimatized colonials' from Canada. The humorist devoted one of his regular *Punch* columns to an encounter with a very forward 'Canadian' squirrel while sitting on a Broad Walk bench. Nibbling peanuts from Guthrie's bag, the squirrel chatters away:

> Yes, we're pretty numerous here. When we first arrived, all the most desirable residences were occupied by brown squirrels. Mighty condescending they were to us. Said they were superior to colour prejudice, and if we *did* chance to be born grey, we were nevertheless squirrels and brothers. Told us we were welcome to any branches or nuts they'd no use for ... Offered to show us around. But I guess we showed *them* around. There was no *enterprise* about those squirrels ... Too stuck-up and stand-offish. And as for hustling – why, they spent more'n half the winter asleep! It was get on or get out.

The natives had emphatically failed to get on; and so there were no more 'brown' squirrels. As the acclimatized colonial saw it, they only had themselves to blame. His success and that of his fellow squirrels showed that 'even in an old country like this' there was still plenty of opportunity for a creature with get up and go.[115]

By 1913 London's new squirrel was equally well established in the southwest, at Kew, in hundreds of acres of woodland at the Royal Botanic Gardens.[116] Some locals disapproved, as red squirrels were reportedly 'still plentiful' thereabouts.[117] Cecil Brown, who lived close to the main entrance, considered them 'undesirable aliens' (he blamed the Duke of Bedford) that wrecked plants and trees and pilfered nests (acts of vandalism for which humans would be blamed and penalized).[118] Other locals warmed to the new arrivals, mirroring an American trend. Certain Central Park squirrels with big, entertaining personalities became visitor favourites, notably 'Bunny'.[119] During the First World War, American Red Cross nurses stationed in London at St Katharine's Lodge Hospital delighted in feeding the squirrels that visited from adjacent Regent's Park. A photograph caption explained that the 'great pets of the nurses' were 'as domesticated as kittens[,] play about the lawns and even enter the wards and help

Feeding 'Bunny' in Central Park, New York, 1904, photograph by William H. Rau.

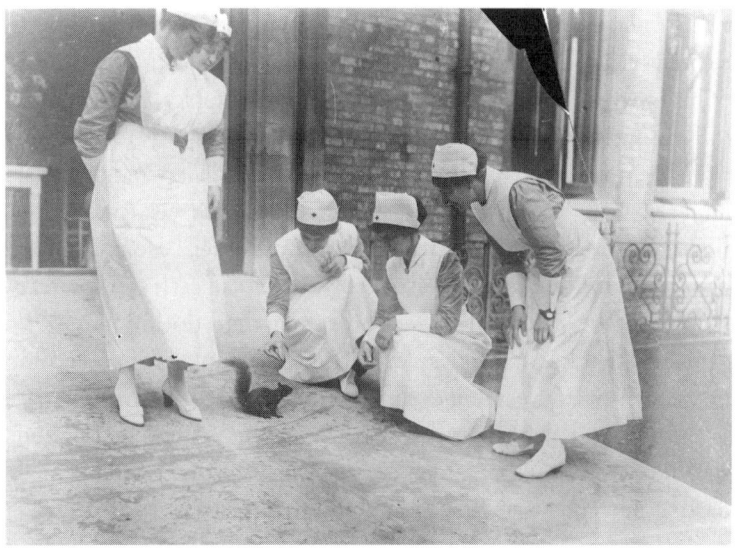

A nurses' pet at St Katharine's Lodge Hospital, near Regent's Park, London, 1918.

themselves to dainties from the patient's trays'. But an American (perhaps the photographer) noted that the squirrels 'don't know what you mean when you say "Here Bunnie"', because 'in England squirrels are called "Cuthbert".'

Neither of London's two best-known squirrels at the time were actually called Cuthbert (in fact, I have yet to come across a squirrel in Britain called Cuthbert). 'Tiny' held court around the Peter Pan statue in Kensington Gardens in around 1918. 'Jimmy' was a 'fat' grey living in a beech near Kew's Great Pagoda. His hybrid status, wild-cum-tame, was the best of both worlds. Jimmy liked nothing better than to squat encircled by youngsters who fed him nuts. A journalist characterized this ritual as a *levee*: a formal reception for invitees that a royal personage holds shortly after rising in the morning, to demonstrate a ruler's accessibility and convey intimacy – a ceremony that reached its apex in Louis XIV's court.[120] After eating his fill, Jimmy climbed onto a child's shoulder and then upward to a woman's, ending up perched atop her hat, tantalized by its waving plume.[121]

Greys were the biggest hit with young Londoners. In Kensington Gardens, recent transplants from Regent's Park rivalled the peacocks and Japanese geese of the Serpentine area as a prime attraction. In 1917 Lovat Lionel Gordon-Stables sent photographs to *Country Life* of his five-year-old son, 'Master' Brian Gordon-Stables, with a squirrel in Kensington Gardens. Brian had trained the squirrels to 'take nuts from his mouth and run all over his body'.[122] Gardeners and flowerbed admirers were less enamoured of the friskier squirrels. As Gordon-Stables observed, they 'unearth the bulbs wholesale, devour carnation buds and hawthorn fruit, and are deadly enemies of cacti and saxifrage'. Adopting the stance of grown-up rather than doting parent, he explained that the stringencies of wartime interrupted regular feeding by visitors, giving freer rein to squirrels' 'natural predatory instincts'.[123]

Other grey behaviours proved irresistible. Visitors to London's squirrel playgrounds, regardless of age, were smitten by the animals' darting movements as they flashed through the trees – a masterclass of daredevilry. Those interested in the grey's mode of travel beyond Regent's Park also noted its lithe adventurousness. Within a couple of weeks of the first release into Regent's Park in 1906, an advance guard had travelled nearly 6.5 kilometres (4 mi.) to the southwest. A resident of South Kensington, noting a small grey's appearance in his and his neighbours' gardens, wondered whether it was one of those that had been 'recently liberated from the Zoo'.[124]

This particular garden owner did not speculate about the squirrel's means of travel. But a sighting in Hendon, Middlesex, 8 kilometres (5 mi.) northwest of Regent's Park, in October 1906, prompted inquiry. In the United States, telephone and electric power lines facilitated migration from sites of introduction in cities and incursion into leafy suburbs. Based on observations of greys' movements in the grounds of the Statehouse (state capitol) in Columbus, Ohio, a central London resident thought they might also have travelled up to Hendon along telegraph wires, trapeze artist-style.[125] Central Park's streetwise population was the reference point, however, for a finding that Regent's Park's greys, the source of 'emigrants', had made it to northern Middlesex by January 1909. More audacious than their 'British red' counterparts, crossing the park's busy perimeter roads did not faze them.[126]

The 'little strangers' were largely sympathetically received. They lacked the native's cutely tufted ears and their heads were more rat-like, but the 'magnificent tail more than compensate[d]' for these deficiencies.[127] William Beach Thomas – who encountered his first grey on the New England estate of the English-born nature writer Ernest Thompson Seton – agreed.[128] Though 'our own squirrel' was 'the prettier, and more engaging', and also less destructive, Thomas found the 'attractive tameness' of the 'imported' squirrel that had expanded from its beachhead in Regent's Park irresistible.[129] And at least one London-based journalist was unfazed at the prospect of the 'dapper, friendly, little grey American squirrels' spreading across the entire country.[130]

At the same time, a 'curious point' was noted: wherever the grey appeared, the red disappeared, though there was no evidence of physical combat. (Direct confrontation was the way in which laypersons approached competition between grey and red at the time.)[131] The future looked bright for greys in London on the eve of the First World War. At large in the zoo and the park, Regent's Park squirrels were 'breeding freely'.[132] Its greys were apparently the target of thieves: 150 were allegedly stolen between April and September 1912. One thief that was caught (armed with a 'walnut with a hook attached') was reputedly fined £5 (over £300 in 2017's money). But the denizens of 'the squirrels' trees' enclosure represented a protected source of replenishment for zoo, park and beyond.[133]

Constance Innes Pocock, wife of the ZSL's superintendent, addressed grey appeal in an anecdotal book about the zoo. After covering familiar, 'captive' parts of the collection such as hippos, elephants, kangaroos and lemurs – and characters such as 'George, the Mandrill' (reputed to snack on greys) – she switched to 'non-captive'. This category consisted exclusively of the 'beautiful little' North American squirrels gifted by ZSL's president, the Duke of Bedford. Captivated by the grey's antics, zoo visitors were keen to establish their own 'colonies', initiatives that ZSL strongly discouraged. Within the zoo, Pocock explained, this 'charmingly innocent-looking' creature excelled even the 'Cockney rat' in terrorizing birds' nests. And were the red's 'more pugnacious American cousin' to be released in 'our country retreats', 'it would be good-bye to the little English red squirrel with which we all are so familiar'.[134]

The Grey Peril

Britain's leading magazines of rural life and outdoor pursuits, *The Field* and *Country Life*, echoed Pocock's anxiety about releases beyond London. In 1912 *The Field* chronicled the sudden decline in red numbers, noting massive reductions in Dorset, Somerset, Surrey and West Norfolk. In Surrey, scarcity was attributed to the unusually harsh 'continuous Arctic' winter and spring of 1895; in West Norfolk, the paucity of acorns and other nuts in the winter of 1910–11 was held responsible.[135] An outbreak of mange was a suggested explanation for Somerset's sudden decreases, and tapeworms were mentioned as a cause of declines elsewhere.[136] For parts of Surrey close to London, a correspondent identified a different agent. A Kingston-upon-Thames resident attributed the absence of reds from his garden (and Richmond Park) to an introduction of the 'North American grey squirrel' locally in the autumn of 1909. *The Field*'s editor interjected that this did not explain the dearth of reds in Dorset, West Norfolk and Somerset – nor in Nottinghamshire. No greys had been released in any of these counties. He also cited lack of evidence for aggrandizement in North America, where, according to Hornaday, the red was in fact the aggressor.[137] The absence of firm evidence continued to trouble the editor a month after the First World War broke out, when *The Field* reported the widespread belief that the 'imported grey' would 'probably drive ... away' 'our common red squirrel'.[138]

The readers of *Country Life* also closely followed the grey's progress. Sir Albert Kaye Rollit, a businessmen and former Conservative MP for a north London constituency, had frequently encountered greys in North America. He was also acquainted with the semi-tame variety of Central Park, which, he noted, provided similar habitat to Regent's Park. Though happy to see greys over here, Rollit cautioned that the 'brown variety' (pretty but damaging to certain conifers) vanished whenever it showed up. Coexistence was unlikely in his view.[139] Given Rollit's reservations, a resident of rural Essex ('M. C.') sought advice on 'turning out' the grey as 'ornament'. Was there risk of injury to trees, gardens or crops? Would it stay put or wander off? One thing M. C. did not have to worry about was the grey 'driving out the common brown squirrel'; in his part of Essex, it was 'practically extinct'.[140]

A respondent to M. C.'s request for advice was William Herbert St Quintin, a gentleman naturalist. He strongly advised against introducing them to a country estate like his own: Scampston Hall, Rillington, North Yorkshire. Though no match for 'our far prettier little red species', their presence in a city park was entirely acceptable to St Quintin, who sat on the council of the Royal Society for the Protection of Birds. But in the kitchen garden, aviary, poultry run, orchard and deciduous woodland, the newcomer was a perfect nuisance. 'As clever and daring as rats' in locating stores of grain and seeds, they were also adept at gnawing through wood to access food.[141] St Quintin was the voice of bitter personal experience. Within two to three months of having rashly 'turned down' thirty greys (from Woburn) at Scampston in 1911, an advance colonist was killed 11 kilometres (7 mi.) away on the far side of a large river, though no bridge led to the new site. Three years of relentless warfare, he hoped, had completely purged his estate.[142]

Though he did not directly name him, another contributor to this thread of letters in *Country Life* had the Duke of Bedford in their sights. The Bedfordshire-based correspondent explained that a landowner 'with a taste for natural history' had let 'this monster of iniquity' loose on his part of 'the English countryside'. Now, a quarter of a century later, in 1914, they were shot and trapped without compunction by the entire community surrounding Woburn, including the chastened introducer himself. And yet, numbers were strong as ever.[143]

Watt undertook to map the grey's presence 'in our country', from Hampstead Heath to Scampston. Shortly after Watt published his preliminary findings, Rollit added one of his addresses, St Ann's Hill, near Chertsey, in Surrey, to Watt's list of places the grey had 'made its home'. For want of solid evidence of its injuriousness (which Watt had now supplied), Rollit, a keen horticulturalist, admitted to complacency while 'the American has been active in its advance'. Rollit confessed that he too had toyed with the idea of releasing greys on his property of Sutherland Grange, in Dedworth, Berkshire. Instead, he had heeded the warning of a 'naturalist friend of high authority' that 'the American' would oust the 'indigenous brown'.[144]

The situation in Kew Gardens focused the attention of naturalists, estate owners, foresters and suburbanites on the contrasting fortunes of grey and red. Captain E. G. Lawton reminisced in 1917 that, when he first moved to Kew, circa 1905, it was customary, on an autumn evening, to see up to a dozen of 'our pretty English red squirrels' frolicking under Kew's beeches. Soon after greys were released, they vanished. A few years later, observing an 'American squirrel' in Richmond Park, Lawton asked a park keeper if he ever saw a red. But, since 'the foreigner' arrived, no 'red English squirrel' had been spotted. For Lawton, the two were incompatible. He favoured confinement to London (where the grey was 'all right in the ... parks'), reserving the countryside for 'our own infinitely prettier and more graceful native'. He did not explain how an exclusion zone would be enforced. Once 'liberated', however, *The Field*'s editor noted that it was virtually impossible to 'keep them within bounds'.[145]

Earlier in 1917, a long, vituperative letter to *The Observer* on 'a coming plague' encapsulated mounting concerns. The correspondent was Sir Frederick Treves, an eminent surgeon credited with saving Edward VII's life in 1902 by operating on his appendix. The king had granted Treves the use of Thatched House Lodge in Richmond Park, where he now lived in retirement. Treves characterized the Regent's Park release as the 'brainless act of some muddler in natural history'. That children liked them was the only redeeming feature of 'these rat-like' animals'. Because of 'the foreigner', the 'British squirrel' was 'extinct' in Richmond Park (where a hundred greys may have been released in 1902[146]). Having found the emaciated (but not injured or diseased) corpses of two British squirrels in his garden in the spring of 1908, Treves surmised that greys had 'robbed' their food caches. Wanton destructiveness, in his view, was their signature trait, wrecking 'twenty times more than they eat' of buds, shoots, peas, gooseberries, pears, apples and plums. Exceedingly fond of strawberries, they reportedly bit a single chunk out of twelve rather than eating one in its entirety.[147]

Treves's purpose was to fire a warning shot, specifically to alert owners of orchards, fruit gardens and plantations in Surrey and Kent.

The debate he stirred up was picked up by the London correspondent of *The Press*, New Zealand's main daily, which noted that 'the famous surgeon, author, and physician to the King' was in the vanguard of criticism, sounding his alarm from a prime site of 'foul work' by the 'American invaders'. Treves set the tone by dubbing Richmond Park's greys 'German-minded invaders' and *The Press* made tongue-in-cheek references to another peril Britain faced on top of U-boats and Zeppelins. By appropriating and spoiling food supplies, greys were exacerbating the shortages that marauding submarines were already inflicting – not to mention killing off reds ('pretty and nice to have around'). Riffing off racial fears that peoples from East Asia, specifically China and Japan, threatened white supremacy and Western nations' interests, the reporter identified a 'grey peril' to compound the 'yellow peril'.[148]

Pointed references to Germany in 1917 joined established if innocuous or strictly descriptive mentions of the grey's American nationality. In addition to Treves's letter, the New Zealand paper cited the string of squirrel letters that *The Observer* subsequently printed. *The Press* quoted from a letter by Arnold Mathew, the Old Roman Catholic Archbishop of Great Britain and Ireland. Mathew observed that the American squirrel was 'more wary and . . . hardier' than its British counterpart and 'longer lived and more prolific'. Dissatisfied, though, with generalized arguments about the advantages of being bigger, tougher and more fecund, he highlighted the mechanism of direct elimination: greys slaughter the 'indigenous British red' and 'devour its young'.[149] Casting around for human analogies involving Britain and the United States, *The Press* reversed the roles of colonizer and colonized. Imagining Treves as instigator of American-style anti-colonial insurrection, the reporter conjured up 'Patriot' Paul Revere's momentous 'midnight' ride to Lexington, Massachusetts, on 18 April 1775. 'Other patriots' were joining the anti-grey cause 'as the minute-men of Lexington and Concord rallied to the clarion call of Paul Revere'.[150]

Treves's letter blamed ZSL. A week later, ZSL counter-attacked. Emphasizing the multiplicity of distribution points, 'it is a mistake, the authorities of the Zoological Garden say, to suppose that all the grey squirrels that are found in various parts of England come from Regent's Park.' Moreover, ZSL did not view greys as a serious wartime threat to national food production. The zoo was also confident that, if, when and where greys did become a problem, economically or for reds, they were controllable, like other 'vermin' such as stoats and weasels, by trapping or shooting.[151]

The Press contrasted Treves's and fellow patriots' indignation with ZSL's more relaxed (and, the reporter thought, better-informed) position. The New Zealander had interviewed Pocock, the ZSL's superintendent, who explained that the Regent's Park squirrels were extremely popular. And within the zoo, Pocock suspected (echoing his wife's views), 'most visitors enjoy feeding the squirrels more than they do seeing our other animals.' Pocock granted that greys had 'driven' out reds where they still existed, but reiterated that greys were manageable. Now that Pocock had allayed fears, the New Zealander felt that 'Britain can breathe more freely, and concentrate on the U-boats!'[152]

The red's champions could not take wider support for granted. A *Country Life* editorial ('A Plea for the Squirrel', 1912) entreated owners of private parks and woodlands not to 'persecute one of the few characteristically wild creatures that belong to England' (control should be left to its 'natural enemy', the pine marten).[153] Some who wrote to *Country Life* insisted that greys were uniquely guilty of eating eggs and fledglings and of harassing adult birds.[154] Others were less convinced. Objecting to a letter writer who gave the red the benefit of the doubt, a Lake District resident had witnessed one suck an entire nest of blackbird eggs in their garden.[155]

In 1917 – when its presence in southwest England was noted, following a release in Rougemont Gardens, Exeter[156] – attitudes to greys remained fluid. Watt (in his survey's final report) observed that the 'sweeping condemnation of the grey squirrel's habits of life is often accompanied by a superfluous sympathy for our native squirrel based on the assumption that the last-named is being displaced, while so far as destructiveness goes the two species are in the same category'. One of Watt's informants (from Bedfordshire) mentioned the 'very popular fallacy' that the grey has 'turned out' the red and noted that 'the keeper's gun' has 'ever warred against the red squirrel'.¹⁵⁷ But for Kew resident Lawton, the cases of Kew Gardens and Richmond Park, where resident reds were never persecuted by the 'keeper's gun', clinched the grey's responsibility for their demise.¹⁵⁸

Like Watt, Oldfield Thomas, curator of mammals at the British Museum of Natural History, canvassed opinion and reported in *The Field*. Thomas had corresponded with North America's leading sciurologist, Edward W. Nelson, chief of the United States Bureau of Biological Survey. Nelson freely conceded that, like all squirrels, the grey caused some damage on its home grounds. But his Biological Survey had received just ten letters of complaint against squirrels in 1916 – mostly regarding the filching of eggs and fledglings – and not one complainant singled out greys. Nelson drew on the insights of his colleague Vernon Bailey, an expert on prairie (Richardson's) ground squirrels. Bailey kept a nesting box for greys in a large oak in his Washington, DC, back garden and reported that none of his trees and plants suffered injuries.¹⁵⁹ Based on the grey squirrel's record of conduct in North America, Oldfield Thomas felt that the 'violent denunciations [in Britain] of the grey species as a present pest, and a possible future uncontrollable plague', were 'exaggerated'. Nelson's advice to British park keepers, conveyed through Thomas, was to lure any nuisance-causing greys into traps. In fact, assuming they were as tame as their counterparts in American parks, a 'landing net' would suffice.¹⁶⁰

The grey's presence *in* Britain did not necessarily mean it was *of* Britain. Archibald Thorburn left it out of his two-volume *British Mammals* (1920), despite his purported coverage of mammals that 'inhabit or visit our islands'.¹⁶¹ Whether he excluded it judgmentally, to highlight the distinction between mere presence and rightful presence, or simply because he thought numbers were insignificant is hard to judge – despite the fact that the walrus and hooded seal, occasional visitors, merited inclusion. Introducing a 1974 reproduction of the original text and drawings, Sir David Attenborough explained Thorburn's omission of a species so numerous half a century later. Back in 1920, greys were 'so rare that [Thorburn] could ignore them'.¹⁶² By contrast, Thorburn considered the 'indigenous' 'Common Squirrel' still 'common'. However, those who wrote letters to magazines and newspapers about the growing grey problem before 1920 would disagree with Thorburn (and Attenborough). A reviewer in 1920 was certainly somewhat surprised by its absence, venturing that Thorburn 'may have to add the grey squirrel to his next edition, for this adaptable American animal' had 'taken full advantage of the hospitality given it here'. 'Fearlessly' following canal banks and roads out of London, it had 'now got far out into the country'.[163]

Pitchforks and Guns

The titbits London's parkgoers liberally dispensed demonstrates that greys benefited from an emotional resource in abundant urban supply: public affection. 'Seeing one of them sitting on its hind legs nibbling at a nut held in its forelegs', a reporter contended in the early 1920s, counted among the capital's 'loveliest sights'.[164] Borrowing a slogan from the Forestry Commission 25 years later, zoologist Maurice Burton presented the divide in British sentiment as a split between town and country: 'In the parks a pet, in the countryside a pest'.[165] But Burton's juxtaposition obscured the complexity of attitudes to greys. Moreover, individual attitudes, regardless of place, could vacillate between approval and disapproval, especially given the complication of a pest often described as pretty.[166] Being pretty did not guarantee the salvation of the 'pretty pests' that raided the fruit trees of a resident of Putney, west London, targeting 'the choicest ones first, Cox's orange apples and the best pears'. In the late summer of 1919, the frustrated Putneyite sent *The Field*'s editor a small apple plucked from the mouth of 'the raider' he had shot, 'caught in the act'.[167]

But when the grey had first featured in parliamentary discussion a decade earlier, it was in terms of its value as quasi-public pet. Julius Bertram, Liberal Party MP for Hitchin, Hertfordshire – a keen fox hunter with a strong interest in dogs – wanted to know why, along the Broad Walk in Regent's Park, only dogs on leads were permitted. Lewis Vernon Harcourt, who, as First Commissioner of Works, was responsible for London's public parks, explained that this restriction

Squirrel illustration from Archibald Thorburn, *British Mammals*, vol. 1 (1920).

followed representations by locals on behalf of the 'semi-tame squirrels which have become so charming a feature'. Harcourt was confident dog owners would tolerate this minor inconvenience 'to contribute to the pleasure of those who are too poor to possess pets of their own, but are able to enjoy those which they now find in the park'.[168]

Urban sympathies frustrated and angered the pretty pest's rural critics. No matter how zealously they pursued control in the countryside, slashing overall numbers was pretty much impossible, they contended, so long as city parks and leafy gardens operated as sanctuaries and reservoirs. However pampered and half-domesticated, these populations fuelled spontaneous recolonization of countryside areas painstakingly cleared. The editor of a London weekly newspaper dismissed the grey as a 'devil entertained unawares' that fed off 'public sentimentality'. To headline a piece about the irrepressibility of a trio of non-native species – the little owl and muskrat in addition to the grey – the editor chose the first two words of a famous saying by Horace: *Naturam expellas furca, tamen usque recurret et mala perrumpet furtim fastidia victrix*: 'Drive Nature Out With a Pitchfork, She'll Come Right Back, Victorious Over Your Ignorant Confident Scorn'. Letting 'the aliens in – human or animal', was 'easy enough'.[169] Driving out the grey would require more than a pitchfork – or homeowners willing to defend their Orange Pippins with shotguns.

4
AMERICAN HUSTLE, *c.* 1919-39

OWNERS, occupiers and managers of private land – like the Putney resident defending his apples – had been shooting and trapping grey squirrels since the releases at Henbury and Woburn established colonies that started to spread too far for some locals' liking. Foresters spearheaded calls for government to add its weight to 'persecution' by 'private people' and what *Country Life* dubbed a 'war conducted by private enterprise'.[1] World war exposed Britain's dependence on imported timber and, in 1917, a forester demanded action, 'legislative if necessary', for the 'extermination' of the 'troublesome' grey ('ten times worse' than its 'pretty little ... brown' counterpart) that jeopardized post-war afforestation.[2] Central government, though, did not yet share these concerns. Responding to a landowner's entreaties in the House of Commons on behalf of the Board of Agriculture and Fisheries, its parliamentary secretary, Sir Richard Winfrey, conveyed that the board (which became the Ministry of Agriculture and Fisheries (MAF) in 1919) was unaware of any 'exceptional damage' to plantations.[3]

During a debate in the Lords on setting up the Forestry Commission (FC), a non-ministerial department under MAF's auspices, a major talking point was the disadvantages and advantages of squirrels. Squirrels – red by default – arose in connection with the Forestry Commission's powers to destroy 'rabbits or vermin' in the proposed government-owned forests and also on private lands beyond. Subsection 4 of Clause 4 ('Prevention of damage by rabbits and vermin') read as follows: 'For the purposes of this section the expression "vermin" includes squirrels.' And so Lord Buckmaster, a champion of squirrels, proposed an amendment to afford some protection to a creature that 'must many times have endeared himself to your Lordships' affections'. His reasons were threefold. First, the squirrel's reputation as woodland pest was undeserved. Second, it was 'a most delightful animal to watch'. Lastly, the squirrel merited exemption from the 'vermin' category because it 'incorporates more than any other animal the very spirit of English woodland life'. As such, the (native) squirrel should be particularly welcome in a 'Government forest', which, being created from scratch, would be devoid of animal life – the antithesis of the national sylvan spirit. Buckmaster's amendment was withdrawn due to lack of support. (So was Viscount Haldane's proposal to replace the reference to 'squirrels' in subsection 4 with 'such animals or birds as the Commissioners may from time to time determine'.) At least there was something on which proponents and opponents of protecting the native squirrel could agree, noted Buckmaster: that the 'alien' grey squirrel merited 'no mercy'.[4]

The Great British Squirrel Debate: Embracement and Rejection

In the early 1920s, pestiferous reds still peppered debates in the Lords and the pages of *The Field*.[5] Growing numbers of Britons, though, were more willing to overlook or downgrade their villainy to misdemeanour. At the same time, affection for greys, however grudging, persisted. Though indubitably an 'expert thief' with a 'voracious appetite', a Londoner with a soft spot for Regent's Park's 'little Grey Brother' acknowledged that 'even a burglar has a human side.'[6] Yet criticism hardened – and in cities too. For every article about a charming encounter with a grey frolicking among autumnal leaves in the 'misted parks' there was one enquiring 'Is the Grey Squirrel a Bad Lot?' (*Daily Mail*, 31 March 1922).[7] Reporting from a 'favourite haunt' of the 'brown' squirrel – a coastal pine grove on the Isle of Wight – William Beach Thomas noted the growing presence, elsewhere in England, of the 'less furtive' yet 'very comely' and 'very engaging' grey cousin (children, particularly, were 'captives to his charm'). Thomas posed a question – are we to welcome the grey? – but quickly realized it was redundant. For 'he makes himself welcome.'⁸ Explanations of grey popularity pivoted on the red hole in town dwellers' lives. Since urbanites, especially Londoners, seldom, if ever, saw a red, they warmly welcomed the grey hole-filler that made itself so welcome.

Nonetheless, the first place in Britain to see concerted action was the capital. Cavalry commander H. Howard-Vyse railed in 1921 against 'misguided' releases near his 121-hectare (300 ac) Stoke Place, Buckinghamshire, of an 'alien' species 'stronger and more hardy than our English squirrel' whose competitive strategy, he alleged, included killing the red's youngsters. As nearby Stoke Common and Burnham Beeches were the grey's 'stronghold', he beseeched their owner, the Corporation of London, to authorize destruction locally.⁹ But the first official fields of engagement where a public authority battled greys were not the leafy shires but Corporation-owned green spaces in London or those in the capital under royal park authority.

'Too many grey squirrels in London', complained Harold John Massingham, a London-based ruralist writer, in 1922. Massingham, who embraced the 'precious' red as 'our native, handsomer and more engaging' squirrel, considered greys even worse for birds than cats. To protect against the bird-eating 'Attila' ('an undesirable alien if ever there was one'), he advocated sanctuaries in parks.¹⁰ Within six months of Massingham's suggestion, an internal report of the Office of Works (a government department overseeing London's royal parks) recommended reducing numbers 'as far as possible' in Kensington Gardens and other royal parks in tandem with designation of bird sanctuaries. The defences were fences too light for cats to scale and with outward-leaning tops, which (presumably) also made them squirrel-proof.¹¹ In the Corporation's green spaces (among them Hampstead Heath), culling had begun in 1917. And in 1924, the Corporation's park committee delivered a decisive verdict of grey guilt based on the latest evidence of depredations on birds in Highgate Wood.¹²

Massingham compared the impact of a transplanted 'stranger' on a country's 'native fauna' to the appearance of an 'unscrupulous trader among a tribe of primitive aborigines'. Yet – articulating the sentiments of what has become known as nature therapy – he recognized the grey's appeal: 'he comes to them out of a green world, which . . .

partially appeases their hunger for nature's self.' Moreover, 'his acute parasitism makes him the friendliest of beasties.'[13] For every Londoner like himself, who resented the country gent 'foisting pure foreigners upon us', there were thousands who considered greys a 'source of joy and delight' (the sentiments if not the language of cultural ecosystem services). And so, Massingham calculated, park keepers 'positively dare not make war upon him so long as he is under the protection of the public'.[14] A Regent's Park official reminded a journalist covering Britain's first 'squirrel war' that they wanted to keep numbers down, not wage all-out war. By the end of 1925, officials evaluated the size of London parks' population as 'tolerable'. Reflecting on the culling campaign, the Office of Works reiterated that 'we never intended a war of extermination.'[15]

Bigger 'squirrel wars' were fought elsewhere against a 'plague' of greys.[16] In 1924, for instance, the National Farmers' Union (NFU) coordinated a 'campaign of slaughter' against the 'little alien of American origin' in Kent's strawberry fields and apple and nut (cob) orchards.[17] Others were more intent on defending the native squirrel, such as Edward Max Nicholson, future director general of the Nature Conservancy (1952–66) and co-founder of the World Wildlife Fund (1961). In 1925, when studying history at the University of Oxford, Nicholson saw a Manichaean, zero-sum confrontation between 'two absolutely irreconcilable species'. First, there was only one squirrel. Now there are two. But Britain could eventually return to just one; and that would be the grey, he feared. Reds, as well as enjoying ancestral rights, were 'handsomer in colour' and 'more graceful'. The odds were stacked in favour of the 'hateful alien', however. Red and grey were 'fighting to the death with such an inequality of stamina that the foreigner almost invariably triumphs'. Bringing the grey over did not augment Britain's faunal richness, as the acclimatizers claimed. This was rank substitution of one squirrel by another in the Home Counties. The only way to prevent further expansion westward, beyond Berkshire, in Nicholson's view, was to establish a cordon sanitaire. The time had come to choose between a native of 'a fairly respectable character' and an incorrigible and irredeemable 'interloper' – to 'rouse ourselves and stamp out the grey enemy which is quickly overcoming him'.[18]

Between Nicholson's call to arms in 1925 and Middleton's landmark survey in 1929–30, newspapers and magazines were replete with first sightings and reports of misdeeds in yet another neck of the woods.[19] The customary time lag between a grey's initial appearance and when it was first noticed made it harder to protect gardens and woods. As 'M. P.' explained in 1930:

> One awakes one morning to see three or four on one's lawn, or the gardener appears and reports that suddenly most of the ripened plums or peaches in the garden repose now on wall tops and only partly bitten, to realise that the horde is there – and the ones seen merely the outer fringe of the main body.

'M. P.' doubted whether, during the early stages of an invasion, more than 1 per cent were seen. At Highclere Park, near Newbury, Berkshire, renowned for its beeches and firs, no greys were recorded as recently as 1927. In 1930, more than 1,800 were shot.[20]

Aside from greys, the English countryside faced various interwar challenges. In the late 1920s, many middle-class Britons – especially residents of southern England – felt wistful for a rural lifeway and landscape, confronted by the disintegration of traditional patterns of landownership and the corrosive forces of modernity. Those who founded, in 1926, and joined the Council for the Preservation of Rural England bemoaned the encroachments of petrol stations, electricity pylons, advertisement hoardings, council houses, bungalows, new roads and ribbon or 'pepperpot' developments. These defacements – the infiltrating infrastructure of modernity – violated the countryside so intrinsic to their sense of identity, personal, class and national.

For someone deeply conservative like Thomas, the grey squirrel was among a panoply of threats contaminating tradition-nurturing home soil. Prominent among uglifying innovations was Great Plains-style 'prairie farming' on 'immense' holdings that he blamed for collapsing rural networks, a population drain and the death of the quasi-paradisaical English village.[21] On one occasion, he lamented the 'vanishing labourer', reflected, in Gloucestershire, in the decline of cheese-making.[22] Within this worldview, the grey squirrel was another major despoiler – not just a symbol of unwanted change but itself a source of unwanted change.

If the countryside was the real England and a common good, the shared inheritance of English people, then the native squirrel also had a rightful claim on an inviolate rural patrimony.[23] Part of this veneration of the English countryside was the trumpeting of its scenic virtues (qualities Thomas wanted to protect by creating national parks). In 1921, Francis Younghusband, the president of the Royal Geographical Society, who had spent much of his adult life abroad in British imperial service but recently returned, celebrated what he called 'home beauty'. 'The Englishman,' he claimed, 'though he loves the Alps and the Himalaya, is touched by nothing so deeply as by a Devonshire lane with its banks of primroses and violets.'[24]

Younghusband's affection for home beauty (after returning, he lived in Kent and Dorset until his death in 1942) fits into a tradition of nature patriotism stretching back to William Wordsworth. In the 1820s, the poet – as synonymous with the Lake District as Beatrix Potter – lambasted non-native timber trees. Profit-seeking incomers from Manchester and London were 'thrusting every other tree out of the way to make room for their favourite, the larch', whose plantations were 'over-running the hill-sides'. Wordsworth also railed against ornamentals such as copper beech (native to southern England and southeast Wales), 'avenues of fir-trees' and lime (linden) that industrialists planted in the grounds of their new lakeside summer villas, thus leaving what made the Lakes special 'disfigured'.[25]

Wordsworth also promoted his native ground's scenic appeal comparatively within Europe. Lakeland's mountains, waters and woods were 'diminutive in comparison', but 'the superiority of the Alps is by no means so great as might hastily be inferred.'[26] Younghusband never compared invidiously but was fond of 'intimate' as a descriptive term for 'home beauty'. Intimacy (small is beautiful) was also a central scenic ingredient for Wordsworth. Though grander in scale, the Alps were often less pretty, subtle and accessible.[27]

Thomas's view that 'size is not essential to grandeur' dovetailed with the celebration of the intimate. The place he showcased was

Cheddar Gorge, Somerset. The juxtaposition of quality and quantity and of refinement against bigness was an established convention in British landscape aesthetics. While clearly dwarfed by 'American canyons', Cheddar 'inspires wonder and admiration in a like degree'.[28] This championing of things English – smaller but more attractive – shaped his views on 'our' native squirrel's superiority. Rural England would be further impoverished if, as already seemed likely by 1925, greys became a 'common object of the countryside throughout England'.[29]

In 1928 the weekly magazine *Country Life*, a leading defender (since 1897) of England's rurality and countryside, hosted a squirrel debate – which ran until 1931. Vernon James Watney, a brewery magnate and MP, sparked it off with a letter about the absence of reds, since 1925, in the ancient woodland of Wychwood on his Oxfordshire estate, Cornbury Park. 'These big foreigners' ('more like very powerful stoats than squirrels') that reached breeding age early had 'exterminated' them. Watney had responded immediately to this existential threat to the 'most valued things of English country life'. The kill rate at Cornbury Park was 329 in 1926 and 742 in 1927. During the first half of 1928, 306 had already been eliminated. He wrote to 'enlist' wider support for a crusade to 'destroy' the grey 'scourge'.[30]

Watney's letter generated a flurry of correspondence. The Secretary of the Royal Society for the Protection of Birds (RSPB), Linda Gardiner, alluded to the defence of the grey's presence in London a few years earlier by Oldfield Thomas, curator of mammals at the British Museum of Natural History, who queried the Office of Works' claim that the grey had evicted the 'red British species' from the capital. Gardiner conceded that their occupancy of London's parks and woodlands might be acceptable; reds were already dwindling in urban settings. Keeping the feisty grey in London was the problem, as it 'cannot be tied within the railings of a public resort'. However, for those who valued songbirds above other forms of native fauna, 'extermination' of the red was 'the least of his crimes'.[31]

Major Portal was also more concerned about birds than 'our red one' – though they were birds of a different feather. As 'no coop of chickens or pheasants is safe', he warned, near a woodland with greys, he advocated shooting 'every one seen on sight'.[32] As a template for effective concerted action in southern England, the Secretary of the Forestry Commission and a member of the University of Oxford's Imperial Forestry Institute cited the Highland Squirrel Club's crackdown on reds.[33]

Not all respondents to Watney agreed with Gardiner, Portal and those representing forestry. Refocusing on London, Sir Peter Chalmers Mitchell, ZSL's secretary, repeated the standard defence: city dwellers enjoyed greys. He discouraged further introductions or internal redistributions, but denied that they evicted, let alone killed reds. 'Our native squirrel' was melting away because of a difference in temperament. Whereas greys were 'fond of human society', the red 'shuns human neighbourhood', averse to urban encroachment ('extension of building'). Mitchell also emphasized that reds' numbers fluctuated, citing Scotland, where recent multiplication presented a 'menace to forests'. While accusing Watney of unvarnished prejudice ('because [he] dislikes grey squirrels, he assigns every kind of mischief to them'), Mitchell recognized his own bias ('I like them and may be a partial witness').[34]

Americans noticed the clash between Mitchell and Gardiner. The *New York Times*, reacting to a Forestry Commission investigation initiated in July 1928, ran the headline 'American Squirrel on Trial for His Life in England: He is Charged with Killing His English Cousins and With Destroying Bird Life' (11 November 1928). The article's opening gambit was the 'sparrow war' that Americans had waged previously against a pestiferous bird from Britain. Now, Britain had an equivalent problem species. The reporter quoted Mitchell's conviction that nothing indicated that the 'pet' of numerous American parks (the reporter's phrase) ousted or slaughtered its 'smaller English brother', but gave equal time to Gardiner's warnings.

By 1930, pressure on central government to commit to the fight was mounting. Moore Hogarth, founder of the College of Pestology (London), demanded prompt action to exterminate a pest threatening to 'over-run' England just like rabbits in Australia.[35] 'Every other week or so', remarked William Beach Thomas, 'some new crime is brought home to the grey squirrel.'[36] 'I had thought that my dossier of the crimes ... was complete,' he commented in 1930.[37] Yet fresh infamies and transgressions kept surfacing.

Allegations that reds were dying out because greys were interbreeding flew around.[38] An incredulous American journalist reported a Hertfordshire resident's theory that the American squirrel was guilty of making love, not warring against reds. And so, the misbehaving greys were not pure American greys but cross-breeds 'combining the vices of both races with the impudence of *Carolinensis*'.[39] The most lurid charge, that greys slaughtered and mutilated reds, proved remarkably tenacious. Humbert Wolfe, a high-ranking civil servant who was also the best-selling British poet of the 1920s, featured this murderousness in a poem, 'The Grey Squirrel', which was included in a children's book about Kensington Gardens:

> Like a small grey
> Coffee-pot
> Sits the squirrel
> He is not
> All he should be
> Kills by dozens
> Trees, eats
> His red-brown cousins[40]

A British correspondent recounted how his son, seeing a grey exit a red's drey at his home near Etchingham, Sussex, climbed the tree to find it contained a 'warm and bleeding' red. A high-ranking lawyer, he offered this eyewitness account as 'very near to proof of murder'.[41] A reviewer of Margaret Holden's *Near Neighbours* (1930), a book about the birds in her garden, shared the King's Counsel's outrage, disagreeing with Holden's 'championship of the alien grey squirrel'. Once an admirer of its 'cheeky courage' (sauciness), the reviewer was now alive to 'unpleasant' habits such as 'evicted baby red squirrels lying crumpled at the foot of a tree'.[42]

The Middleton Report and the National Anti-Grey Squirrel Campaign

For forty years, the anecdotal, often sensationalistic reports of individuals constituted the main source of information about Britain's second squirrel and its relations with the first. A critical mass of credibly empirical, scientist-generated data arrived in 1930–31. In March 1929, in *Country Life* and *The Times*, A. D. Middleton announced his investigation on behalf of a research group at Oxford University led by pioneering British animal ecologist and invasive species biologist Charles Elton.[43] Middleton's premise was that the 'oft-repeated assertion that the grey has driven out the red cannot be so true as is generally supposed, for in many districts where red squirrels are now unaccountably scarce there are not, nor ever have been, any greys'. He sought grassroots, micro-level information on years of red squirrel abundance and scarcity; the date and circumstances of the grey's first appearance; the extent of its increase and current distribution; its impact on native fauna; and disease incidence among squirrels of either sort.[44] A year later, Middleton solicited similar information from readers of the *Daily Mail*, *Manchester Guardian* and *Observer*.[45]

Middleton conceded that the grey 'does make life very unpleasant' for reds. He noted reports of 'actual combat' between them, including an example from Yorkshire akin to the mauling of red kits in Sussex that the K.C.'s son recounted.[46] Middleton also mentioned the rumour that the male grey castrated its red counterpart by biting off its testicles. But he emphasized – just as American biologists pooh-poohed reports that American reds emasculated greys in North America – that the practice was wholly unsubstantiated.[47] As for the allegation of extinction through interbreeding, he had unearthed 'no shred of authentic evidence'. (Greys and reds cannot interbreed; as different species, they are reproductively incompatible.[48]) Reds with grey pelage, Middleton explained, simply exhibited a seasonal stage, as did greys with rust-coloured streaks on heads, backs and tails.[49] At times, reds were not particularly red and greys were not particularly grey.

These were just some of the findings in Middleton's report of 1930, fleshed out in a later book, *The Grey Squirrel* (1931).[50] Despite wielding the customary lay terminology of 'the aliens', 'the foreigner', 'the enemy' and 'a plague', Middleton's objective was an 'unprejudiced view of all that is known concerning the grey squirrel in its new home'.[51] And though he offered a sobering, even exonerating corrective to fantastical beliefs about what it did to the red, Middleton substantiated, meticulously, that the grey inflicted considerable economic injuries: greys ate the buds and shoots of eighteen species, and the bark, nuts and seeds of eleven, as well as raided twelve varieties of cultivated fruits. Middleton compared their wantonness to that of a fox in a henhouse. Greys nibbled on wheat, maize, beans and peas. They took the eggs of fifteen bird species, including pheasant, partridge and domestic fowl, and preyed on the chicks of ten.[52]

While attributing avian predations to 'rogue' individuals, Middleton accepted that the species' general traits meant it was likely to become a throbbing headache for horticulture and forestry by 1950.[53] He also confirmed that it already occupied most of England and Wales; only England's far north remained grey-free. If the rate of multiplication and spread established over the 1910s and '20s was allowed to continue, there was 'every reason to believe that these aliens will quickly become an unmitigated pest of a hitherto unknown character throughout the country'.[54] And so, especially as its natural predators the wild cat and pine marten were virtually gone from England, Middleton urged prohibitive legislation. Moreover, as the grey was set to become 'as formidable a pest' in English woodlands as reds already were among Scotland's conifers, he suggested a 'national squirrel club' and other outfits at regional and county level, akin to the Highland Squirrel Club, which had racked up a body count of 80,000 reds so far.[55]

Middleton's questioning of greys' violent behaviour surprised some commentators.[56] The violence he found was of an entirely different sort – and it was confined to British reactions. The *Manchester Guardian* noted the 'great many violent outbursts by patriotic naturalists, agriculturists, and foresters'.[57] Americans certainly leapt to their squirrel's defence in response. The *New York Times* homed in on Middleton's finding that parasitic disease was the likeliest cause of the 'smaller and less powerful' reds' die-offs hundreds of miles from any grey 'centres'.⁵⁸ A follow-up piece informed Americans that the 'sturdy emigrants' had 'acquired a bad name' in Britain and that 'casual mention of their wicked ways in a London newspaper brings forth eager confirmation from all parts of England', including reports of plants deflowered 'as if in pure malice'.⁵⁹

Nevertheless, Middleton remained optimistic about the native squirrel's long-term future, confident that large-scale post-1919 afforestation, with conifers maturing in the 1950s, would provide some permanent sanctuaries 'from serious competition'.⁶⁰ And once bedded down as an 'established member of the community', he saw no reason why the 'new-comer' would be immune to the fluctuations characterizing red squirrel demography.⁶¹ Also on a brighter note, Middleton confirmed the grey-free status of a region whose natural heritage value was arguably unparalleled among England's localities: the Lake District.

Among the happiest recipients of this good news item about Lakeland was Thomas Alfred Coward, the *Manchester Guardian*'s country diarist, who placed the grey's invasion within a wider cultural and commercial context. 'The complaint that we are becoming Americanised is not without foundation,' he observed. 'We see it in our kinemas and fiction, in our business methods and publicity stunts, in a section of the press, and in our speech, which "Punch" calls "the American slanguage".' And to top it all, 'even our fauna is threatened' by an 'arboreal rat' likely to 'overrun the country' if not tackled immediately.⁶²

Having established such a commanding foothold, greys no longer required the inadvertent assistance of disease to continue their northward trajectory towards the land of Wordsworth, Potter and Nutkin. Some Britons were resigned to this eventuality, the grey's turpitudinous foothold in about a dozen counties suggesting that occupation of the entire UK was just a matter of time.⁶³ In 1925 Edward Max Nicholson had urged 'patriotic' landowners to organize regular shoots and to take responsibility for purging their estates. At the same time, he realized that the actions of private individuals were nowhere near enough. Thinking transnationally and comparatively – 'America curses the "English sparrow" and England the American pond-weed and the rest to an extent of patriotic indignation decidedly unpropitious to cordial relations' – he suggested the League of Nations do something to combat species invasions.[64]

Five years later, Northern Irish journalist and cultural critic Robert Wilson Lynd wound up a discussion of the dangers of intercontinental transplantation of ostensibly harmless flora and fauna with a concrete proposal. Noting that countries around the world were 'becoming as suspicious of a strange root or a strange animal as of a human alien', he foresaw 'more and more restrictions [at national level] of the kind that prevented a poor Irish widow last year from taking her aspidistra [a popular houseplant] with her to Canada'. The British government should bar species introductions such as the grey squirrel.[65]

William Beach Thomas – who had suspected, a decade earlier, that 'now and again, we shall need a Pied Piper to whistle the excessive multitude [of greys] away' – also lamented the absence of restrictions.[66] In 'our well-balanced country', he lectured, in 1931, 'no insect, bird, or mammal ... enjoys a licence free enough to interfere with the liberties of others'. Comparing the dreys high in the elms lining the main approach to London Zoo through Regent's Park to 'wigwams', he felt that ZSL (and the Duke of Bedford) should not have been permitted to allow one of their animals to 'overflow into the depths of the country'.[67]

The press couched demands for action as a declaration of war in the nation's fields, gardens, woodlands and plantations.[68] The government remained unwilling to intervene directly, but MAF gave its blessing to an extra-governmental organization established under the auspices of *The Field*: the National Anti-Grey Squirrel Campaign (NAGSC).[69] For a campaign launch in February 1931, its organizer and honorary secretary, Laurance (L. W.) Swainson, explained via radio broadcast that the organization (which only lasted a few years) was motivated by 'a love of the countryside and by consideration for farmers and allotment holders'. Reporters – who credited the campaign with 're-christening' the grey as the 'American tree rat' – presented this latest clampdown as a revenge of sorts for the grey's 'victory in his private war' against the native squirrel.[70] But the campaign was not anti-grey simply because it was pro-red; in fact, it did not even equate grey expansion with red contraction. 'We have no proof that the Grey kills the Red,' Swainson explained. 'He is quite bad enough on other counts.'[71] NAGSC's main aim was to 'protect our song-birds from the ravishing' of greys.[72] The grey was 'an enemy of all British song birds, and wherever you find [it] the woods are silent'.[73]

Nonetheless, the press presented NAGSC as the 'waging of a systematic warfare' by MAF and its allies against the 'supplanter of our beautiful [and 'comparatively harmless'] native squirrel' and 'menace' of places that grew food and timber.[74] At NFU's urging, MAF convened in May 1931 a conference on grey squirrels for landowning and land-using organizations, including the National Trust, Royal Horticultural Society and RSPB, to devise a national control strategy.[75] Interviewed by the Atlantic edition of the *Daily Mail* (sold on liners between Southampton and New York City), Swainson explained that the Minister of Agriculture (Christopher Addison) had 'realised the enormous amount of harm that the country is suffering . . . and the importance of gaining the help of everybody to stamp out the pest'. Wildly exaggerating the population of Britain's supreme pest, Swainson cited 30 million, with densities as high, in places such as Burnham Beeches, as six per acre.[76] Without prompt and concerted national action, Britain risked being completely 'overrun'.[77]

Realizing that greys were equally at home in town and country, attendees at MAF's squirrel summit urged park authorities to join them in implementing controls. After the Office of Works revived efforts to regulate numbers in London's royal parks in 1930, London County Council's Parks and Open Spaces Committee resumed parallel measures in the autumn of 1931 on sites under its jurisdiction, prompting a journalist to refer to 'our doomed alien friend of London parks'.[78] Grey control in London was hardly a fringe activity. Reflecting on his spell as First Commissioner of Works and Public Buildings in the Office of Works (1929–31), Labour MP George Lansbury recalled that during parliamentary discussion of supplementary estimates (additional departmental funds) members of the opposition expected him to know 'all about the bird sanctuary' in Hyde Park 'and whether I took proper precautions to deal with the grey squirrel'.[79] In 1930–31, each London park employed a bird warden who reported annually on the squirrel 'menace'.[80] In its 1931 report, the Committee on Bird Sanctuaries in the Royal Parks held the grey accountable, among other things, for woodpecker decline in Richmond Park.[81] When Labour MP Peter Freeman sought an exemption from destruction for greys in public parks, the minister of agriculture, Addison, reminded him of the recent campaign of elimination in Kew Gardens (the only green space in London over which he, as agriculture minister, had jurisdiction).[82] Between 1932 and 1937, 2,100 greys were shot in Richmond Park alone.[83] And in London County Council's parks between 1931 and 1939, the annual kill averaged between three hundred and four hundred.[84]

Given the status of the 'English gray' as prime attraction, London County Council's culling policy puzzled the *New York Times*.[85] Park patrons also found it hard to shift perceptions from 'pet' to 'plague'.[86] Leonard Robert Brightwell, an illustrator and author who specialized in writing stories about animals, found it difficult to resist 'an animal that can pacify a crying baby by giving a trapeze act on the pram handle, or cheer a down-and-out by simply perching on his knee'.[87] This lively likeability – this provision of what we would now call nature therapy and non-human emotional support – was the formidable obstacle authorities faced when trying to win public support for 'control'. The 'American intruder' generated 'antipathy' everywhere, noted Richard Fitter, expert on London's wildlife, except in London's larger parks.[88]

A letter writer insisted, however, that London County Council and the 'Crown authorities' must tackle greys in London, where well-entrenched park populations generated a spreadable surplus. He had witnessed a squirrel 'jump from a tree in the Regent's Park to the upper deck of a No. 121 bus and travel [north] to Mill Hill, there to jump off unmolested'. And in the early hours of various mornings, he had watched 'parties' migrate along the parapets of Vauxhall, Chelsea and Battersea bridges, bound for leafy southern London suburbs and the Home Counties. Without 'complete destruction in all London parks and squares, as well as in private gardens', NAGSC's valiant efforts in the countryside would be in vain.[89]

The Grey Squirrels Order of 1937

'Someday', speculated, hopefully, an article titled 'Bolsheviks in Fur and Feather' (1932), 'we shall exact a licence for the introduction of strange species – as we now do, after long delays, in the case of, say, Bolsheviks.'[90] The author's main bugbear was 'that rather pretty rat-like creature', the grey squirrel, which, *urbi et orbi* (to the city and to the world), had embarked on a three-pronged campaign against birds, grain and 'our real and delightful brown squirrel', now largely displaced 'from its own home'. Legislation of the sort that the writer (and Lynd) had in mind soon followed: the Destructive Imported Animals Act (DIAA) of 1932. DIAA's sole target, though, was the muskrat, a recently introduced furbearer that had escaped from fur farms. DIAA, however, contained a provision for adding 'other destructive mammals not native to this country and not established in a wild state in Great Britain or which had become so established only in the last fifty years'.

Five years elapsed before MAF listed the grey squirrel under DIAA.[91] MAF's main action after the May 1931 summit was to distribute a leaflet outlining available control methods ('rigorous shooting at all seasons of the year' was recommended) and exhorting landowners to act.[92] Though grey numbers nationwide had fallen substantially in 1932–3, disease rather than shooting was the likely cause, and numbers were recovering by 1935. MAF secretary Walter Elliot planned to renew publicity efforts, including a broadcast and a new pamphlet, but reiterated that MAF could only do so much, lacking authority to act directly.[93] The most proactive branch of government was the Forestry Commission. Between November 1936 and the end of October 1937, the organization 'destroyed' more than 1,500 greys, 'systematically', on its plantations.[94]

Meanwhile, private landowners continued to wage micro-campaigns. In deciduous woodlands, shooting was only feasible in winter, when trees were bare. One method was to spray a drey with shot, hoping it was occupied. In the late 1940s, in Micheldever Wood, Hampshire, the FC's forester for southeast England reported that 'every drey to be seen was blasted with shot – many squirrels were killed in this way.'[95] This practice was discouraged, however: the invisible occupant might be a native squirrel. The most efficacious method required a minimum party of three – plus dog, should the quarry jump or fall to the ground. Using an extendable pole up to 18 metres (60 ft) long, a drey was poked to dislodge any grey occupants.[96] A dislodged grey that spotted a gun would move to the opposite side of the branch or trunk. While one gun circled the tree, following the squirrel, the second shooter remained stationary. The dog helped push the squirrel towards the guns ('marking') and the standing gun usually shot it.[97]

In a section of Burnham Beeches covering 200 hectares (500 ac), 4,270 squirrels were shot – not necessarily after drey poking – between 1920 and April 1930, peaking at 1,011 in 1929.[98] In Windsor Park, the body count for May–June 1930 was 1,370. On a single, modest-sized estate in Surrey – where, according to the owner, 'they swarm, and preservation of fruit is impossible' – the annual toll was six hundred to seven hundred.[99] Yet the impact on overall numbers seemed negligible. To make more of a dent, this landowner implored county councils in the most heavily invaded counties to offer an incentive of 2*d*. per tail to check further invasion in areas where it was present but not yet 'in hordes'.[100] A veteran gamekeeper agreed that countryfolk, with already enough pestiferous creatures on their plates (rats, moles, crows), needed a payment to encourage pursuit of this 'new vermin': 'What good are they ... in terms of cash?' As the value of the 'impudent and objectionable aliens' as food or fur was low, they had to be 'worth killing'.[101]

MAF was unenthusiastic about rewarding such culling with money, not least because incentivization had proved ineffective and open to abuse.[102] Nineteenth-century 'vermin bounties' handed out for squirrels, pine martens, wild cats, polecats and assorted raptors in Scotland illustrated these drawbacks. On the Cawdor estate, for instance, between 1862 and 1878, bounty hunters collected payments for reds in numbers far exceeding the total population within a 160-kilometre (100 mi.) radius.[103] In 1937, however – as sightings of greys from northern England started trickling in, and Middleton appealed for information – MAF's Grey Squirrels (Prohibition of Importation and Keeping) Order added them to DIAA.[104]

Opinion on the MAF order was evenly divided between representatives of urban and rural constituencies, though not always along party lines. Edmund Radford, a Conservative, represented inner city Manchester but lived in the adjacent county of Cheshire, where greys abounded. He wanted stronger evidence that they committed 'crimes' like frightening off 'the old English brown squirrels'. Otherwise, he felt unable to back a 'sentence of death' on a creature that, however mischievous, was endearing and 'picturesque'.[105] Labour MP David Adams, who represented an area of northeast England without greys (Consett, Co. Durham), concurred. General complaints about 'an American introduction' and 'non-indigenous' creature that reproduced faster than its red counterpart – which was not quite as harmless as the grey's opponents alleged – were insufficient to justify scheduling.[106]

Other MPs had already made up their minds. It was crunch time. Should the grey be allowed to become a permanent fixture in England and Wales? Or should its expansion be arrested and even rolled back? For Herwald Ramsbotham (Conservative – Lancaster) the 'undesirable alien' was a 'bitter enemy' of agriculture, horticulture, silviculture and birds.[107] The parliamentary secretary to MAF, the Earl of Feversham, who submitted the 'Grey Squirrels Order' for approval in the House of Lords, owned a large estate on the North Yorkshire Moors. For him, the order was 'an opening step in a campaign of propaganda and publicity which it is hoped will encourage the owners of land to exterminate the grey squirrel by their own efforts'.[108]

Among those welcoming the order (the greys 'at last condemned by the authorities') was Phyllis Kelway, a renowned author of animal stories for children. At her home near Huddersfield, Yorkshire, she kept specimens of many of Britain's smaller mammals for study and breeding, including both squirrel species, which she unsuccessfully encouraged to fraternize.[109] In a book about Britain's arboreal denizens, Kelway prefaced her chapter 'Red and Grey' – which followed one on the red alone – with a 'Note' explaining that it was written earlier that year, in April 1937. Back then, the grey squirrel remained a 'free citizen' of Britain, though, because of American origins, 'his naturalization papers had perhaps never been granted.' Alluding cryptically to the imminent order (effective from 31 July), Kelway noted there was 'worse in store for the grey squirrel than nine days' quarantine on Ellis Island'. By the time her book was published, 'he would be officially outlawed.'[110]

Despite press talk of a collective death sentence, the order's bark was worse than its bite. Importations had already ceased. The last recorded introduction point (Needwood, Staffordshire) was in 1929 and it was a secondary transplantation, as all releases had been for nearly two decades. The last batch brought directly over from North America had been in 1910, in Frimley, Surrey.[111] Moreover, many of Britain's greys were not being kept in the manner that the order defined as 'keeping'. Simply having grey squirrels on your land, of which you might well be unaware, did not constitute 'keeping'. The order applied only to those who knowingly kept greys and induced them to remain on their property, whether as pets or at large (thereby incurring the £5 fine Nick Clegg later ridiculed).[112] And if a landowner/user was not directly responsible for a squirrel's presence – it may have entered from a neighbouring property – then they were not legally responsible for removal. The owner of a garden next to Bagley Wood, Oxfordshire, the source of his greys, wondered whether, 'in the event of a prosecution for failing to deal with our squirrels', he could 'refer the summons to the Wardens of Bagley Wood'. If the adjacent property was publicly owned, by the FC, for instance, then the legal position was even more complicated, as the Forestry Commissioners' authority was delegated from MAF.[113] From an American perspective, however, these were niceties. The order was tantamount to initiating 'vigorous war' to 'stamp them out'.[114]

To secure widespread support, the order's proponents were at pains to underline the limited liability of a squirrel 'keeper'. MAF's Feversham sought to reassure that despite the statutory power of entry conferred, this authority would only be invoked should private (that is, voluntary) efforts at control prove unforthcoming or ineffective. For the grey's most ardent foes, this was a major weakness. To maximize the order's clout, they wanted anyone with greys on their land to be held accountable.[115] But MAF was loath to undertake shooting and trapping on private lands.

Grey opponents assigned FC plantations a role comparable to London's parklands: that of sanctuary, reservoir and springboard for infiltration of private property. Because Commission lands surrounded his Duncombe estate in North Yorkshire, Feversham explained, 'as soon as one kills 300 or 400 ... in any given month a similar number come the next month and take their places.'[116] The vulnerability of his lands did not make him an advocate of enhanced MAF powers, however. The ministry would offer every form of help to private landowners short of direct financial assistance or manpower. In practice, the sort of help Feversham envisaged was strictly educational, specifically a pamphlet advising on methods of extermination.[117]

Shortly before the order became operational, George Courthope, the Forestry Commission's parliamentary representative, reassured critics concerned about economic damage and the rapid extermination of the 'indigenous and harmless red' that the Commission was already 'systematically' killing greys on its lands.[118] However, MAF could and would not intervene directly. Fitter summarized the ambiguity and complexity of MAF's position in light of the order. On the one hand, it was MAF's 'declared policy ... to exterminate this recent addition to our fauna'. On the other, there was 'at present no intention of waging an official war on it'.[119]

Some of the beleaguered red's champions might have dreamed of rounding up all greys and shipping them back to North America. Fantasies such as repatriation could be indulged, though, in children's fiction. Helen Vaughan Williams's *Squirrel War; or, The Fight for the Doll's House* (1936) – a book tinged with what might be called squirrel racism and eco-xenophobia – tells the story of interspecies struggle through a contested object, a doll's house. Rollo is a red squirrel whose troubles begin with the arrival of 'an unknown cousin' from America, 'a big grey fellow'. Rollo 'didn't like the look of him' but was willing to show him 'the best trees for berries and for buds in the spring'. Rollo even introduced him to his friends: 'And how did he repay me? Drove me out of my home.' Pine Tree Cottage in Flibberty Wood had been in Rollo's family for generations. Yet he could do nothing about it, the 'crafty', cruel, rude and deceitful grey being so big and strong, greedily eating all of Rollo's carefully stored food.[120]

Red squirrel soft toy, National Trust gift shop.

Eastern grey squirrel (*Sciurus carolinensis*), hand-coloured lithograph from John James Audubon and John Bachman, *The Viviparous Quadrupeds of North America*, vol. 1 (1845).

Red squirrel with hazelnut at Alverstone Mead Nature Reserve, Isle of Wight.

Grey squirrel carrying pine cone.

Hans Hoffmann, [Eurasian] *Red Squirrel*, 1578, watercolour and gouache over traces of graphite on vellum.

An encounter between grey and red on a stump in Dodd Wood, Bassenthwaite Lake, northern Lake District. The difference between the two species is most visible here in the contrasting shapes of the ears. Whereas the grey's are rounded and tuft-less, the red's are pointed and tufted.

Grey squirrel campaign: 'Destroy the grey squirrel: 1/- for this tail', 1955, poster.

'Happy birthday from Tufty', n.d., card depicting Tufty holding the hand of his 'mummy', Mrs Fluffytail.

Red squirrel mural, Zase and Dekor, Cottrell Road, Bristol, September 2017.

Illustration by R. J. [Reginald James] Lloyd for Ted Hughes's broadside poem 'Squirrel' (1986).

Rollo thinks his new home in the doll's house (a Christmas gift) is impregnable. But his nameless adversary, the generic Grey Squirrel, lays siege and starves him out. Then Rollo's wildest dream comes true. Grey Squirrel is sent back to where he belongs, in a ship on which a parrot that visits Flibberty Wood is due to travel with his owner. The reds drug Grey Squirrel and slip him aboard in the parrot's cage. The trick works. On reaching American shores, the parrot's owner releases Grey Squirrel into 'the forests that were once his home, to greet again a thousand friends and relations'.[121] Meanwhile, back in Flibberty, the reds and other native creatures break into celebratory song:

Hooray, hooray,
He's gone away,
Our crafty foe, Grey Squirrel!
We'll live at ease
Amid the trees
And do exactly as we please,
Extremely gay
Because we're so
Delighted that he's had to go.
Hooray, hooray,
We've chased away
Our crafty foe,
Grey Squirrel!

In the chapter 'We Meet an Enemy', the narrator explains that an 'intruder' disturbs the harmony of the woods. Not the usual suspects, the jay or fox, but a 'grey one', sitting up on its hind legs with a cherry in its forepaws. Between bites 'he screamed the rudest possible things to Father and Mother in the squirrel language.' At first, the narrator does not realize the 'strange' grey one is also a squirrel, though, as the parents announce after chasing him off, 'An enemy! One of your wicked grey cousins!'[122] The kids want to know why Mother calls the grey cousin an enemy and wicked. She explains how, once upon a time, the only squirrels in the woods were red, who thought their closest cousins were dormice. Then, one day, an ancestor, in a forest further south, had the same encounter with a grey, whose 'queer' colour he attributed to illness – a diagnosis quickly dismissed when he saw the queer squirrel tucking into *his* cherries. The ancestor knows he must chase him away but faces fierce resistance. This cherry-stealing grey, it turns out, is not alone. Fights break out all over the woods, the greys usually winning and often killing vanquished reds.

Surviving reds head north. But even in Flibberty Woods, greys have started showing up. Sanctuary-seeking reds might have to retreat even further north. A city sparrow summering in the countryside tells 'great-grandfather' where greys came from: a London park.[123] Mother tells her kits that their human allies are getting 'angrier and angrier', even shooting greys. One day, she hopes they can return south to their ancestral homes. The family then goes to see a grey shot by Tim the Tinker for trespassing in his garden, its body 'swinging stiffly' on the washing line behind Tim's caravan. Tim's body count climbs to five after five days and the helpless reds are enormously grateful to humans like Tim for trying to 'keep the woods safe for us'.[124]

The grey as aggressor also provided a useful analogy to highlight human besiegement in an article published shortly after the Battle of Cable Street (4 October 1936), precipitated by a proposed march by the British Union of Fascists (Blackshirts) through an East End neighbourhood with a sizeable Jewish population. Harry Roberts, a doctor in his mid-sixties with thirty years' experience working in London's East End, noted the growth and spread of the Jewish population stemming from late nineteenth- and early twentieth-century immigrants from Russia and elsewhere in eastern Europe. After settling initially in Whitechapel and Aldgate, they had expanded into Mile End, where Roberts's surgery was located, 'on the edge of the advancing Jewish wave' (according to a biographer[125]). He characterized them as clannish, exclusive and reluctant to intermarry, but flourishing compared to existing residents because they were 'more "pushful"' and 'more persistently industrious, than most of those with whom they compete – and often as a result replace'. The 'ways of the pushful, unsporting Jewish minority' provoked a backlash from displaced Mile Enders, who no longer felt safe in an area rightfully belonging to them. 'An attitude of racial dislike', Roberts surmised, 'is apt to be indiscriminate in its target.' The 'smouldering anti-Semitic resentment', Roberts reckoned, was 'much resembling the feeling our native squirrel might have towards the grey interloper'.[126]

An exchange of letters in the *Manchester Guardian* in 1936–7 about an interloper's displacement of indigenes highlighted differences in attitude to native squirrels between Scotland and England. There were still enough in Scotland for them to be saddled with the transgressions for which greys bore the brunt of blame south of the border. 'E.E.M' of Southport, Lancashire, just north of the red stronghold in the pinewoods at Formby, Freshfield and Ainsdale, was highly critical of the Highland Squirrel Club's large-scale culling of reds. Arnold W. Boyd, a *Manchester Guardian* 'country diary' columnist, conceded that this slaughter might seem anomalous in English eyes, given reds' scarcity in England. But 'E.E.M' had overlooked the threat they posed to plantations (and fledglings). Boyd agreed, though, that nothing good could be said about its 'foreign' replacement, save for its 'tameness and engaging appearance'.[127]

Next up in the *Guardian* letter thread was a lover of squirrels, red and grey, for whom the Highland Squirrel Club's exploits were a reminder that the red's popularity was largely confined to England. Both species inhabited Watnall Hall, Nottinghamshire, the estate where Lady Rolleston, the social reformer, lived with Sir Humphry Rolleston. A pair of greys arrived at Watnall in 1931 and her reds were rescue squirrels bought from the RSPCA. Neither type had injured a single tree in Watnall's extensive plantations or among the long avenues of red and white chestnuts strung between entrance lodges and the main house. Completing the picture of interspecies harmony, she reported that her greys and reds shared feeding tables with local birds.[128]

The Reverend Basil Viney, a London-based animal welfare proponent, endorsed Rolleston's happy vision. 'Every Londoner', wagered the author of *The Revelation of God in Nature and Humanity* (1931), 'will agree that the cheerful, friendly little grey squirrel is well worth the occasional bulb, fledgling, or even bifurcated larch!' He deplored the compliance of London County Council's parks officials with MAF's demands. He acknowledged that greys were not good for reds. In

London's open spaces, though, there were none to displace. Remove greys and the capital would be squirrel-less.[129]

Rolleston followed up with another pro-grey letter in 1937, a few weeks before the order came into force. Greys at the Rolleston property in Leicester respected its trees and were 'friends of the birds' there too.[130] Viney wrote again as well, stressing that neither squirrel was completely innocent, the red's impact on firs cancelling out the grey's on market gardens. Together, they were 'the most delightful wild mammals in Britain'. He struck a defiant tone on behalf of fellow Londoners, who 'will never forgive the officials who robbed him of his new ... friends – best of all the denizens of the parks'.[131]

Directing attention to the imminent order, a resident of Hale, greater Manchester, rejected Rolleston's grey-friendliness, propounding the pest's 'complete elimination' by extending the order's definition of 'keeping' to legal responsibility for measures against squirrels on your land.[132] Harper Cory, a Canadian author of popular books about animals, highlighted the selectiveness of Rolleston's evidence: 'That the two squirrels [she] observed ... knew how to behave does not rule out the fact that some squirrels misbehave.' Cory, who lived in southeast London near Crystal Palace Park, was no friend of any squirrel. The zealous pursuit of greys (the 'destruction order'), he predicted, would backfire, bringing 'epidemics' of reds. He was particularly dismissive of Britons who wanted to wipe out the grey 'because he is an alien.' 'Such an argument', he reckoned, 'comes strangely from a race noted for its world-wide expansion.'[133]

Naturalist F. Howard Lancum pulled rank in his response to Rolleston. The *Manchester Guardian* billed his letter as 'the evidence of a field naturalist' and he cited his first-hand experience of a 'tree rat' snatching four fledglings from the nest of a hedge sparrow (dunnock) and of others dining on young thrushes. Living in a nut-growing area, in Sidcup, Kent, he had also seen walnut trees stripped of fruit.[134] A correspondent from East Sussex later that autumn (1937) reiterated the need for constant vigilance in his hazel-nut grove adjacent to woodland dense with greys. In 1936 he had enjoyed his first decent crop after destroying all dreys and shooting a dozen the previous winter. But having neglected to 'de-squirrel' the following winter, they reappeared.[135] The choice boiled down to nuts or squirrels.

Other contributors to the *Manchester Guardian*'s discussion prioritized the threat to reds. A danger to timber trees, admittedly, the red squirrel nonetheless had a 'claim to consideration as an indigenous species' and, 'as a native of the country', was 'entitled to its share of the fruits of the earth'. Turning to the grey, the correspondent warned against distortion of judgement by emotion. 'Like a jury condemning a handsome ... criminal', his advice was to 'harden our hearts' for the sake of 'biological, or ... zoological nationalism'. The defence of British fauna against a 'dangerous, hardy alien' that 'inevitably upsets the balance of nature' demanded these new varieties of nationalism.[136]

Explanations of Grey Success: Pioneers on a New Frontier

Hardiness was just one reason offered for the grey's ascendancy. The commonest explanation was that 'the foreigners' were larger and more powerful, which allowed them to 'dominate their smaller and weaker cousins and to turn them out of their rightful homes and breeding quarters'.[137] These explanations of success focused on their invasiveness as a species: the propensity to invade owing to attributes such as size, strength and hardiness. Yet the invasibility of the recipient environment – its susceptibility to an introduced species' naturalization and expansion – must be considered too.[138] A newcomer enjoys maximum success when high levels of invasiveness and invasibility combine. A species' observable attributes, among them body size, competitive capacity, life history and foraging strategies, are called phenotypic traits.[139] Since the early 1900s, British commentators had attributed grey invasiveness to four phenotypic traits: body size, strength (competitive strategy), prolificity (life history) and adaptability (foraging strategy).

The advantage of bigness was most transparent when greys threw their weight around. In 1937 an essayist witnessed the 'little brutes' interrupting a red pair's courtship and evicting them from their homes. And sometimes, he added, mobs of the 'newcomer' killed pairs of reds.[140] This belief persisted into the 1970s. Lord Wise's explanation for the ousting of reds from the mixed woodlands on his Norfolk farm was simple: greys 'fight and kill the red'. Though open, by 1998, to additional explanations of red demise, Wise still suspected that 'marauding' greys might have slaughtered young reds in their dreys.[141]

Chasing away, injuring or slaying a native congener is certainly one way for an introduced species to root itself in new soil. These kinds of visceral confrontations constitute direct competition. Nonetheless, as Middleton and other professional science-informed voices realized in the 1930s and '40s, however strong it might be compared to the red, the grey was not that kind of evictor. Even William Beach Thomas accepted that, while the grey might 'banish' the red, it did not destroy it, literally.[142] Rather than just bigger and stronger, the grey was a more effective indirect competitor. Red population declines (shaped to a lesser and greater extent, respectively, by food availability and epidemics) were not catastrophic over the longer term provided their habitat niche remained available. Young greys 'preempt' the best habitat, however, forcing red counterparts into areas of poorer quality. And once greys had replaced reds the latter's chances of regaining former territories were remote.[143]

Shortly before the First World War, a ZSL member explained the grey's superior competitiveness as a function of adaptability. This attribute reflected the grey's membership in a community of animals known as commensals. Reeling off a list of others – the house sparrow, house mouse, black rat, brown rat and cockroach – the member defined a commensal as a species that readily adjusts to the 'artificial conditions created by man'. First-hand encounters with the 'American' and 'Canadian' squirrels released at London Zoo in 1905–6 informed the correspondent's views. Marvelling at their tameness, fearlessness and hardiness, the ZSL member pinpointed a 'radical difference in temperament' between greys and reds. In fact, the red was the opposite of a commensal. Even in captivity, the grey was tougher than other squirrels, not only the zoo's European (red) squirrel but the Malacca Prevost's squirrel and African ground squirrel. Not least, greys were less prone to paralysis of the hindquarters.[144]

These constitutional and temperamental advantages, allied to more obvious assets of size and strength, allowed greys to settle into places reds had vacated and to secure the 'upper hand' in woodlands where reds still resided.¹⁴⁵ Nature writer Hugh Wallwyn (H. W.) Shepheard-Walwyn feared for the future of a 'fascinating rascal' that barked trees, nipped off buds and snatched hatchlings (that is, the red) if the 'hardy alien', tolerable in London's parks, where it provided free entertainment for 'babies and nursemaids', became naturalized in rural woodlands.¹⁴⁶

Kelway (no fan of the 'jocular cosmopolitan tree-rat') identified a similar portfolio of assets and advantages, incorporating her experience of greys and reds kept as pets. Greys were 'more robust and forceful'; highly adaptable and unfussy; 'rough and hearty'; 'voracious' and coarse; worldly and 'audacious'. Not only were their appetites bigger, in keeping with their size. They were also greedier, not to mention 'presumptuous and insolent'. By contrast, reds ('imps' at worst) were polite, cautious, gentle and fastidious, 'dainty' feeders with 'microscopic' appetites, whose 'somewhat delicate' constitutions were more susceptible to disease and intolerant of disruption.¹⁴⁷

Greys maximized their competitive advantage through a superior foraging strategy. That the acorn forms the red's staple is a persisting misbelief. Yet, as early twentieth-century British commentators realized, reds prefer hazelnuts.¹⁴⁸ That did not mean, though, that the fussier red enjoyed unrestricted access. A Buckinghamshire farmer protested that greys ripped down unripe hazelnuts in July and August.¹⁴⁹ The ability to digest hazelnuts as soon as kernels begin to form by neutralizing their toxic tannins deprived reds of a vital food source later in the year. Greys then double their advantage by switching to acorns – a staple in North America – whose toxins reds have difficulty digesting (later research indicated), even when acorns are ripe.¹⁵⁰ As squirrels do not hibernate, this capacity to eat unripe nuts (and fruits) allows greys to put on fat earlier, equipping them better to survive a hard winter.¹⁵¹

Red squirrels were never associated with sturdiness, brawn and stamina. Conservative MP and future prime minister Harold Macmillan told future Labour MP Richard Crossman in 1943 that Americans (like Romans from the ancient Greek perspective) were 'great big, vulgar, bustling people, more vigorous than we are'.¹⁵²

When a grey squirrel popped up in their neighbourhood, British commentators identified qualities associated with Americans, American life or the United States, interpreting grey success as part of the larger phenomenon of American triumph. Americans, in turn, often saw their squirrel's achievements as another sign of their nation's entrepreneurial vigour.

These attributes of Americanism were readily linked to pushiness and hustle.[153] Particularly in u.s. English, to hustle means to act quickly and energetically. As a noun, hustle denotes vigorous, loud, even forceful activity that might entail pushing, jostling and crowding others out. Interwar commentators on both sides of the Atlantic detected tremendous hustle and bustle in the grey's activities in Britain. The British government's so-called declaration of war against it in 1931 (namely, the MAF-convened squirrel summit and MAF-endorsed National Anti-Grey Squirrel Campaign) inspired a British journalist writing for the *Daily Mail*'s Atlantic edition. The reporter explained to American passengers eastbound on Cunard's RMS *Aquitania* that 'American "hustle"' was 'driving out' the 'home variety' and that 'the alien' was far more adaptable than 'our native red'. As its fortunes did not depend on finding an exact match to its homeland's thick hardwood forests, the reporter predicted its spread would continue unabated to cover most of England and Wales by 1981.[154]

The red's inferior hustle and hardiness were widely recognized by the grey's opponents and proponents alike.[155] Alert to parallels between human and non-human colonists, Coward observed that, faced with an invader, the native, 'as a rule, withdraws or declines in numbers or power'. Nature 'only provides enough accommodation and food for a limited number of inhabitants, and if a new-comer is powerful enough to obtain what is provided the native must quit or go under'. He insisted that there was a 'squirrel country' for reds and a 'squirrel country' for greys. And as the latter usurped the former's 'ancient rights', there was increasingly only one kind of squirrel country. The red squirrel had 'nowhere to colonise afresh'.[156] Rural England's squirrel future looked grey.

The fate of 'weaker', previously secluded tribes retreating in the face of stronger incomers was also on the mind of a reviewer of three books about peoples forced to seek sanctuary in the mountains between India and China. These marginalized Himalayan groups were likely to disappear, like the aboriginal Tasmanians, taking their places alongside aurochs and dodo. For the reviewer, greater economic muscle and more powerful weaponry were insufficient to explain how one 'race' prevailed over another. To be factored into the equation as well were biological considerations of expansion and contraction, 'akin to the laws of animals and plants, which decree that the grey squirrel shall supplant the red'.[157]

Instead of dwelling on the manifest destiny of biology fitness, nature writer Frances Pitt singled out an attribute associated with Americans. Identifying competition for food and dwelling places as the crucial battlegrounds in red–grey rivalry, American 'go aheadness' delivered success. For all its 'fairy grace', 'our native' lacked the 'stamina of the foreigner', which, if 'less spritely', was 'larger' and 'more solid'. This 'greater energy', allied to its 'particularly sturdy and virile' nature, explained the grey's power to 'supplant'.[158] These success-underpinning qualities of boldness and initiative were specifically associated in British minds with the white American male's pioneering instinct. According to a despatch from a frontline of the 'squirrel war' of the mid-1920s, 'pioneer' greys, swarming on London's outskirts, such as Highgate Wood, had gone on to 'capture' and vandalize red neighbourhoods as far west as Newbury in Berkshire (80 kilometres (50 mi.) from Regent's Park).[159] A MAF poster distributed in the early 1930s ('You Versus Pests: The Grey Squirrel Menace') pinpointed the grey's 'propensity for exploring', with trailblazing pairs found up to 32 kilometres (20 mi.) from existing colonies of infestation.[160]

Pioneering on a newly opened frontier was a theme further developed by Brightwell, a well-known artist of London Zoo scenes as well as a writer. Also a Fellow of ZSL, his story 'The Grey Raiders' (1930) – subtitled 'The Romance of Some Squirrels that Invaded Rural England' – was a semi-fictional account of that zoo's release of greys into Regent's Park. The three pairs, which spent nine months in a cage, may have resembled prisoners but were actually 'alien conquerors . . . waiting for the day when they should make the woods for a hundred miles around their own past all dispute'. From the start, they displayed the 'endearing' qualities of the 'born conqueror': 'active, gay, and fearless, consumed with that burning curiosity which marks the pioneer'. In the story, a map next to their enclosure shows their historic distribution 'from southern Canada to Guatemala', which a seer might have complemented with a map showing their future distribution from 'Kew to Fifeshire and from Colchester to Polperro'. These 'chubby' squirrels fought among themselves, accumulated food supplies, tripled their numbers within two years and generally flourished in the 'clement climate' of this 'most tolerant of countries'. And when the zoo strung up a rope between a tree in their enclosure and one outside (a safety valve for their swelling population) they crossed into a new land of freedom.[161]

As adept, adaptable and inquisitive as the rat, sparrow and cockroach, Brightwell relates how the 'roystering little ruffians' explore every available park nook and object. A pioneering pair soon strikes out into the wider world, Brightwell succinctly summarizing their trajectory: 'From distant Canada to cage, from cage to public park, from park to private estate'. The first couple occupies the nest of an evicted red squirrel. Might triumphs over right, the English population soon in 'full retreat'.[162] The 'hardened marauders' benefit from the absence of predators in their acquired home and run amok, akin to the Canadian (also known as American) pondweed that has taken over the fresh waters of 'half the world'. To secure a new residence, they eject young reds from their drey and slay the returning mother. Wherever the 'Western cousin' appears, incumbent reds are 'exterminated'. The 'nomads of the Western wild', 'undaunted' by efforts to reduce their numbers, are simply unbeatable, dug into 'British soil for good'.[163]

Early twentieth-century explanations of success often weighed up both fitness for invasion and the hospitable properties of the host environment that Brightwell believed the grey guests had abused. Invasibility (according to a recent interpretation) depends on the 'structure of the resident ecological network and is defined as the total width of an opportunity niche in the trait space susceptible to invasion'.[164] William Beach Thomas, who had observed greys in Albany, upstate New York, and in the grounds of New York City's Bronx Zoo, did not use the language of opportunity niche and trait space. Nonetheless, he felt they had found 'what botanists call their optimum in England', enjoying their adopted home so much that 'their prosperity may interfere with the freedom of others.'[165] (So much for Coward's claim that the grey squirrel was 'not in its right setting in Britain'.[166]) Rapidly increasing numbers between 1921 and 1931 indicated to a kindly disposed correspondent of Thomas that the creature had 'undoubtedly made the best of the opportunity given it to hold its own in a new country'.[167]

One of the reasons greys had taken 'so kindly ... to an English environment', Thomas explained, was their capacity as generalists to make maximum use of available foods: 'Was ever so omnivorous a pest?'[168] MAF subsequently likened its broad diet to the rat's 'very catholic' tastes.[169] Without discussing diet, specifically, Pitt agreed that the English climate and countryside suited it very nicely.[170] That the offspring of a single pair had colonized a 52-square-kilometre (20 sq. mi.) area of west-central Scotland near Loch Long and Loch Lomond in under thirty years underlined for the *New York Times* how Britain's new squirrel 'thrives in his new environment'.[171]

An opportunity niche – a new squirrel frontier – consists of absences as well as presences. Left behind in North America were customary competitors and 'natural enemies'. The grey's evolution in a more competitive squirrel environment, particularly with the sympatric American red squirrel and the fox squirrel, prompted 'overbreeding', whereas British reds, innocent of interspecific competition, rarely produced more than two or three kits per litter when they bred (normally twice a year). Greys bred no more frequently, but each litter consisted of between four and five, with up to nine recorded.[172] Moreover, though the enemy release hypothesis was not formally articulated until 2002 (and remains contested), explainers of grey success in the 1930s were effectively subscribers.[173] The grey 'made itself at home in its promised land', Coward remarked, not least because, back home, 'natural checks' (predators, pathogens and parasites) curbed numbers.[174] What did the rabbit and prickly pear in Australia have in common? And, in England, the brown rat, little owl, American waterweed and grey squirrel? British ornithologist Arnold

Boyd posed this rhetorical question in the mid-1930s. His answer: their new homes were largely predator-free.[175]

By 1939, greys were a common feature of English (and Welsh) landscapes and a fixture within Britain's mammalian community. Yet membership of those landscapes and community was not something many Britons were prepared to grant readily. Many felt about red squirrels just as Wordsworth had felt about the Lake District's 'usurped' native trees: they were superior to non-native forms and had more right to be there. The native squirrel, like the indigenous oak, ash, birch, holly, elm and alder, was integral to the spirit of place and inseparable from a sense of belonging and heritage. And like those trees, the red was threatened by an alien form that, in Wordsworth's language for 'exotic' trees, was 'discordant' and a 'deformity' within the broader landscape.[176] In 1942 Frank Newbould sought to harness Britons' patriotic spirit for the war effort by evoking images of the soft, southern English countryside. Yet that countryside had already been invaded, in a fashion. If the War Office had asked the poster artist to mobilize the public for the defence of Britain's red, all Newbould would have had to do was modify his slogan slightly to read: 'Your Squirrel – Fight for it Now'.

5
WAGING WAR ON THE 'GREY PERIL', *c.* 1939-73

TALK of (grey-initiated) warfare between Britain's two squirrels had been routine since 1914, with perhaps the first reference to 'the squirrel war' just over a decade later.[1] The Second World War accentuated this martial discourse. Evolutionary biologist Julian Huxley, recently appointed the first director of UNESCO (1945), brought up warring squirrels in 1946 while discussing whether war was 'instinctive in man'. For Huxley, fighting between individuals or small groups – whether humans or other animals were involved – even to the death, did not constitute engagement in war. This was simply competition writ large. Huxley agreed with many fellow Britons that greys had 'ousted' reds. But he disagreed that warfare was its means. References to war against pests, he also maintained, were strictly metaphorical.

War, he insisted, 'means only one thing – organized [physical] conflict between groups of the same (or closely related) species, aimed at the imposing of the will of one group on the other'. According to this definition, just three species waged war: humans, certain bees and ants. Huxley happily characterized as war territorial 'battles' between hive-bee colonies and between rival ants' nests. Any other form of conflict within the animal kingdom, such as individuals fighting over food, mating rights or habitat, fell short of 'true war'.[2]

Britons acting against the grey squirrel disregarded such distinctions. As they saw it, the grey was fighting and winning a war against the native squirrel. Those trying to help the red (and confronting the grey's other affronts) characterized their actions as righteous defence of the defenceless native squirrel (and various human interests). According to a MAF leaflet in 1953, those who brought over the grey thought it would be 'an attractive addition to our native wild life'. 'Little did they know', though, 'what trouble they were laying up for later generations!'[3] MAF had just introduced a bounty to incentivize the killing of a creature of which it was now 'easier to say where it is not found than to say where it is'.[4]

Central government expressions of disquiet in the early 1900s did not necessarily translate into firm measures, or leadership of efforts to fight the 'menace' various interest groups had pinpointed by 1914. And government action, when it materialized in the 1930s, was motivated more by economic damage than concern for the red's prospects. Insofar as branches of the state worried about impacts on native wildlife, the at-risk species identified were 'insect-eating' birds beneficial to crop yields. At no stage did MAF embark on an open-ended commitment to reduce, let alone try to eliminate what pest controllers referred to as infestations (defined, in the early 1940s, as densities of three or more per acre).[5] An unlimited commitment to control was never attainable, financially or politically.

Prior to the Grey Squirrels Order of 1937 ('an official declaration of war', according to *Country Life*), state-level measures were restricted to those of municipal authorities such as the Parks and Open Spaces Committee of the Corporation of London.[6] Alongside the Office of Works, this Committee conducted a 'squirrel war' on a limited scale in the early 1920s.[7] Between 1937 and 1973, when MAFF issued a Grey Squirrels (Warfarin) Order, central government induced, supported or led various campaigns. Yet most government officials, politicians and scientists too, accepted that eradication was infeasible. Even rolling back the advance was considered unrealistic. Containment represented the most successful outcome. Whatever the measure of victory, no weapon proved particularly effective. According to the Forestry Commission's chairperson, Sir David Montgomery, 'the best place for a grey squirrel is in a pot.'[8] But the tactic of beating the enemy by eating it made few inroads either.

Squirrels in the Second World War

The Second World War sharpened the edge of accounts of inter-squirrel warfare, non-fictional and fictional. Richard Church's *A Squirrel Called Rufus* (1941) was a viciously partisan children's story about the struggle for mastery between Red Tail and the savage 'foreigners' (Grey Tail) who 'demand so much living space'.[9] ('One cannot help feeling', a reviewer noted, 'that the grey squirrel hordes, with their "pincer movements" are described with other grey-clad armies in mind.'[10]) For Church, the red was the 'true English breed... native to these woods'. As a red called Scurry announces, there is 'only one kind of squirrel, our own kind, the English squirrel, the red squirrel'. Unfortunately, these 'true possessors of the woods' took their sole dominion and rightful ancestral rights for granted, growing complacent and oblivious to the 'grey shadow' haunting their woods. For too long, reds are in denial.[11] Scurry's friend Rufus, the emerging hero, gives him a brief history lesson: 'They came from thousands of miles away, over the ocean. My father says they came here quietly enough and settled down in the woods without upsetting anybody.' Gradually, they got to know 'every path, every hiding-place, every larder' in 'our English woods' and 'somehow or other' the rank opportunists, greedy and pushy, 'are always first on the spot'. Scurry agrees: 'my father says they get the best of everything, and make the most of everything.'[12]

The enemy's leader, Grey Gleam, a rough-mannered, bullying brute, heads an army of invaders that, according to an elderly red (Red Warden), are bigger and have sharper teeth, longer claws and larger appetites. As a self-proclaimed superior 'race' of squirrel, they feel entitled to take what they want. At this point, Church turns his criticism onto the natives as well, identifying how they have hastened their own demise. The greys' stronger work ethic exposes the reds' creeping laziness (failure to secure winter food and housing). Eventually, the reds form a Defence League led by Rufus, who finally slays Grey Gleam. The grey army has been decapitated, but, at the novel's end, the war is only half-won.[13]

The usurpations of Church's fictional greys were confined to woodlands. They did not impinge on the fields and orchards whose unimpaired productivity became imperative during warfare between humans. In a bid to maximize food output, MAF targeted various species through its County War Agricultural Executive Committees (WAECs), which provided farmers with a 'pest-destruction service'.[14] The grey squirrel loomed large on MAF's blacklist, outranked only by rat and rabbit.[15] The Grey Squirrels Order of 1943 required owners and occupiers of land to destroy greys on pain of prosecution and authorized County Agricultural Executive Committee officials to enter non-compliant properties to execute the order and recover costs from the defaulter.[16]

An encouraging sign for MAF was that 'grey squirrel shooting clubs' operating on private lands sprang up along the lines of traditional rat and sparrow clubs in response to the grey's brisk northward advance during the 1930s. In addition, established vermin clubs such as Elham and District Rat and Sparrow Club (Kent) added greys to their lists.[17] After 1939, though, these anti-pest activities were hampered (as during the First World War) by a shortage of cartridges and shooters.[18] Stepping into the breach, WAECs dispensed free cartridges to registered squirrel clubs – the government's first attempt to promote grey squirrel control directly, according to MAF.[19]

Casseroling the Enemy

The squirrel carcasses that the clubs racked up had no monetary value. MAF's verdict (1943) that 'nothing can be said in favour of the grey squirrel while it is alive' applied to dead squirrels too. Red squirrel fur, in medieval times, lined coats, cloaks and hoods.[20] But in the 1940s their fur had little commercial value and the tail's sole monetary worth (fetching half a penny apiece) was as hair for paintbrushes.[21] Under wartime stringencies, however, government officials identified and promoted a largely untapped value: as meat that tasted like a combination of rabbit and chicken.[22] This was an early expression of an approach to invasive species management summed up by the catchphrase 'if you can't beat them, eat them.' Given the near-extinction of the wild cat and pine marten, A. D. Middleton had noted that the 'post of squirrel-eater' was 'vacant'.[23] Whether humans could fill that niche was a question that took on fresh meaning when meat rationing was instigated in March 1940. Board of Education and Ministry of Food (MOF) recipe pamphlets featured sheep's head broth and sheep's heart pie and cuts of cow such as clod and shin.[24] In that context, a MAF official expressed 'no doubt whatsoever that grey squirrel (or for that matter any squirrels) would make excellent eating'. He then offered a suggestion: 'Why not procure some and invite the Scientific advisors of the Ministry of Food to the banquet?'[25]

The edible squirrel was a more venerable tradition in the grey's homeland. At the 23rd Annual Game Dinner at the Grand Pacific Hotel in Chicago, in 1878, among the boiled meats were fox squirrel, grey squirrel and black squirrel, supplemented, in 1901, by grey squirrel ragout.[26] Moreover, an Irishman who joined a grey squirrel hunt in upstate New York in 1891 reported that the animal's flesh (white when cooked) was 'considered excellent for the table'.[27] The early twentieth-century American wildlife conservationist William T. Hornaday was less enamoured. Dismissing its meat as '"gamey" and rank', he observed that 'Americans are the only white men on earth' who ate squirrel. 'An Englishman', he insisted, 'would as readily eat a rat!'[28]

Though Hornaday did not identify any Englishmen, his point had been borne out by the reaction of Thomas U. Brocklehurst, some twenty years earlier. In 1881, when Brocklehurst was served up a dish in Montgomery, Alabama, he 'supposed [it] to be a salmi [ragout] of prairie hen, or some birds, and [I] helped myself a second time; on being told it was squirrel, I had to rush to the bar at the end of the room for a *petit verre* [quick drink]'.[29] Two British zoologists agreed with Hornaday that most Britons considered eating squirrel outlandish. They disagreed, though, regarding its edibility (until he realized what he had eaten, Brocklehurst found it tasty enough to have seconds). In view of the dwindling numbers of reds by 1918, 'it is perhaps fortunate for our species that the excellence of its rabbit-like flesh seems to be unknown.' For the 'attraction provided by a destructive animal with ... a palatable carcase might well have proved fatal'.[30]

Despite the impression Hornaday and Brocklehurst gave, squirrel consumption in Britain was not unprecedented. Squirrel pie was a favourite dish, for example, in the New Forest. Prior to 1914, the squirrel that was eaten – or considered inedible – was red by default. Then, in 1917, red and grey featured together among the 'edible mammals' in a wild foods guide – though the author, a field sports enthusiast, judged the often 'very numerous' red tastier (roasted, and stuffed with pounded beech mast or sweet chestnuts).[31] Pointing out that squirrels were regularly eaten in western North America, a letter writer recommended putting Britain's troublesome grey on the menu.[32] The 'little patriotic experiment' a regular *Country Life* contributor had in mind at the time of Middleton's survey was shooting and cooking a grey.[33] When squirrel pie, roast squirrel and squirrel curry were suggested thereafter, the squirrel was explicitly grey.[34]

Persuading significant numbers of Britons of the grey squirrel's palatability was a tall order. British aversion was contrasted with American enthusiasm. A few months before the United States joined the Allies in December 1941, a British journalist regretted the 'irrational prejudices' that prevented Britons eating an animal 'widely enjoyed in their homeland'.[35] Shortly after war broke out in Europe, William Beach Thomas recalled seeing a butcher's in upstate New York 'as abundantly hung with the bodies of grey squirrels as any [British] game-shop with pheasants'.[36] The *Manchester Guardian*'s assistant editor ('Lucio') noted that grey squirrel, fried or casseroled, was a prized American dish though there was no shortage of livestock meat stateside, during peacetime or the present conflict.[37]

British proponents of consumption faced a considerable obstacle: the off-putting soubriquet of 'tree rat'.[38] 'The average countryman', a correspondent explained, 'cannot overcome a distaste for "squirrel pie". He has been told that the grey squirrel is a "tree rat", and would not dream of eating one.'[39] Trying to undo this reputational damage, 'Lucio' commended greys as 'clean, almost entirely vegetarian feeders'. Tastier and tenderer than rabbit (prominent in MOF cookbooks), they were also surprisingly meaty.[40] Among the recipes for stewed rooks, roast starlings and hedgehog pâté in an early foraging text (*They Can't Ration These*) were squirrel-tail soup and grilled squirrel.[41] Had the grey's edibility been more widely appreciated during the war, Thomas reflected in 1945, greys would have 'vanished like the rabbits' (whose dearth returning servicemen widely noted).[42]

Coverage in the *New York Times* of MAF's initiative found Middleton's earlier observations on British dietary foibles eminently quotable. A devotee of grey squirrel meat, Middleton had remarked that 'for some extraordinary psychological reason', 'many [British] people who will readily enjoy jugged hare and swallow oysters with gusto are positively revolted by the thought of eating a grey squirrel *en casserole*.' And so, the American reporter explained, the British government had to tell 48 million Britons something self-evident for centuries to rural residents of the eastern United States (pioneers and planters alike): grey squirrels were 'quite good eating'.[43]

Some rural residents were receptive. Eating greys augmented domestic meat supplies and protected birdlife.[44] But it is unclear whether MAF's or MOF's entreaties made much difference. Evidence is scattered and anecdotal. Data gathered by Mass-Observation hint at rising consumption. A diary entry from an elderly woman in Watford, Hertfordshire, read: 'Have just heard of Squirrels being eaten for dinner. Am told they taste good, a cross between rabbit and hare. One makes two meals.'[45] By far the biggest group of consumers, though, were American military personnel. They may have been over-fed, but GIs from Appalachia still craved an authentic taste of home.[46] MAF officials identified the U.S. military as a sizeable market for what was a prime game species in states like Kentucky.[47] 'They're a great delicacy in the States,' a U.S. Embassy official explained, adding, 'I'd like some right now.'[48] At its conference on wartime pests, MAF acknowledged this unusual form of American assistance in the war against Britain's third most damaging pest. 'Whenever possible', the press reported, American troops 'eat squirrel pie'.[49] Monica Shorten underlined this market's significance. Between August 1944 and December 1945, she conducted a nationwide Agricultural Research Council-funded survey of grey squirrel distribution.[50] In her subsequent book about squirrels for a general readership, Shorten reported that homesick American and Canadian servicemen paid up to five shillings (almost £9 in 2017's money) per squirrel.[51]

Shorten's book, published in 1954, included recipes for roast squirrel, squirrel pie and squirrel curry – and she disclosed that her unwitting dinner guests usually mistook squirrel meat for rabbit or chicken.[52] While most Britons recoiled, some considered squirrel an eminently fit dish for their dogs. Food shortages were a major argument for the pre-emptive, precautionary euthanizing of pet dogs both when war broke out in 1939 and again during the Blitz.[53] But the plenitude of greys across the southeast represented a vast untapped reservoir of petfood. Civil servant Stephen Tallents, who worked at MOF during the First World War, and at the Ministry of Information during the second global conflict, reported that a neighbour's dogs 'enjoyed' eating boiled-up squirrels. Reassured by this information ('that was good enough for me'), Tallents shot a squirrel on his lawn at Sutton-at-Hone, Kent, then instructed his cook to turn it into a casserole. He praised the result as an 'appetizing dish of good white meat, at once more delicate and more interesting than rabbit'.[54]

After Tallents told this story in *The Times*, 'inquiries poured in for [his cook's] grey squirrel recipe'. Also, a gift of six greys from Sussex, trussed up with string, appeared on his doorstep. And an enthusiastic squirrel eater from Alabama wrote to him about his squirrel-hunting dog. Not least, Tallents was deluged with recipes, some from top gastronomes. One casseroled squirrel made little difference to their numbers at Tallents's Kent home, though. Soon afterwards, one of them polished off 'the entire crop of my favourite young walnut' in a single morning. Tongue firmly in cheek, Tallents announced he would not be satisfied until the government arranged for the repatriation, even extradition, of every grey.[55]

With food rationing continuing into peacetime, and to protect emerging crops in the spring of 1946, MAF extended wartime distribution of free cartridges to members of grey squirrel clubs authorized by County Agricultural Executive Committees. The number of accredited clubs rose to 150 in November 1945, then climbed to 320 (across 27 counties) by the end of January 1947; by the close of 1947, they had increased to 450. During the winter and spring of 1945–6, 60,000 greys were shot at a cost of circa £1,000 (£35,000 in 2017's money) to the taxpayer.[56]

In austerity Britain, explained the American weekly *Time*, 'food was so short that . . . the Ministry of Food hawked tasty recipes for squirrel pies.'[57] Squirrel hotpot and squirrel pie – Tallents characterized such dishes as 'an uncouponed extension of our meat ration' – also featured in a department store's restaurant in London's West End.[58] Otherwise, MOF (and MAF) efforts were largely in vain. As grey squirrel was a 'delicacy . . . not yet appreciated', the *Times of India* observed that specimens selling for three shillings apiece on poulterers' stalls at London's Leadenhall Market found few buyers (though a pound of runner beans was priced at 45 shillings).[59]

The post-war explanation for this indifference, even antipathy towards the edible grey, was the same as that offered interwar and during wartime. In 1943 MAF observed that 'in view of their being described as tree rats it is very doubtful whether the public would consume them.'[60] Otherization – the stripping of the grey's credentials as squirrel and reinvention as rat – was counterproductive here. A resident of Hertfordshire was horrified that the minister of agriculture considered them 'fit for human consumption'. He proceeded to set the minister right on the grey's true nature: 'as bad or even worse than rats, for they will rob bird's nests of eggs and young, will kill and eat the Red squirrel, and even devour one another if caught in a trap'. He claimed, preposterously, that a single pair could produce up to four hundred offspring a year (government biologists set annual productivity at seven, maximum), which threatened to render them as ubiquitous as the common rat, which was 'not recommended for human consumption'.[61]

Characterization as tree rat scuppered efforts to put greys on plates. Among those responding to the Grey Squirrels Order of 1943 and the news that the government was touting greys as food, was Viscount Hailsham. A former secretary of state for war from 1931 to 1935, Hailsham believed that the edible grey was manna from heaven for Joseph Goebbels, Germany's propaganda minister. Reacting to an item in the *Daily Express* about how squirrel meat could compensate for ever barer fishmongers' slabs, Hailsham remarked that 'if Goebbels got hold of the article, it would afford him fresh rat propaganda' for headlines such as 'U-Boat Campaign Compels British to Eat Rats'.[62] A British journalist – for whom the grey's faults were simply those of its equally ratty red counterpart 'magnified by numbers and over-boldness' – agreed that 'the rat *motif* has been overdone.'[63]

The grey's unappreciated culinary merits continued to give MAF food for thought in the early 1950s. Officials shared advice that had been received from acquaintances with large estates who had sampled the goods. Lord Dorchester, a keen sportsman and popularizer of neglected domestic food sources, who had joined squirrel shoots stateside, informed a government contact that Americans liked squirrel pie. And one Major M.E.B. Portal was partial to squirrel stew.[64] The Forestry Commission's Richard Cotterell, who chaired the FC's Grey Squirrel Committee, conceded that he had not yet sampled squirrel meat. But in the spring of 1953, the FC belatedly joined MOF and MAF, issuing four recipes for its 'worst enemy'.[65]

Bob a Brush

The FC threw most of its weight, though, behind a scheme long advocated by rural interests and boosted by magazines such as *Gun and Game*: a bounty payment.[66] There had been a successful precedent set in South Australia. In its capital, Adelaide, where greys were released in 1917, a bounty instigated in 1920 of two shillings and sixpence wiped out the local population (admittedly considerably smaller than England's) in just two years, by 1922.[67] In Britain, three decades later, there was a growing conviction in government circles that the anti-grey cause was foundering. The *New York Times* reported on 'recent intelligence from London' suggesting that Britain's 'squirrel feud' had reached 'a pretty pass'.[68] The prompt for this American despatch was a BBC Home Service story of 1952 about greys driving reds out of Epping Forest, one of the few localities around London where they persisted. And Epping Forest lay within the parliamentary constituency of Prime Minister Winston Churchill.

To help Americans grasp the gravity of the situation, the U.S. reporter portrayed it as 'a crisis' of 'British patriotism and pride', for the red was 'deeply cherished by British nature lovers' and 'known to be British to the core of its long, curly, beautiful tail'. By contrast, Britons considered greys 'thoroughly misprized and suspect'. Not only were they 'aggressive bullies', they were foreigners, and 'American foreigners' to boot. The reporter was particularly struck by the British tendency to remove greys' squirrel identity by attaching the 'loathsome epithet of American Tree Rat'.[69] Noting that London's park keepers had already waged 'patriotic' war against the 'intruder from abroad', on behalf, not least, of the 'littler, more loveable native' squirrel, the *New York Times* wondered whether Londoners were interested in greys back home. Over in New England, the tables were turned. The American red was evicting the grey ('big cowards'). And so, perhaps the only hope for Epping's 'true-blue' reds was to draft in American reds to sort out the greys that had snatched British resident status.[70]

The *New York Times* reported that, according to 'the B.B.C. man', those who should undertake the systematic shooting required so reds could 'live in peace' were the nation's gamekeepers. As it turned out, the FC assumed command of efforts to unsettle the usurping grey. The commission was a non-ministerial government department under MAF whose land acquisition since 1919 for reforestation and afforestation made it one of Britain's biggest landowners by 1939. Of late, the FC's stance on the grey had hardened considerably. In 1937 its chairperson had pointed out that the species was 'of far less importance than at least half a dozen other wild animals', including the red, which had inflicted substantial damage on conifer plantations, especially northeast Scotland's pines. As yet, there were no greys in the New Forest, he claimed.[71] By 1950, however, levels of damage were creeping up across southern England.[72] The body counts squirrel clubs racked up looked impressive: 7,105 and 8,449 in the New Forest and Gloucestershire's Forest of Dean, respectively, during 1952.[73] Yet populations were rebounding.

The FC operated a payment-by-results scheme (ninepence per tail) in the late 1940s among employees on commission plantations such as the Forest of Dean. The FC's director-general, Arthur Gosling, recommended rolling this out more widely as an alternative to free cartridges – a scheme, he suspected, that was open to abuse as dead squirrels did not have to be submitted.[74] The 'shilling a tail' ('bob a brush') programme was specifically designed to tackle infiltration of FC plantations from adjacent private estates by incentivizing control in borderlands.[75] Members of MAF-registered clubs could choose (until 1955) between two free cartridges and the new cash payment, which MAF county finance departments paid out after certification of dead squirrels by a county pest officer.[76] The anti-grey campaign, some commentators reckoned, was accelerating into an 'all-out' 'national drive'.[77]

'Tail-money' advocates cited the previous bounty for eastern greys in colonial North America. Others warned of potential misuse: if pest killing becomes lucrative, a vested interest emerges.[78] Why dry up an income stream by eliminating all breeding stock? An MP for the Kent constituency of Tonbridge, Gerald Williams, wondered if the minister of agriculture had heard of 'boys capturing grey squirrels, cutting off their tails and releasing them to breed more grey squirrels and obtain more tails?'[79]

Despite reservations, 'bob a tail' (as 'bob a brush' was also known), initially a two-year scheme, got off to a flying start. MAF's information officer even enlisted BBC radio's daily soap series *The Archers*, 'an everyday story of country folk' set in the fictional village of Ambridge in fictional Borsetshire. *The Archers* – which had attracted 9 million listeners since its launch on 1 January 1951 – broadcast MAF's self-styled 'propaganda' in a February 1953 episode. Tom Forrest, the local gamekeeper, forms an Ambridge branch of the Grey Squirrel Shooting Club to organize Saturday morning shoots with free cartridges dispensed by Borsetshire's County Agricultural Executive Committee.[80]

During March and April 1953, the members of non-fictive squirrel clubs killed 60,785 in England and Wales, with a bounty paid on 23,751. (The total body count for the previous March and April, pre-bounty, was 34,653).[81] Seven months into the scheme, FC officials and their MAF colleagues were largely satisfied. Between September 1952 and September 1953 – a period that included six months of 'bob a brush' – 258,644 squirrels were killed across England and Wales, an increase of 90,606 on the September 1951–September 1952 figure (168,038).[82]

Killings were concentrated in southeast and south-central England. (London's West End restaurants catering for American diners reportedly bought tail-less squirrels for a shilling apiece.[83]) Kill rates were lowest in Cornwall, Durham, Lancashire, Lincolnshire,

Pembrokeshire and Merionethshire. The zero counts for Anglesey, Caernarvonshire, Cardiganshire, Cumberland, Norfolk, Northumberland, Suffolk and Westmorland suggested these counties were still mainly, if not entirely, grey-free.[84] Natural barriers such as large rivers, mountains and treeless lowlands such as the fens (where there were never any reds either), with the additional obstacles of extensive industrial areas, were credited with keeping the grey out of England's northwest and northeast extremities.[85]

Alongside incremental northward expansion, a fresh challenge emerged in the early 1950s. Reporters and sciurologists noted that southern England's greys were occupying residential gardens, suburban parks, woodlands on urban edges, leafy cemeteries and churchyards, and tree-fringed playing fields and golf courses.[86] Monica Vizoso (née Shorten) characterized this wide spectrum of private, public and municipal green spaces, where shooting was very difficult, as 'danger places'.[87] In London's parklands, for instance, numbers had recovered during wartime, after the pre-war vice grip had been relaxed.[88]

FC officials were concerned that these 'pockets' would function as reservoirs, propelling 'reinfestation' of cleared areas.[89] But numbers declined again in London's squirrelscapes during the bounty period (1953–8). A three-year survey by the London Natural History Society's Ecological Section yielded a single sighting in central London. Possibly an escaped pet, it was 'being chased by a black Labrador dog in Kensington'. In suburban areas, however, greys were found 'more or less wherever there are trees'.[90] Bearing out the commission's reinfestation fears, greys re-established themselves in Greenwich Park by 1958. And in 1959 and 1960, the survey reported fresh sightings in Bushy Park, Hampstead Heath, Highgate Wood, Richmond Park and Wimbledon Common.[91] In 1961, greys were also logged in Clapham Common, Peckham Rye Park and Ruskin Park. 'Firmly established in the suburban garden belt', observed the *London Naturalist*, greys continued 'to advance' towards central London from north and south.[92] The sighting of a grey in Regent's Park in 1961 – the first for thirty years – was confirmed by reports of dreys in 1962.[93]

As greys reinvested London parks, old habits, squirrel and human, returned. In Greenwich Park, recorders noted in 1959, some 'suburban' (inward-migrated) squirrels had become 'bold enough to make overtures towards the public'.[94] Treats such as peanuts were liberally dispensed, despite being 'officially unwelcome'.[95] Walking in Greenwich Park's Wilderness and Flower Garden, on 1 January 1961, the density of numbers reminded a proto-citizen scientist of Regent's Park's Broad Walk in the 1920s.[96]

Though greys were being fed instead of shot in London, when the bounty scheme's two years expired in the spring of 1955, the total body count since March 1953 had piled up to just over 750,000, with 590,653 presented for reward in England, Wales and Scotland for a total payment of circa £29,500.[97] For the likes of Cotterell, the fall in numbers between 1 April and 30 September 1955 – 72,606 compared to 164,000 for the previous six months, a drop of 159,000 compared to the corresponding period in 1954 (227,714) – was 'sensational'. While conceding that breeding conditions in early 1955 were poor, he and other FC officials interpreted this decline in the body count as proof that the bounty was working.[98] Reduced kills reflected low populations.[99]

Among the sceptics was Monica Vizoso of MAFF's Infestation Control Division (ICD) who attended meetings of the FC's Committee on Grey Squirrels (CGS). Converting kill rates into a percentage of the total population was impossible, she cautioned, because nobody had even a rough idea of how many greys there were, before or during the bounty scheme. Reliable estimates were unattainable because solid figures for acreage of woodland occupied were unavailable. Greys were also living, and perhaps in greater numbers, in gardens and parklands. Complicating matters further, squirrel populations, grey and red, underwent regular downturns. The numbers recorded depended on when, precisely, in the cycle of fluctuations the counting exercise was conducted.[100]

Vizoso attributed the decline in greys to a quadruple whammy of environmental factors unrelated to the bounty. Poor beechmast and acorn harvests in 1954 resulted in squirrels' substantial weight loss that heavily impacted reproductive success; greys died, in other words, before attaining sexual maturity. Enfeebled populations were hit again by a severe cold snap during the late winter/early springtime breeding season of 1954–5. The third blow was summer drought in 1955. Finally, predation by foxes and stoats had increased, reflecting a rabbit shortage brought on by an outbreak of myxomatosis.[101] Numbers would have fallen regardless of the bounty. They would have risen again, regardless.

Nonetheless, the FC persuaded the Treasury that the bounty was working just fine. In fact, a complete wipe-out might be possible in some areas, the commission contended. But an added incentive was required because the few that remained would be harder to catch.[102] There was also the danger that all the effort and expense of killing greys over the past two years might be in vain: with anticipated bumper crops of beechmast and acorns in the autumn of 1956, populations could quickly revive. And so, in what a journalist described as a 'national call-to-arms', the commission doubled the reward to two shillings for a year only, from the start of 1956.[103]

The doubling of the bounty was widely covered in the United States.[104] After three months at the higher rate, the *New York Times* reckoned 'native British' reds were finally seeing some gains in their sixty-year war against the 'alien American', thanks to 'Government help'. (In unfettered competition, greys were bound to prevail.) More than a million greys had been despatched, the article claimed, since the bounty was first launched in 1953.[105] Trying to seal its victory, the FC extended the doubled bounty until 7 March 1958.[106]

Squirrel Knowledge: Unofficial and Official

Verdicts on the bounty's success fell into two camps. As data from scientific studies increasingly informed understandings of grey squirrels, tensions rose between the findings generated by experts and the first-hand experience of laypersons ('countryfolk', in contemporary parlance). The post-mortem by government biologists on 'bob a brush' noted that game hunters shot 2 to 3 million greys annually in North Carolina, and this had negligible impact on overall numbers. They also noted that the 'practical man' sometimes failed to appreciate how the 'unsystematic killing of a few hundred thousand grey squirrels, in Britain, may do no more than help to keep the present population stable and healthy'. During a largely forgotten bounty scheme for reds that operated between 1903 and 1917, annual kill rates varied between 2,601 and 7,199, though the figures for the first and final years were broadly similar at 4,640 and 3,998, respectively. They urged rural interests to accept reality: the grey was 'here to stay' and 'no more likely' to be extirpated than 'any other well-adapted vertebrate'.[107]

Withdrawing the bounty in March 1958, the FC trusted that 'despite the absence of a monetary reward everyone concerned with the well-being of agriculture and forestry will continue, wherever possible, to put down the pest.'[108] Many farmers, foresters and other countrypersons, also their allies in Parliament, regretted its discontinuation. Attributing the recent sharp increase in grey numbers in 1959 to cessation of payments, Viscount Mersey urged MAFF to reinstate the bonus.[109] A prominent landowner-politician who represented this position was Harold Macmillan. While Prime Minister from 1957 to 1963, he characterized the clashing bodies of squirrel knowledge as 'official' and 'unofficial'. At his 485-hectare (1,200 ac) Birch Grove estate in East Sussex (the Macmillan family home since 1896), Macmillan – who made his name in publishing – pursued the lifestyle of an Edwardian country gentleman (his wife, Dorothy, was a duke's daughter). Like many landed gents, Macmillan was hostile to greys – a hostility so ingrained he did not need to explain it. He loved a pheasant (and a grouse) shoot, and shooters saw greys as a major threat to gamebird chicks. Even during the Second World War, while serving as British Minister Resident in North Africa, Macmillan had greys on his mind. As he wrote to his wife, referring to their thirteen-year-old daughter, Sarah, who showed 'prowess with a gun': 'Perhaps one day she will shoot a grey squirrel, which will be useful as well as exciting.'[110]

In early January 1960, the prime minister opened his campaign with a 'personal minute' to MAFF minister John Hare: 'What is happening about grey squirrels?' (Macmillan's position showed that to be anti-grey was not necessarily to be pro-red. He never raised the red's plight with MAFF ministers.) His experience at Birch Grove indicated that incentivization had been a roaring success: 'in my part of the country we had got rid of them all, thanks to your subsidy.' Without the bonus, they were now 'swarming again' across his estate. Macmillan thought the 'subsidy' should have been doubled to four shillings, not 'abolished'.[111]

Hare replied that his scientific advisers (echoing Vizoso's position) indicated that 'fluctuations in natural conditions', such as food availability, were the critical influence on population levels. The decline in 1955 reflected a 'bad breeding year', not the availability of a bonus – a finding confirmed by the subsequent rise following a good breeding season, regardless of the doubled bonus. Numbers would plummet again 'once we get a hard winter and a shortage of acorns, beechnuts, etc'. Hare reassured Macmillan that research into control methods continued, though nothing promising had emerged.[112]

'What are you doing about grey squirrels?', the far-from-placated prime minister fired back, emphasizing their resurgence not just in East Sussex but across southern England. 'I am afraid I have again been compelled to accept the scientists' view,' Hare responded, citing a thorough investigation of grey squirrel population dynamics. Even clearer in March 1960 than in January was that the 'controlling factor' in population size was food supply. Hare assured Macmillan that 'every possible line of research' would be pursued in the search for viable control measures. Deeply unimpressed by so-called expertise, the prime minister scribbled on Hare's minute: 'What do scientists know about squirrels?' Trusting in his own observations at Birch Grove and privileging fellow countrymen's convictions, he added: 'What I say is true and known to every [game]keeper in Britain.'[113]

In the summer of 1960, Macmillan accosted Hare's successor, Christopher Soames. Grey squirrel numbers were 'getting up again; in some areas quite seriously'. 'The official view', Macmillan explained, 'is that this is just the ordinary cycle.' The 'unofficial' view – one he shared – was that the bounty's 'abolition' explained the rise. He put his minister on the spot: 'What do you think?' A fellow southern English country gent (and Winston Churchill's son-in-law), Soames was no grey squirrel fan: his reply included the mollifying, handwritten statement, 'I shot two yesterday.' But, like his predecessor, Hare, Soames told Macmillan he was 'looking into the grey squirrel problem and the matter of the bounty'. Knowing full well how exercised

Macmillan was about greys, Soames's private secretary, A. J. Phelps, sent, on 30 September 1960, a minute to a civil servant at MAFF, 'just to warn you that when he returns from America... the Prime Minister may well ask what is happening about grey squirrels'.[114] Because eastern greys abounded at Camp David, the presidential retreat in the Maryland woods, perhaps there had been squirrel talk with President Eisenhower at their previous meetings there in March 1959 and March 1960. The two leaders probably did not have a chance to swap squirrel stories when they met on the latest trip (27 September 1960), Macmillan's main purpose being to deliver a speech at the United Nations.

The president had his own squirrel headaches. In the midst of the bounty scheme, in the spring of 1955, squirrel consciousness was high enough in Britain for a British newspaper to run a *New York Herald Tribune* story about the grey's impudence at home. (A couple of months later, John Steinbeck updated the readers of *Punch*.)[115] Eastern greys had inhabited Washington, DC's treescapes since the early 1900s, including the White House grounds.[116] A member of the administration feeding the White House squirrels before or after a cabinet meeting provided an irresistible photo opportunity during the Warren Harding administration of the early 1920s. 'Pete' was a particular favourite of cabinet officers, who often 'attended' press conferences and was viewed as 'Laddie Boy's rival' (Laddie Boy was Harding's Airedale terrier).[117] Not all the presidential squirrels' antics met with approval: 'Piggy' allegedly ruined a large bed of crocus bulbs in springtime 1923.[118] Post-1945, Harry Truman took a particular shine to the White House's furry residents. He appointed an 'official feeder', sought relaxation by hand-feeding them himself and was often accompanied on his strolls by another 'Pete' the squirrel.[119]

When the U.S. Golfing Association installed a putting green on the South Lawn in 1953, the previously welcome residents took liberties. Disfiguring the immaculate turf by burying nuts, they drove President Eisenhower to distraction. 'Operation Squirrel' began in mid-March 1955 with an attempt to drive them away using high-pitched sounds. Equally unsuccessful was a tape recording of apparently unbearable sound effects.[120] Groundskeepers then resorted

Waging War on the 'Grey Peril', c. 1939–73

Secretary of the Navy Edwin Denby feeds 'Pete', the White House squirrel, 1922.

to a traditional method, trapping, and the 'spoilsports' caught in walnut-baited box traps were relocated (possibly by Secret Service agents) to an undisclosed location (probably nearby Rock Creek Park, though more distant Arlington Cemetery was mentioned too).[121] (How many the presidential dog, Heidi, a three-year-old Weimaraner, bagged is not recorded, though a journalist observed that the White House squirrels 'are getting very wary of Heidi'.[122])

Meanwhile, dreys were removed from White House trees. Then, during National Wildlife Week in 1955, freshman Democrat senator

Richard Neuberger of Oregon, a renowned conservationist and animal welfare proponent, gave a speech in the Senate in which he exhumed an arcane 1906 law prohibiting the trapping, catching, killing, injuring or pursuit of 'any squirrel or any chipmunk' in the District of Columbia (the penalty was $5 and/or imprisonment in the workhouse for up to thirty days, for each squirrel or chipmunk). Neuberger called their presence at the White House a 'tradition and an institution ... for over a century'.[123] The weight of public opinion was firmly behind the uprooted squirrels; Neuberger was inundated with packs and even sacks of nuts to distribute to the evicted.[124]

Though the 1906 law rendered 'the greater part of the District in effect a game preserve', the White House press secretary insisted on the legality of trapping. As they belonged to the federal National Park system, the White House grounds were exempt from DC codes. Nonetheless, trapping had already ceased days before Neuberger's speech on 25 March, with a modest total of three squirrels apprehended.[125] Regardless of public indignation, there was the matter of efficacy. The National Wildlife Federation had predicted that White House staff would struggle to catch the miscreants as fast as they could breed.[126]

The squirrels that irked Macmillan were no more easily despatched than Eisenhower's irritants. Soames's reply to Phelps's missive alerting him that squirrels were likely to be a priority on the prime minister's return from the United States contained no fresh information about control measures. Still, an unusually appreciative Macmillan annotated Soames's note with 'Thank you. Keep after the squirrels.' A few months later, on 21 December 1960, 'in case this should come up at your end', a top civil servant at MAFF, D. H. Andrews, updated Phelps on the squirrel situation.

Andrews reiterated that more accurate data on the grey's depredations were needed before MAFF and FC officials could offer recommendations to Soames. A colleague of Phelps was unimpressed, commenting, on 22 December, 'It is a little disheartening to find that in 1960 we still apparently have no reliable information on the extent of damage caused by grey squirrels when the taxpayer was paying at first a shilling and then two shillings a tail ... between 1953

Dead young grey squirrel, Failand, Somerset.

and 1958.' Had this cash been 'well spent'? Or was it 'money down the drain'?[127]

The following spring, after Macmillan repeated his call to revive the bounty, Soames reiterated that since the weather was the overriding influence on population levels, determining food supplies, breeding conditions and the survival rate of kits, reinstatement would be a waste of public money.[128] Systematic trapping, he thought, offered the best chance of denting numbers (though, with six traps per hundred acres required to make a difference, he was reluctant to impose this cost on the taxpayer). Stepping back from the squirrel war's frontline, the government should 'leave it to the individual woodland owner to deal with [them] in his own woods'.

Macmillan remained unconvinced about the weather's governing role and unpersuaded that there had been a dip in numbers post-bounty: 'I don't believe this is true.' Soames replied on 15 May 1961 that since grey numbers at the scheme's end were practically the same as when the bounty was brought in five years earlier, £81,000 had been shelled out 'to little effect'. Only in a few discrete areas was the overall population big enough to cause 'serious trouble'. As back-up, Soames enclosed an 'interesting' recent *Country Life* article in which nature writer Garth Christian stressed that the bounty represented poor value for money.[129] If Macmillan found the time to read it, what would really have seized his attention, I suspect, was Christian's reference to large numbers of the 'North American invaders' 'emigrating to pastures new' with a 'powerful sense of purpose'.[130]

The prime minister remained silent on squirrels for a few years. Then, in the spring of 1963, despite the distractions of sex and spy scandals (the Vassall Affair and the Profumo Affair), he contacted the FC anew, to learn what it intended 'to do about the grey squirrels at Birch Grove'. This intervention prompted plans for the experts at both the FC and MAFF to confer, though Macmillan had resigned (due to ill-health) by the time they eventually met in October.[131]

Warfare with Warfarin

While Macmillan and his kind wallowed in bounty nostalgia, the FC's pest experts and their MAFF counterparts looked beyond shooting and trapping. Around the time of Macmillan's resignation on 18 October 1963, the experts selected poison as a promising alternative. Warfarin, an odourless and tasteless anti-coagulant, had been dispensed to rats across Britain since the early 1950s.[132] But there was a legal obstacle to administering warfarin to 'tree rats' in England and Wales (following field trials in England).[133] Under the Protection of Animals Act of 1911, it was only lawful to poison 'rats, mice or other small ground vermin' in England and Wales.[134] In Scotland, use of warfarin against squirrels was already permitted because the corresponding animal-protection legislation of 1912 omitted the qualifying adjectives 'small' and 'ground', and left it open as to whether or not the category of 'vermin' included or excluded squirrels of any colour. Nonetheless, warfarin was rarely, if ever, used on squirrels in Scotland for two reasons. Scotland had relatively few greys, not least because broadleaved habitat was scarcer than coniferous cover.[135] The risk to the still common red also restricted deployment.[136]

Politicians representing farmers and foresters amended the Agriculture (Miscellaneous Provisions) Bill of 1971 to add greys to the list of 'vermin' permissible to poison under the Protection of Animals Act. In 1927 the Act had been amended for England and Wales to legalize the poisoning of designated pest species, namely, 'insects and other invertebrates, rats, mice and other small ground vermin'. The latest accepted amendment read 'other small ground vermin including grey squirrels'.[137] The amendment's mover in the House of Lords, Lord Dulverton of Batsford Park, Gloucestershire, was particularly interested in afforestation and timber resources. 'Woodbine Willy' (he belonged to the Wills tobacco family) explained that when the Protection of Animals Act was drawn up, there was no grey squirrel 'problem'. Since 1911, however, it had 'ousted' its British counterpart, 'by cold-blooded murder' – disembowelment, to be precise.[138]

A significant difference between Britain's two squirrels, often remarked on in the early 1970s, was that greys spent much more time on the ground. From a legal standpoint, though, squirrels were tree-dwellers. But for Dulverton and other pro-poisoners, the grey's arboreal traits, such as nesting aloft, did not exempt it from the ranks of 'small ground vermin'.[139] Lord Inglewood strongly supported Dulverton's position. Inglewood was vice-president of the Royal Forestry Society and a landowner in a corner of England, just east of the Lake District National Park, that greys had not yet penetrated. His Hutton-in-the-Forest estate, near Penrith, lay not far from the Scottish border. He thought it absurd that a grey squirrel 40 kilometres (25 mi.) to the north in Dumfriesshire could be legally poisoned, whereas one that showed up in his county of Cumberland could not.[140]

Other amendment-backers highlighted the nationality of squirrels. Britons were much fonder of reds than of 'the foreign North American animal', according to the Earl of Lonsdale, president of the Timber Growers' Association, whose family seat was also in the Lakes. Dulverton pointed out that reds were 'far more beautiful and native'.[141] On the question of belonging in Britain, Lord Leatherland, the journalist and Labour Party peer, took a different slant. Neither a farmer nor owner of woodlands, he lived in Essex, so the squirrels that brought him joy each morning when he surveyed the garden from his bedroom window were grey. If killing them to protect trees was absolutely necessary, then they should be shot, not left 'writhing in agony' after ingesting poison.[142]

Control advocates might deplore but could not ignore the grey's urban status as quasi-public pet. A solution was offered by Birmingham MP Sir Sydney Chapman, who had recently instigated, on 28 July 1971, what became a government-sponsored initiative (*Plant a Tree in '73*) to compensate for the millions of trees stricken by the latest outbreak of Dutch elm disease. In all official business and correspondence, he suggested, MAFF should henceforth refer to the 'unwanted immigrant pest' as the tree rat. Chapman's request for formal eradication of its identity as squirrel gave the newly appointed parliamentary secretary to MAFF, Peter Mills, the opportunity to demonstrate the differences between the *Sciuridae* and *Muridae* families (the latter includes rats, mice, hamsters and gerbils): 'I have it on good scientific authority that this pest is not a rat.'[143]

Tree rat or squirrel, pest or pet, government scientists recognized that shooting had not only failed to curb population growth but may have backfired spectacularly by boosting fertility.[144] Facing parliamentary inquiries about reintroduction of financial incentives for control – prompted by the forthcoming National Tree Planting Year (1973) – MAFF re-emphasized the cash bonus's ineffectuality.[145] The momentum had shifted to warfarin-laced grain bait administered in early summer when trees were most vulnerable to injury.[146] To protect England's and Wales's remaining native squirrels, the Grey Squirrels (Warfarin) Order 1973 proscribed outdoor use of the poison bait in sixteen counties that still harboured reds.[147]

Twenty years earlier, MAF had regretted that 'still far too many people' regarded the grey as 'an amenity rather than a menace'.[148] This amenity value was readily apparent in 1979, during parliamentary debate about dispensing warfarin in Scotland as well as England and Wales, triggered by the Scottish Woodland Owners' Association's concerns that damage from greys to privately owned commercial hardwoods had gone from negligible to severe in under a decade (their northward advance had reached Perthshire). Lord McCluskey, Solicitor-General for Scotland, alluded to the recent scaling back of the adult grey seal cull on the Orkneys. Enthusiasts for more liberal use of poison confronted the same obstacle that seal cullers faced: growing public opposition to the 'control' of an animal that millions of urbanites and suburbanites enjoyed.[149] Not only were greys too entrenched to dislodge; too many people liked them.

In February 1981, the FC advised against further official action: grey populations rose and fell irrespective of intervention. This was met with dismay by peers with timber interests, who continued to grumble about governmental defeatism. Lord Gibson-Watt, a forestry commissioner, reiterated the need to wage perpetual war 'with every weapon at your disposal', including warfarin. The Earl of Cranbrook agreed, rather reluctantly given the grey squirrel's attractive appearance and urban appeal, that Britons should take all measures, but short of poisoning, to 'rid our shores of this unwelcome pest'.[150]

The Last Red Squirrel

Despite 'persecution' of greys, a *Guardian* country diarist (J.K.A.) noted on 3 July 1963 that 'there always seem to be others to take their places.' 'If the rate of spread shown over the past ten years were to be maintained,' wagered Monica Shorten in 1964, 'the squirrel might celebrate the centenary of its arrival [1876] by colonizing the entire land surface of England and Wales.'[151] She was close to the mark. In 1959, 81 per cent of the areas greys occupied were red-less.[152] At the end of the 1960s, only England's far north, the southwestern extremity of Cornwall and the fenlands of East Anglia and Lincolnshire remained grey-free.[153] The year after Shorten's prediction, countryside writer John Cecil Moore took her scenario of total conquest a step further – as far as the red's extinction. Moore, like William Beach Thomas, defended the rural status quo against modernity's intrusions. The main human characters in his novel *The Waters Under the Earth* (1965) are Janet and Ferdo Seldon, who live at Doddington Manor, in the heart of the English countryside. But a deepening malaise afflicts their cherished world, with dry rot inside the manor and plans for an invasive motorway beyond. To cap it all, greys ('them') are ousting reds ('us').

Part one, 'The Day of the Last Red Squirrel', opens with the lines: '*Saw red squirrel*, wrote Janet Seldon in her extraordinary diary... *Rare as dodo now!* The date was 31st July 1950.' Janet asks Ferdo when he last saw a red at Doddington. Perhaps as much as a year ago, he replies, explaining that 'they've been steadily getting fewer ever since those grey bastards came' in the late 1930s. (Janet associates them with the impending war and the physical appearance of Neville Chamberlain.) A 'misguided rich man' brought them over from North Carolina. They evicted reds 'from the woods where they belonged' and flourish among the oaks Ferdo's ancestors planted centuries ago.[154]

Over time, the fortunes of (red) squirrels and squires became entangled. Everything would be hunky-dory provided there were '*Doddington squirrels and Doddington squires*'. But beware the day '*Should squirrels be gone and oak trees fall – Then down go Seldons and down goes all.*' And so reds, symbols of a happier time before the clouds of imminent war blocked the sun, were forgiven for pinching the walnuts. Waiting for war, life goes on. The gamekeeper, Old Northover, campaigns against the 'newly-arrived' greys. He shoots the first pair that shows up – just as he shot anything 'unusual or strange' – but to no avail. Soon the place is crawling with them and Old Northover, 'driven frantic by xenophobia', has his work cut out: 'furtive, hardy, proliferous; intrusive and persistent... the upstart rodents remained, nested, brought forth.' Ferdo and his gamekeeper just about keep up with them until war breaks out and Ferdo goes off to fight. Old Northover manages to continue, for a bit longer, his 'private war against the grey squirrels which seemed to him a more real and immediate foe than the Germans across the Channel'.[155]

When Ferdo returns in 1945, the war on the home front is lost. Greys have 'colonised Doddington, spinney by spinney and copse by copse' and reds have virtually vanished. Janet writes '*Saw red squirrel ... Rare as dodo!*' and Moore interjects: '[t]hey couldn't know that it was the last red squirrel they would ever see at Doddington: the very last.'[156] Though, as Shorten observed, the grey's 'eastern frontier' advanced slower than its western frontier – large parts of southern and central England had seen their 'very last' reds by 1965.[157] As numbers shrank further, expanding numbers of human allies rallied to their side.

6
WANTED: RED AND ALIVE

WHEN warfarin's use was discontinued in 2015, among those urging for its reinstatement was the Earl of Kinnoull. A peer who farmed in Perthshire, Kinnoull chaired the Red Squirrel Survival Trust (RSST), launched by its patron, Prince Charles, in the Lake District in 2009. Kinnoull also chaired the United Kingdom Squirrel Accord (UKSA), a partnership of 33 organizations invited by Charles to sign up in 2014. UKSA consisted of a mix of governmental agencies, companies such as the Killgerm Group, a warfarin manufacturer, and NGOs such as the National Trust and the Woodland Trust.[1] UKSA represented the crest of a wave of organized concern that had been building since 31 July 1996, when the Biodiversity Action Plan (BAP) for the red squirrel – second only to the water vole as Britain's fastest-declining mammal – was released. Known as the UK Strategy for Red Squirrel Conservation, it was led by the UK Red Squirrel Group (UKRSG), an umbrella organization including representatives of the UK's statutory national conservation agencies.[2] The inaugural Red Squirrel Week, an annual series of consciousness-raising activities organized by Red Alert North East and the Wildlife Trusts, quickly followed the BAP.[3]

Voluntary groups, private/public partnerships and consortiums that antedated RSST included Red Alert North East (1992), the European Squirrel Initiative (2002) and the Red Squirrel Protection Partnership (2006). In 2010 they were joined by Red Squirrels Northern England, a collaboration between Natural England, the Forestry Commission, the Wildlife Trusts and RSST tasked to coordinate grey squirrel control in Cumbria, Northumberland and other red strongholds. In 2001, Petronella Wyatt, deputy editor of *The Spectator*, referred to the red's human champions as 'red-squirrelians'.[4] Collectively, these groups constitute the regiments of the red-squirrelian defence force.

A *Guardian* country diarist described Red Alert as a kind of neighbourhood watch scheme.[5] A training ground for citizen scientists is another illuminating characterization. Those who sign up as red defenders log sightings of reds and greys; monitor feeders; set up remote, thermal imaging trail cameras; and collect hair samples from sticky strips in feeder boxes.[6] Volunteers trained by consortiums such as Red Squirrels United also host live capture traps in their gardens and sometimes kill greys. Those who have culled for groups like Red Squirrels Northern England include an artist, musician, newsagent, mechanic, cleaner, carer, accountant, retired police officer and magistrate.[7] Since 2012, RSST has recognized these contributions with a Volunteer of the Year Award. That inaugural year, RSST also gave out a Lifetime Achievement Award. The 78-year-old recipient, David Stapleford, was a member of the National Red Squirrel Captive Breeding Programme (East Anglian branch).[8]

RSST also organizes an annual Red Squirrel Appreciation Day (21 January).[9] 'I take enormous pleasure in having them around – and in – the house when I am at home in Scotland,' wrote Prince Charles to mark the day in 2021 – home being Birkhall on the Balmoral estate. 'They have even been known', he disclosed, 'to hunt down a few of their favourite [hazel]nuts left out in an unguarded jacket pocket!'[10] Charles's extreme fondness for reds – Prince William calls it an infatuation – epitomizes the sanctification and petification of reds (Charles gives his Birkhall squirrels names, according to William) that can be traced back to *The Tale of Squirrel Nutkin*.[11]

The Sanctification of Squirrel Nutkin

As numbers shrank ever smaller, the red's acclaim swelled inversely. A nostalgic *Guardian* country diarist's entry from Somerset, in 1960, told of how 'the children excitedly found a hole in the beech tree where a squirrel had hidden his winter store of nuts.' 'How I wish', the diarist continued, 'it could be the home of the red squirrel which, alas, has now become very rare in this part of the country.'[12] Another plaintive voice from 1960 was John Betjeman's. Then in his mid-fifties, the poet recalled his Edwardian boyhood haunts around Highgate Hill and Hampstead Heath, where greys arrived in 1908. In the early 1900s, 'here on the southern slope of Highgate Hill Red squirrels leap the hornbeams.'[13] Now, he recalled in an article heralding publication of his autobiographical musings in verse, *Summoned by Bells* (1960):

> Still I can hear the trains from Gospel Oak,
> Electric, though, not steam; and still the squirrels
> Leap on the boughs around, but grey, not red;[14]

Grey squirrels and electrification represented the onslaught of unwanted modernity.

Rarity and esteem are closely linked: the fewer, the more highly esteemed. In 1977, the Royal Mail validated the red's status within a select group of 'prized' British wildlife by issuing a set of stamps depicting the red squirrel, hedgehog, hare, otter and badger.[15] A quarter of a century later, when the Royal Mail advertised its special set of 'Woodland Animals' stamps, which included the red, its headline posed a question: 'the nearest you'll get to these elusive woodland animals?'[16]

Key ingredients of sanctification are iconization (icon formation) and valorization (creation of surplus value). In 1998 Lord Inglewood, who resided on the eastern edge of the Lakes, proclaimed that 'in the personification of Squirrel Nutkin', the red was 'almost the icon of the Lake District'.[17] Journalist Patrick Barkham described the home of Julie Bailey, a member of the Penrith and District Red Squirrel Group who trapped and shot greys, as a 'shrine to the red squirrel':

> The time was told by a red squirrel clock, the woodburner was adorned with cast-iron squirrels. Bailey's study walls and carpet were squirrel-red; there were ornamental squirrels

made by a local sculptor, a red squirrel jigsaw, goblet, boot brush, paperweight and piggybank. We drank tea from red squirrel mugs.[18]

Bailey took no pleasure in killing greys but felt an obligation 'to undo the damage the Victorians did by bringing them here in the first place'.[19] In 2009 an American reporter explained that Red Alert Northern England's Save Our Squirrels group (SOS, 2006) wanted Britons to eat 'the ones in our backyards and attics' to save 'the cute one with the tufted ears featured in Beatrix Potter books'.[20]

Conflation of reds with Squirrel Nutkin, indispensable to their iconization and valorization, was already evident in the Royal Photographic Society's exhibition of Nature Photography in 1957, which included 'Young Squirrel' by Ronald Thompson. The caption explained how it 'looks as if it had stepped from a page in one of Beatrix Potter's books'.[21] A quarter of a century later, Nutkin served as a powerful tool for the red's champions. 'We do not want the red squirrel just to remain as a character of fiction in . . . Potter books,' cautioned Labour's Baroness David, opposition spokesperson on the environment.[22] When Potter published her story, reds were a common sight, explained actor Andrew Sachs, narrating an episode of ITV London's *Survival* series on the red's decline.[23] *Daily Mail* coverage of a campaign to radio-collar reds and greys in the borderlands between Lancashire and Cumbria, using the data to bias woodland management towards reds, was headlined: 'Help Is at Hand for Nutkin's Last Stand' (19 July 1990). In 1980, a journalist – ignoring centuries of mixed feelings at best towards the red and downright hostility at worst – proclaimed that 'we all loved that little animal.'[24] Just as the grey could do no right (guilty by default, pre-disgraced), the red, it seemed, could do no wrong. Red proponents regarded any criticism of what is an arch-example of nationalized charismatic mini-fauna as unpatriotic badmouthing. To sanctify, for instance, was to downplay flesh-eating. Victorian and Edwardian commentators had acknowledged the red's carnivorousness. Increasingly, though, its champions promoted it as a 'natural vegetarian'.[25] Instances of dietary misbehaviour were aberrant.

Road safety action song written for the Tufty Club, 'Tufty Fluffytail', words and music by E. M. Blair and Ken Keith, 1962.

The Royal Society for the Prevention of Accidents (ROSPA) gave the red squirrel's reputational capital its biggest boost since *Squirrel Nutkin*. In 1953, when the FC launched 'bob a brush', Elsie Mills, head of ROSPA's 'under-fives' department, wrote a series of books illustrated by Marcia Lane Foster. One for each month, they dispensed road safety advice for infants through animal characters who lived in an actual place, East Cheam, a south London commuter suburb. The youngsters (the 'furryfolk') were Minnie Mole, Willy Weasel, Harry Hare, Fiona and Freddy Fieldmouse, Bobbie Brown Rabbit and Griselda Grey Rabbit. Mrs Owl was their teacher; Mr Horace Hedgehog was the 'lollipop man'; and Sergeant Badger was the local bobby. But the lead character – one of the 'furryfolk' – was a five-year-old red squirrel named Tufty Fluffytail.

Greys also have fluffy tails. But Tufty Fluffytail's first name denotes his species' signature ear tufts, a distinctive wintertime trait. Grey ears are round rather than pencilled, and tuft-less. Mills later

Wanted: Red and Alive

explained that she had stumbled on this 'suitable vehicle' for winning young hearts and minds during a walk near her Kent home. After crossing a busy road with difficulty, she entered a tranquil woodland. When she spotted a 'little' red squirrel under an oak – and a larger one she took to be its mother – it dawned on her that 'this was exactly the type of character' to transmit her road safety message.[26]

Mills did not explain exactly why she considered the red squirrel (seldom seen in Kent by the early 1950s) the most appropriate vehicle for her mission. Nutkin was hardly a role model for children – and, as a supreme risk-taker, might well have been the jaywalking furry-folk's ringleader. What she probably had in mind was Nutkin's enormous value as a symbolic creature. Tufty (unlike Nutkin) was a 'red squirrel who always tries to do the right thing'.[27] In total, 242 children under five years old died in road accidents in 1960 and a further 2,344 were seriously injured; the age group's fatalities for 1961 rose to 304.[28] And so the most important things to do right were not to run into the road; to cross it safely holding an adult's hand; and to memorize the kerb drill. After British youngsters embraced Tufty, he lent his name to a new kind of squirrel club.[29]

'Stop, look, listen',
Tufty Club button, n.d.
(after 1993).

The Tufty Club for under-fives, launched in December 1961, was obviously a very different animal from the clubs whose adult members shot squirrels of the other colour. This club was a nationwide network of road safety groups that attracted 53,000 members within a single year. A year later, membership was expanded to include any children under the age of eight.[30] A string of eight-minute Tufty adventures on Rediffusion TV's children's series, *Small Time*, and further stories of him and his 'furryfolk' friends, extended Tufty's renown.[31] Figures for registered local clubs vary from 24,000 in 1970, 10,000 in 1973 and 22,000 in 1978.[32] Between 1961 and 1978, nearly 4 million badge-wearing, club song-singing children enrolled.[33] The first film, *Furryfolk on Holiday*, was made in 1967, sponsored by Butlin's. In the early 1970s, the government's Central Office of Information (Department of the Environment) commissioned a series of animated films for TV ('shorts') featuring Tufty and his pals. In 2019 the parliamentary under-secretary of state at Defra, Thérèse Coffey, who grew up in Formby, a place renowned for its reds, observed that she was 'just about old enough [born in 1971] to remember Tufty from the road safety films... shown in the '70s'.[34]

The 'safety conscious red rodent with the bushy tail and cute face' was so successful as a promotional device that he quickly rivalled Nutkin as journalistic shorthand for red squirrel.[35] Like Potter, Mills

'Play Where It Is Safe': Tufty on the slide in a playground with the rest of the Furryfolk, 1960s. The stronghold strategy that the UK government adopted in 2005 was designed to ensure that reds could continue to play safely in their last bastions.

'How Tufty crosses the road', n.d.

appreciated the power of fictional animal characters to become real in young imaginations. And like Nutkin, Tufty inspired a thicket of merchandise, including board games, jigsaw puzzles, colouring books, Christmas cards, 'bendy' toys, mugs, *Good Housekeeping* knitting patterns, ceramic figures, puppets, handkerchiefs, toothbrushes and flame-resistant pyjamas.[36]

As Tufty approached his 25th anniversary, some adults criticized Tufty Club publications as unrealistic. They were turned off by the absence from visual educational materials (including films) of dwellings such as high-rise flats, pointing out that not all children had secure private gardens to play in – and parks within spitting distance.[37] In 1981 *Child Education* magazine referred to 'twee little squirrels'.[38] Then, in 1984, Lambeth Council's safety subcommittee urged ROSPA to confront Tufty's white, middle-class and detached-home-in-the-suburbs identity. The voices in the Tufty films, for example, use expressions such as 'jolly' and 'super'. The subcommittee's report urged ROSPA to revisit its books, cartoon strips, posters and films. The subcommittee's report also identified gender stereotyping; the grown-up's hand that Tufty holds is that of his 'mummy', who does not go out to work. They wanted Tufty to be more relevant to children living in multiracial, multicultural inner London. The leader of the Conservatives on Lambeth's Labour-controlled borough council retorted that Tufty was a 'perfectly innocent brown squirrel', who 'might be allowed to stay if he was grey and black' instead.[39]

In conjunction with the business Adams Childrenswear, Tufty was 'modernised' in 1993.[40] Despite the makeover (jeans and T-shirt), the brand remained essentially unaltered. And yet, the revamped Tufty never regained the heights of popularity enjoyed during his first two decades.[41] Two years after the relaunch, health educationist Margaret Collins published the results of a survey on animal characters that primary school age children encountered on TV and radio and in books and comics. Screen animals – such as Bambi, Sonic the Hedgehog and Minnie Mouse – rather than creatures from books and comics were the ones that engrossed young minds. In fact, no child mentioned Tufty – or Nutkin.[42]

Collins followed up with a survey specifically about Tufty. Teachers and parents in their twenties and thirties had varying levels of awareness. But everyone who remembered him knew he was a squirrel. In the 1960s, '70s and '80s, ROSPA invariably referred to Tufty as a red squirrel; reflections marking the 25th birthday of the 'fully clothed, upstanding woodland rodent' noted that this lifespan 'far exceeds the expectation of any ordinary red squirrel'.[43] By the early 1990s, Tufty's

'Red squirrels. Please drive slowly,' sign near Loweswater, Cumbria.

colour had faded. As for the kids, only two or three in a reception class of 35 had heard of Tufty. As for what he was, only one child knew he was a squirrel (colour unknown).[44] But as a journalistic reference point for reds (however glib), Tufty's influence did not wane. 'Tufty's Last Stand' (1998), an article based on a visit to the 'real-life Tufty' at Formby, reported that trailblazing greys had been spotted here for the first time, also in the Keswick area of Cumbria, the heart of Nutkin country.[45] 'The Tufty Club's Safe Haven' was how the *Daily Mail* on 22 September 2001 styled Brownsea Island, in Poole Harbour, Dorset, with its population of thirty people and 250 reds.

When the roles of humans and squirrels were eventually reversed in terms of who was helping who to cross the road, Tufty was the go-to celebrity. A rope bridge in County Durham was headlined 'Tufty Gets a Helping Hand Crossing the Road'.[46] Squirrel researcher Peter Lurz observed in 2014 that youngsters from southern England 'cannot help but wonder why an animal that introduces road traffic safety to children is so frequently seen dead by the roadside in the North'.[47] Roads in south and west Cumbria now bristle with signs advising motorists to watch their speed: 'Red Squirrels: Please Drive Slowly'; 'Red Squirrels: Slow Down!'; 'Caution: Red Squirrels Crossing'.

Tufty even cropped up in a story about Wayne Rooney in 2006, sparked by the announcement of his book deal with publishing company HarperCollins. As the reporter adopts Rooney's persona, putting words into his mouth, reporter-as-Rooney recalls the brouhaha a few years earlier over his fiancée's engagement ring, allegedly jettisoned among Formby's squirrels. Reporter-as-Rooney's comment on the ring's fate? 'So now it belongs to Mrs Tufty.'[48] (Tufty had a mother, Mrs Fluffytail, but no wife, obviously, given his age.)

A couple of years after Mills created Tufty, the native squirrel as embodiment of Englishness received a further fillip. In 1957, as squirrel-shooting club members were receiving bounty cash, litter-combatting charity Keep Britain Tidy chose the red to front its new anti-litter campaign for the countryside. Reginald Bevins, parliamentary secretary to the Ministry of Housing and Local Government, announced that 'it is a delightful small creature which reminds people of the country.'[49] Unlike ROSPA's squirrel, Keep Britain Tidy's squirrel was depersonalized. Posters depicting a red squirrel against a blue background carried the slogan, 'When in the country, don't forget . . . take your litter home. Keep Britain Tidy.'[50]

Not everyone was convinced. Leonard Harrison Matthews, director of ZSL, explained that the red had been selected because it was 'supposed to lay up stores of food in neat little dumps against the cold weather'. But the connection between hoarding food and keeping Britain tidy eluded him. For, 'pretty as it is', the red squirrel 'alas, is a most careless scatterer of litter', pulling nuts and young leaves off trees and leaving them on the ground, mostly uneaten. In short, the red was 'destructive, wasteful and untidy' – a non-human litter lout. Fifteen years later, in 1972, sociologist Stanley Cohen would identify Teddy Boys as an example of a group that engendered widespread social fear. Casting around for a human equivalent to the delinquent red squirrel to drive home his point, Matthews alighted on this male youth subculture with a flamboyant dress code and reputation for trashing cinemas and other acts of vandalism. Hence 'Teddy boy of the trees'.[51]

When MP G. B. Finlay (who represented Epping Forest) brought this untidiness to Bevins's attention, he retorted: 'If my hon. friend can suggest an animal which gets higher marks for tidiness while being equally photogenic', then the minister of housing and local government would 'gladly consider adopting it' for 1958's anti-litter campaign.[52] Noting that Herefordshire MP David Gibson-Watt had nominated the badger instead as an exemplar of cleanliness and tidiness, a reporter conceded that at least the native squirrel ('whiskery, ear-tufted and a lovely reddish chestnut colour') had been chosen rather than the now much more familiar 'American interloper'.[53] The correspondent acknowledged that red and grey were both wasteful and messy. But it was specifically the grey that red-supporting Britons pounced on as the delinquent, intimidating, arboreal teddy boy.

A Bipartisan Cause

Those in commerce that felt disadvantaged by their competitors' size were keenly aware of the edge the grey possessed over its smaller rival. The stirrings of a tendency that gathered pace towards the end of the 1900s – to treat the grey as a furry embodiment of globalization's downside – were detectable in 1953 during parliamentary discussion on the jewellery and silverware industries. Henry Usborne represented a constituency in Birmingham that contained the city's jewellery sector. Birmingham's jewellers, a collection of small, family-based firms renowned for artisanship, had hit tough times. They struggled to compete with the big outfits, increasingly the norm in the United States, which enjoyed the advantage of large-scale output's economies of scale. 'For some reason or another,' mused Usborne, 'the grey squirrel does not kill the brown squirrel; it frightens it away.' Accordingly, 'the two cannot live in the same territory. Nor can craftsmanship and mass production survive in the same shop.'[54]

Deploying red squirrels for analogic purposes was a far cry from signing up as a red-squirrelian. Politicians representing cities like Birmingham – or other places where they were long gone – showed some allegiance. But the strongest support stemmed from the red's last English strongholds in Lancashire, Cumbria and Northumberland. In early 1980s Scotland, the balance of power between red and grey was the reverse of England's. Lord Mowbray counted himself fortunate that the red was still 'the natural animal at my home in Angus', where 'I have no grey squirrels, happily.'⁵⁵ But Scotland's politicians were by no means complacent.

The storied red's plight packed enough clout to unite politicians and others of all stripes. For Lord Kilbracken, a former naval pilot and intrepid foreign correspondent, whose estate in Ireland's County Leitrim, Killegar, was red-less, protection was a 'completely non-political subject'.⁵⁶ Much of this bipartisan appeal was attributed to Beatrix Potter. As Captain Humphrey Drummond of Megginch Castle, Perthshire (husband of the future Baroness Strange) observed in the early 1980s:

> Without Squirrel Nutkin, most grown-ups would know nothing about them and would care even less. But the red squirrel is now so deeply embedded in the new-found British awareness of wildlife that few people's faces will not crease with pleasure at the sight of a darting russet-brown Nutkin.⁵⁷

No matter how divided Britons were in 1987 after Thatcher's landslide third election victory, reds seemed to offer something to unite around. Viscount Ridley was another landowner-peer 'still fortunate enough to possess red squirrels' on his family seat of Blagdon Hall, a 3,440-hectare (8,500 ac) estate in Northumberland.⁵⁸ Ridley – the older brother of Nicholas Ridley, secretary of state for the environment (1986–9) – chaired the Red Alert North East Partnership, which the Northumberland Wildlife Trust (of which he was president) launched in 1996 to keep England's largest county grey-free.

The Northumberland constituency of Conservative MP Peter Atkinson (Hexham) housed more reds than any other in England. Like Drummond, Atkinson suspected that the British loved them so much because so many had read *Squirrel Nutkin* as youngsters. In 1996 the strength of this love could not be attributed to direct experience. In London and the Home Counties, Atkinson observed, only someone over the age of sixty (like himself) was likely to have seen one in the flesh. That London's children assumed all squirrels were grey saddened him.⁵⁹ Opening a Commons squirrel debate in 1996 – towards the end of John Major's second term – Atkinson remarked on how refreshing it was to discuss a subject of widespread national agreement. Underlining its bipartisan nature, he cited two other staunch friends of the red in the House: fellow Conservative Robert Atkins, whose constituency was in west Lancashire; and Joyce Quin, a Labour MP who represented an urban constituency in Newcastle, 40 kilometres (25 mi.) east of his own.[60]

Atkins, minister of state for environment and countryside from 1994 to 1995, reminisced about his 1950s north London boyhood; reds were a familiar sight on his walk to Highgate School through Highgate Wood. Forty years later, a handful lived in Mere Sands Wood reserve, adjacent to his South Ribble constituency. (Mere Sands, he claimed, was the first place 'on the way north [from London] in which you would bump into red squirrels'.[61])

As for Quin, on 16 January 1996 she had tabled an early day motion: 'Survival of the Red Squirrel in the UK'.

> That this House is deeply concerned about the reduction in numbers and distribution of the native red squirrel throughout the United Kingdom; draws attention in particular to the recent encroachment by the grey squirrel into the North-East of England, one of the few remaining red squirrel strongholds; and calls on the Government urgently to work with and support wildlife organisations and other relevant organisations and authorities in order to promote and safeguard the habitat of the red squirrel and in order to ensure the survival of the species in the short-term and its expansion in the long-term.

All parties were represented among the 166 signatures that Quin's motion attracted, their constituencies stretching the length and breadth of the UK. Signees also spanned the ideological spectrum, from London's Labour left wing (Diane Abbott, Jeremy Corbyn and Ken Livingstone) to Bill Cash on the Conservatives' right flank, which often represented largely rural constituencies.[62]

Though Atkinson, Atkins and Quin were ardently pro-red, red-squirrelians were more numerous in the Lords. How large a dose of empathy hereditary peers felt with another embattled species is hard to say. Based on the amount of time they devoted to the red's predicament and the grey's infamies, squirrels were clearly a topic close to the hearts of the landed gentry that dominated the upper chamber until reform in 1999 removed the automatic right of hereditary peers to a seat. The feeling that this was the last chance to secure the native squirrel's future pervaded a major squirrel debate in the Lords in the spring of 1998, the grey frontier line having reached England's far north. Those who spoke emphasized the need for a big, redoubled push by public, private and voluntary sectors. Though 'driven out of its home' across much of England, Inglewood (the debate's initiator) was grateful that the red still thrived at his family home of Hutton-in-the-Forest, Cumbria: 'Most mornings when we draw the curtains of our bedroom we can see them playing in the trees outside the window.'[63]

Inglewood was patron of Red Alert North West, a group launched in 1993 and sponsored by the National Provident Institution (NPI), a mutual life assurance and pensions company, whose emblem became a squirrel and a clock in 1959. When the demolition and rebuild of NPI's head office in London at 48 Gracechurch Street was completed that year, the building bore a double-facing clock topped by a gold-coloured squirrel. NPI's chairperson, Emlyn Bevan, explained in 1961 that

> we have added for good measure a provident squirrel and a clock to decorate the corner. We hope they will gently remind those who pass by not only of the time but also of the 'time to save' and that this can be providently and profitably arranged at the sign of the Squirrel and the Clock.[64]

Building up an autumnal store of nuts to see it through winter is a squirrel habit associated with saving money for a rainy day: hence the expressions 'save like a squirrel' and 'to squirrel away'. NPI's squirrel became unmistakably red when later adopted as the company logo, featuring ubiquitously on material ranging from annual reports to policy statements. A short TV advert starred a swarm of red squirrels busily sorting a mountain of shell-on hazelnuts into the letters NPI.

Inglewood represented those who dismissed the government's red-saving 'stronghold strategy' – habitat enhancement in the last red redoubts combined with grey-free buffer zones – as pussyfooting. Decisive action, for these aristocrats, meant heavy thinning of grey numbers by any means, including warfarin and bounties. Dismissive of recent studies citing red–grey coexistence at certain Scottish sites, Inglewood considered the belief in longer-term sharing of habitat deluded.[65] A long-standing resident of the Lake District town of Hawkshead told Inglewood that 'when he was a boy, the northern limits of the grey squirrel were Crewe [Cheshire]; now it is at the bottom of his garden.' If nothing was done, Inglewood warned, the 'most lovable and loved of our British native fauna' would soon join the wolf, wild boar and beaver on the extinct natives list.[66]

According to Lord Rowallan, of Rowallan Castle, Ayrshire, reds deserved protection not 'merely because they are beautiful . . . They . . . have another advantage – they are British.'[67] Baroness Farrington (the government's environment spokesperson in the Lords) addressed the distinction between British squirrels (reds, but seldom seen) and squirrels in Britain (mostly greys) by paraphrasing views that the parliamentary under-secretary of state for the environment, James Clappison, had expressed in the Commons two years earlier: 'amongst our school children, I fear, it is an assumption that "British" squirrels are grey.'[68] Lord Plumb, who represented a region where reds were long gone and which greys had 'made their home' (the Midlands), agreed that more should be done to publicize the grey's American origins and to remind Britons that Nutkin was specifically a red squirrel.[69] The nearly two-hour squirrel debate in which Plumb spoke in 2006 was initiated by Earl Peel. A former president of the Yorkshire Wildlife Trust, Peel reiterated that reds are 'an integral part of our woodland landscape . . . immortalised . . . through the charismatic character of Squirrel Nutkin'.[70]

Lady Saltoun of Abernethy also counted herself fortunate to live in a place, Upper Deeside, Aberdeenshire, not yet invaded by greys. Reds practically overran her home on the edge of the Mar Estate, where there were few of the grey-friendly native hardwoods so 'beloved' of what she saw as conifer-hating conservationists. For the

75-year-old chief of the Fraser clan, it was a question of character. The red's demeanour was innately superior, 'rather like quiet, well behaved people, who do not make a nuisance or an exhibition of themselves, or commit crimes, and so do not get themselves into the papers in the vulgar way grey squirrels do'. She implored the government to ignore the benighted masses who had never known the red and 'think, in their innocence' that greys are 'dear little creatures'.[71]

Lord Chorley (Labour) reiterated that the red was 'much more beautiful . . . and, of course, the native type of squirrel'.[72] He had served as honorary secretary of the Council for the Preservation of Rural England for over thirty years (1935–67) and on the National Trust's executive committee for more than forty. A resident of Westmorland County in the southern Lakes, Chorley recalled having been taken to tea with Beatrix Potter in 1941, when he was eleven and she was in her mid-seventies. Doubtless, if still around she would be 'hugely effective' for the red cause.[73] The Earl of Liverpool agreed that Potter would have fought valiantly to keep greys out of the Lakes. He invited colleagues to imagine that, in *Squirrel Nutkin*, 'Twinkleberry had been a nasty, pox-carrying grey cousin [a reference to SQPV] instead of a near and dear brother of Nutkin.' Twinkleberry recast as a grey might have raised awareness of the enormity of the threat the non-native squirrels posed, perhaps nipping the problem in the bud.[74]

Alluding to the importance his wife's grandfather, Stephen George Tallents, placed on concerted action against greys in *Green Thoughts* (1952), Inglewood lambasted the 'squeamishness' of successive governments. Reds badly needed a Churchill, not more appeasing Chamberlains.[75] Inglewood insisted that the only way to really help reds was to hit greys hard, especially pioneers appearing in buffer zones around sanctuaries.[76] One of those buffer zones was the Rede Valley, adjoining Kielder Forest. The largest forestry plantation in England and Wales was home to 12,000 reds, three-quarters of the remaining English population (according to the lowest estimates). Rede Valley was the seat of Lord Redesdale, who became the youngest elected life peer (Liberal Democrat) in 1999. Back then, Redesdale calculated that the grey's northern frontier was advancing at a yearly rate of roughly 8 kilometres (5 mi.).[77] Six years later, in 2005, he witnessed the first greys on his lands (mercifully, as roadkill). Reds were inseparable from his childhood landscape: he spent every summer in the woods around the family cottage.[78] Yet he appreciated the strong attachment to the 'sociable and friendly' greys that were the only squirrels most Britons saw.[79]

Potter's Nutkin was pinned down by Old Brown's talons, but he 'avoided extinction', according to journalist Roger Scott, by managing to pull free. But as greys entered Lakeland's southern fringes and exerted a grip, escaping their predicament was unlikely to be as straightforward for 'the real reds'.[80] Those in Atkinson's Northumberland constituency were not secure for much longer either. The grey frontier crossed the Tyne in November 1995.[81] And before the end of the century, it was becoming clear that grey success had a virological explanation, too.

The *coup de grâce* for remaining reds, their champions feared, might be SQPV. First detected in substantial numbers in the mid-1990s, the virus's symptoms are discharging lesions, weeping pustules, scabs, and swellings that develop on the eyes, mouth, ears, feet, thighs and genitalia, converting soft tissue to sludge.[82] Infected squirrels succumbed so fast – within two weeks – that afflicted live specimens were rarely found. SQPV, which greys probably brought from North America, is a disease-mediated invasion (DMI) because the grey is an asymptomatic immune vector possessing antibodies.[83] In other words, SQPV is 'benign' for a species that evolved with the virus.[84] More specifically, SQPV is a 'spillover DMI', which happens when a non-native directly imports a virus into an area already inhabited by a native congener. And when the affected native species closely matches the 'reservoir host' phylogenetically, a spillover virus is a particularly deadly 'biological weapon'.[85] SQPV accelerated greys' takeover rate by at least seventeenfold, perhaps as much as by 25 times.[86]

Levelling Up: Tools and Tactics

SQPV – which most greys carried – reinforced the widespread conviction that they bore direct responsiblity for the red's demise. Its spread certainly compromised revival measures enacted in compliance with the Convention on Biological Diversity, a multilateral treaty the UK signed up to in 1992. The stronghold strategy of buffer zones and sanctuaries that the likes of Inglewood disdained was one of various measures, short and long term, that red defenders devised and pursued, often in tandem, to even the odds in the red's contest with the grey. One of the most imaginative alternatives to the traditional lethal methods (shooting, trapping and poisoning) that had done little to arrest grey expansion was aggression-management through medication. Joanne Lello, when working as a pre-doctoral researcher at Liverpool John Moores University, floated the idea that drug therapy might reduce the 'anti-social' tendencies that endangered reds.[87]

Nothing came of this quirky approach. By 1996, though, a host of mostly non-lethal options for suppressing greys and revivifying reds were in play. The shorter-term measures with the most legs were red-friendly/grey-hostile food dispensers (feeding boxes); reintroductions in mainland settings; culling on islands as a prelude to reintroduction; and sanctuaries with buffer zones. Longer-term measures involved fertility treatment for greys (slow die-out); breeding of disease-resistant 'super-reds'; and pro-red tree plantings (habitat redesign). More unorthodox was a revival of human consumption under the new banner of 'eat a grey to save a red.' Other red protectors placed greater faith in consumption by a rejuvenated native predator. According to the logic of the proverb that the enemy of my enemy is my friend, the pine marten serves as unwitting ally.

Food Supplementation

Even if an SQPV vaccine was converted from successful prototype to licensed and affordable product, greys would remain a competitive threat. That greys were superior foragers was widely known. But the reasons were better understood in 1993 when researchers provided a phytotoxic explanation for the lower densities and fecundities of red populations, attributing them to acorns' red-specific toxicity.[88] Improving access to digestible food promised quick benefits for reds. Food hoppers (dispensers) turned littleness and lightness (circa 300 versus 550–700 grams (10 vs 19–25 oz)) from disadvantages into advantages.[89] As Clappison explained in 1996, smaller and nimbler 'British reds' could 'out-compete their American cousins' where they were 'better able to access the tiny conifer seeds' at the tips of slender twigs.[90] Only a squirrel heavier than 400 grams (14 oz) could activate the depressible platform of a hopper that brought down a food-blocking screen. Only 'squirrel-lite' could grab goodies such as sunflower seeds.[91] Trapdoors secured by magnets also operated by 'weight discrimination', dumping the heftier grey on the ground.[92] The 'killer hopper' worked in reverse. Fitted with a weighted, hinged flap too heavy for reds to muscle open, it administered warfarin, or a bait infused with a sterility vaccine.[93]

Reintroductions

Different approaches were combined in Thetford Forest, Britain's largest lowland conifer plantation on the borderlands of Norfolk and Suffolk, a post-1919 creation dominated by Scots pine and Corsican pine. After six hundred greys were culled there in the early 1990s, transplants from Cumbria were supposed to reinvigorate the remnant reds; grey control reintensified in the late 1990s, the numbers killed during 1998, 1999 and 2000 totalling 2,209. But the government agency English Nature's hopes that the relocated stock, supported by food dispensers after release from breeding pens, would breed in the wild were dashed. By 2004 Thetford's reinstatement project had officially fizzled out.[94]

To reinstate means to restore to a former position. 'Reintroduction', a word often used synonymously, makes explicit that the species being reintroduced is completely gone from somewhere. Reintroductions in this sense date back to the early 1900s, at least. A 1934 letter to *Country Life* ('Squirrel Nutkin's Tea Party') was accompanied by a photograph of ten 'owners of white waistcoats, red coats and bushy tails' nibbling 'Barcelona' nuts from a basket on a lawn. The letter writer, the daughter of 'Suffolk School' artist Edith Richenda Bland, recalled how her father released two pairs of the 'original British red squirrels' (procured from Leadenhall Market, she disclosed in a follow-up letter) into a patch of red-less woodland within West Norfolk's fruit-growing district. Within five to six years, the original four's progeny had spread to places 'where none had been seen within living memory'. Richenda Bland was prepared to forgive them for biting off unripe peaches, scoring dessert apples and tossing Victoria plums around. These 'amusing antics' offered 'ample compensation'.[95]

Bland's letter prompted requests for advice from *Country Life* readers wanting to start their own colonies or resuscitate flagging numbers. Another private scheme, implemented in 1935 and an early example of reinstatement, transplanted three pairs from the New Forest to Rothbury, Northumberland, where reds had grown sparse.[96] By 1938 a local resident reported that the area's reds numbered over one hundred.[97] Another instigator was ZSL, keen to atone for its role in establishing greys in London. An authority on London Zoo's history called a proposal for the captive breeding of reds the 'Zoo answer to American invasion', the Regent's Park colony having spread into 'open country' where they 'made war on' and 'hunted out' 'our beautiful British squirrels'.[98]

The breeding stock's source was a private collection that novelist and poet Adelaide Mary Champneys (Fellow of ZSL) raised at her Hampstead home. In 1932 ZSL despatched four pairs of young reds descended from Champneys's collection to its recently established Whipsnade Zoo, near Woburn, Bedfordshire, releasing them among the oaks of Bluebell Wood.[99] Whipsnade's greys were trapped and shot as a prelude to reintroduction of a creature 'too long conspicuous by its absence'. Food was provided until they found their feet.[100] Commentators emphasized the proximity of the Woburn estate, where 'this American tree rat . . . first multiplied in large numbers'.[101] Prospects were not encouraging (no sign of reproduction a year after release) but ZSL tried to rekindle the momentum by transplanting all its reds to Whipsnade's woods.[102]

Forty years later – the Whipsnade project having petered out – ZSL undertook a similar venture in its own back garden. ZSL acknowledged that its experiment-cum-pilot scheme (sponsored by NPI) was partly 'sentimental': 'they used to be there, they belong there, and their unnatural absence is widely regretted.'[103] Yet turning the clock back to 1905 and replacing Regent's Park's greys was never the plan. ZSL just wanted to know if reds, last recorded there in 1942, could be re-established in Regent's Park.[104]

In June 1984, fourteen juveniles were trapped in Fife, east central Scotland, where still abundant reds were often 'destroyed' in immature conifer plantations. On arrival at London Zoo, they were acclimatized in an aviary installed in the early 1950s in memory of the 12th Duke of Bedford, the son of Britain's most notorious grey squirrel spreader.[105] During October and November 1984, ten of the twelve surviving specimens were released, radio-collared, into the park itself. Initially, they appeared to be acclimatizing just fine. Many Britons thought reds were incompatible with cities. But ZSL's curator of mammals, Brian Bertram, reminded a journalist that they were a familiar sight in many continental European cities.[106] Shielded from grey foraging and nesting competition by bespoke hoppers with unshelled (that is, bird-proof) hazelnuts and drey boxes inaccessible to greys, there were no signs of conflict (or disease) during their first few months at liberty. Moreover, far from being strangers, genetically speaking, the squirrels from Fife closely matched London's former reds. After all, southeast English stock had replenished Scotland's squirrels in the late 1700s and early 1800s.[107]

Nonetheless, all died before they could breed. Regent's Park is a road-ringed insular space and radio-telemetry data indicated that six healthy specimens became roadkill (a common fate for local greys, too). A feral cat killed another. Two unattached transmitters were found, and an unknown fate befell the tenth.[108] Still, it was a valuable experiment. Places where the two species overlap are rare, so Regent's Park provided the opportunity to evaluate the three standard replacement mechanism hypotheses offered for grey success nationwide: disease, aggression and ecological competition.[109] The official report favoured the second hypothesis; the reds seemed to have access to plenty of food and showed no signs of debilitating disease. Bertram and a fellow researcher logged nearly 1,500 'aggressive interactions', usually greys chasing reds briefly (six to eleven seconds) during the mating and dispersal seasons. But this was inconclusive evidence of negative impact: greys chased each other as well and as often; and reds were never injured during chases.[110]

Island Culling-cum-Reintroduction

Small islands offered the most realistic context for combined eradication and reintroduction. In the late 1990s, efforts centred on Anglesey, an island of 710 square kilometres (274 sq. mi.) in northwest Wales, where the first grey was reported in 1966. By 1998 they numbered circa 3,000; meanwhile, reds were down to under fifty.[111] RSST's chief scientific adviser, Craig Shuttleworth, summed up the project, supported by two hundred volunteers, as 'Nutkin ventured, Nutkin gained.'[112] As coniferous plantations were cleared of greys, transplanted reds began breeding.[113] An interim report prompted this U.S. headline: 'Britain: No More American Gray Squirrels: We Want Our Reds Back' (*Christian Science Monitor*, 5 November 2009). In 2015, after no grey had been sighted for eighteen months, and red numbers stood at roughly seven hundred, Anglesey was declared grey-free. 'Goodbye to Greys as Reds Take Over Island' ran a triumphant local headline on 2 July 2015 in the *Bangor Chronicle*.

National coverage extended to those who regarded the Anglesey project as an engineered local extinction. Diametrically opposed to Shuttleworth's view that greys 'simply shouldn't be here' was Angus Macmillan ('Professor Acorn'), who, with his son, Neil, adopted the grey's persona.[114] They launched a website ('We're As Native As You!') in 2006, maintained a 'factsheet' ('Victimising Grey Squirrels') and started a petition, which attracted 140,000 signatures, to stop the 'barbaric' cull.[115] 'Professor Acorn', who thought the best way to help reds was to plant more conifers and fewer broadleaves, felt that wiping out a sentient species because it was non-native (alien), even to protect another species (so-called killing for conservation), was morally reprehensible.[116]

'Professor Acorn' (sometimes identified as a 'businessman' or 'retired businessman', living in or near Glasgow) grounded his position in research commissioned by the animal welfare group Advocates for Animals. In 2006 the team, led by wildlife biologist Stephen Harris, concluded that killing greys was an ineffectual and costly way to protect reds, identifying Anglesey as an exception to this rule.[117] Those who led the Anglesey cull were fully aware of the fragility of their achievement. Unlike the Isle of Wight, Anglesey is connected to the mainland by a road bridge and a dual level rail/road bridge. Though greys may conceivably have swum across the Menai Straits, these bridges were the most likely invasion routes and cullers were alert to re-immigration potential.[118] During 2014 and 2015, members of the public reported various sightings; three (not pregnant) females were subsequently trapped.[119] Despite a (more modest) cull in mainland Gwynedd to reduce the likelihood of rapid recolonization, Shuttleworth knew that even a more comprehensive mainland cull could not seal Anglesey off forever.[120] Besides, transposing the costs of culling on Anglesey onto Britain as a whole, nearly £851 million (for an eight-year cull) was a conservative estimate.[121]

The Stronghold Strategy

Given the constraints on culling-cum-reintroductions and the grey's capacity for brisk colonization and speedy recolonization, the stronghold strategy exerted substantial appeal. Sanctuaries and buffer zones were at the heart of the North of England Squirrel Conservation Strategy launched in November 2005 by Red Alert North England (RANE), a group spearheaded by the Wildlife Trusts and the Forestry Commission. Each of the seventeen coniferous forests designated as Red Squirrel Refuge Sites was surrounded by a 5-kilometre (3 mi.) zone offering additional protection.[122] MP Trudy Harrison, whose Copeland constituency contained Nutkin's stomping ground, called the refuges a 'ring of steel'.[123]

For the Forestry Commission's Richard Pow, RANE's chairperson, it was time to accept the success of the 'American invasion' in broadleaved woodland and to 'fight where we can win', namely, in the coniferous strongholds.[124] RANE's press release highlighting the red's precipitous decline since 1945 attracted widespread attention in the United States.[125] The 'stronghold' system seemed a 'fitting solution for postcolonial Britain': 'The gray would keep what it had won. The red, like the British themselves, would content itself with a small homeland in return for peace.'[126] (American coverage also noted the grey's appearance on menus, the red's 'special place in Britons' hearts' – thanks to Potter – and how the British fightback channelled the spirit of the Blitz. Typically, though, attention focused on Prince Charles's involvement.[127])

Habitat Redesign

The stronghold strategy pivoted on a distinction between red-friendly non-native conifers (softwoods) and the native broadleaved woodlands (hardwoods) that conservationists preferred, but which were grey-friendly.[128] Though the red's habitat, pre-grey, encompassed both, it was soon clear that reds would struggle to maintain their occupancy of deciduous hardwoods, England's main tree cover, which R.S.R. Fitter identified as the grey's 'natural habitat'.[129] In 1952 the Forestry Commission's press spokesperson, Herbert Edlin, pointed out that the red's most realistic chance of survival lay in conifer expansion.[130] The catch was the slowness of this form of assistance: some conifers only produce seeds after twenty years.

The fit between reds and non-native conifers reflected the Eurasian squirrel's evolution in boreal forests across most of the Palaearctic. That close relationship also reflected marginalization, as greys occupied Britain's deciduous cover. Tracking and computer modelling of future scenarios underscored reds' competitive advantage among conifers, so improvement of existing coniferous habitat was another way to optimize reds' welfare. Peter Lurz suggested that the native Scots pine, with its bigger, year-round seeds, provided better feeding grounds than the dominant Sitka spruce. The FC responded by planting more Scots pine in Kielder Forest, as well as the large-coned but commercially unappealing Norway spruce, which supports higher red densities than other conifers.[131] Plans to diversify Kielder with oaks were dropped.

Red defenders often used the language of greed to impugn greys and cast the native in a superior moral light.[132] The difference in food intake was, of course, simply a matter of physiology. As the grey's higher energy demands require bigger seeded species of tree such as oak, scientists compared planting corridors of oaks to constructing motorways for express grey travel into sanctuary areas. To avoid these self-inflicted wounds in buffer zones, de-coniferization shifted to 'squirrel-neutral' natives such as birch, rowan and willow.[133] The best habitat for reds, the Wildlife Trusts' NPI-sponsored report of 1999 observed, included lodgepole pine, Douglas fir, yew, larch, hawthorn and Sitka in addition to Scots pine and Norwegian spruce – a mix of conifers and broadleaves, non-native and native (yew, hawthorn and Scots pine were the natives).[134]

Another glimmer of hope in the late 1990s was the possibility of red–grey coexistence. Was it always an either/or choice between 'my' place (the red's) and 'your' place (the grey's)? That 'our' place might be an option was raised by the finding that some reds were bucking the usual trend of dying out within fifteen to twenty years of the first grey's arrival. Research sponsored by the FC and Scottish Natural Heritage indicated cohabitation over a forty-year period in Craigvinean forest, a large mixed plantation dating from 1949 near Dunkeld, Perthshire. Explanations included the diversity of trees that offered enough food for both species, learned behaviour on the part of reds and disease-free greys.[135]

This discovery of interspecies harmony was music to the ears of red enthusiasts, among them the *Daily Mail* ('Red Revolution Stirs in the Backwoods', 28 April 1997). As a new century began, many were upbeat about the potential for small victories. Habitat redesign and the stronghold strategy promised to halt the progressive decline in red numbers since the 1920s. 'For the first time', observed Jason Reynolds, a red squirrel conservation officer for the Cumbrian Wildlife Trust, 'we are confident that the species can be saved.' "Squirrel Nutkin is back,' announced the sanguine journalist quoting Reynolds.[136]

Reproductive Control

Other red advocates cautioned against over-optimism. The Wildlife Trusts' director-general, Simon Lyster, acknowledged the benefits of red-friendly plantings, red-only feeders and (as a last resort) the killing of greys. Still, 'the future looks grey.'[137] To ensure that Britain's squirrel future looked at least partly red, squirrel fertility biologists took a different approach by targeting reproductivity, though no one expected this to deliver quick results. Around the world, environmental managers were turning increasingly to fertility interference to bring down 'problem species' – the rabbit and fox in Australia; white-tailed deer,

Canada geese and feral horses in North America.[138] In 1973, searching for a more humane (and more publicly acceptable) alternative to warfarin, the FC started funding research into a contraceptive vaccine.[139]

Investigations centred on chemosterilization to prevent ovulation and render spermatozoa infertile. Yet there were no breakthroughs in the practical administering of lab-developed hormonal reproductive inhibitors, and few advances in this area over subsequent decades.[140] In 1994, though, the FC and MAFF funded a fresh initiative. Reproductive biologist Harry Moore began exploring the potential for female oral immuno-sterilization.[141] When the Lords debated fertility measures for greys that year, some peers smirked: Lord Williams of Elvel enquired 'will there be a government subsidy for squirrel condoms?'[142] By 1998 Moore's project had devised a protein vaccine with monoclonal antibodies that attached to the egg's sperm receptor, preventing fertilization. The species-specific vaccine presented no threat to reds and was dispensable from feeding trays that were inaccessible to reds.[143]

There were two big drawbacks. First, to curb reproduction effectively, 80 per cent of females had to take the bait. Immuno-contraceptive barriers looked promising where populations were small, but efficacy within larger groups and over a wide area was unclear. Second, as sterility wore off and squirrels bred more than once a year, more than one annual treatment was required. Squirrels will not breed in a lab, so the fertility trial in 1999 involving wild-caught females and males previously kept separate took place in large enclosures within a beech plantation. Only 15 per cent of vaccinees became pregnant.[144] A pilot trial in open forest was conducted in the year 2000. Immunity development and the incursion of unvaccinated females between applications presented additional challenges.[145] However, fertility control remains widely regarded as the most publicly palatable and cost-effective method of population reduction.[146]

Recently, a fertility control advance of a different sort emerged at a Scottish genetics laboratory: DIGB (directed inheritance of gender bias) squirrels. Gene-edited males conduct a 'gene drive' that skews the normal 50/50 sex ratio of births towards males. This 'desirable copy' then drives throughout the population, all offspring carrying the transgene and transmitting onwards; eventually, the population crashes because of a dearth of females. DIGB's funder, the European Squirrel Initiative (ESI), sees it as a tool to 'reverse the invasion'.[147]

Disease-Resistant Super-Reds

An earlier ray of hope was a putative 'super-red' at the National Trust's Formby reserve in 1993. Those hailed as a 'true hard nut' that 'puts the tough into Tufty' lived in what was officially called the North Merseyside and West Lancashire Red Squirrel Stronghold. Super-reds had taken on some of the behavioural and physiological characteristics of their grey cousins, from which they were long insulated by the sea to the west and the largely treeless mosses to the east. Reds generally spend much less time feeding on the ground than greys.[148] They also find it difficult to carry body fat into the winter (to maintain their arboreal agility?), whereas greys can put on up to a quarter of their weight in autumn fat. More willing to abandon arboreal security, Formby's reds were less timid than their species' norm. And thanks to ground-accessed foods, they gained weight. These bolder and bulkier super-reds, the reserve's head warden explained, might provide pioneers to restock places from which reds were long gone.[149]

News of a pox-immune super-red arrived shortly after SQPV's ravages were first detected. At the Owl Sanctuary in Ringwood, Hampshire, in 1999, four 'super squirrels' imported from Belgium gave birth to what it was hoped would be the spawn of a more robust, disease-resistant gene pool, shoring up reds around Britain.[150] But the pace of SQPV's spread scuppered any hatching plans. At Formby, an outbreak in November 2007 decimated local reds. The only redeeming feature of this sorry saga was the opportunity that researchers at the nearby University of Liverpool had to study the dynamics of viral spread and the survivors' condition.[151] Between November 2007 and March 2009, numbers plunged from 1,000 to about 100, a survival rate of 8 per cent.[152]

The press latched onto an adult male survivor that, according to the National Trust's Sally Orritt, represented a species that is an 'icon of our entire nation'. Rachel Miller, red squirrel field officer for the Wildlife Trust for Lancashire, Manchester and North Merseyside, explained that this super-resilient squirrel had been named Clark Kent.[153] Unfortunately, despite the occasional red in good shape with SQPV antibodies, most reds that succumb possess them too; and nothing suggests they confer immunity.[154]

Eat a Grey to Save a Red

Some We Love, Some We Hate, Some We Eat is the title of Hal Herzog's book (2010) about our inconsistent and often confused thinking on animals (subtitle: *Why It's So Hard to Think Straight about Animals*). The species illustrating love, hate and food respectively on its cover are the dog (puppy version), the rat and the pig. The book is squirrel-less, but the grey provides a category-confounding example of a species eaten out of dislike – even hatred – or for patriotic purposes. In Australia, eating kangaroo demonstrates the (non-indigenous) eater's Australian credentials.[155] In Britain, eating the squirrel that threatens its treasured native equivalent is a way to express Britishness through solidarity.

The press occasionally noted a call for consumption. Amid sharply rising meat prices brought by early 1970s inflation, Labour MP William Price enquired of MAFF Minister Joseph Godber: 'Are you proud of the fact that the greatest achievement of your [Conservative] Government is to push up consumption of horsemeat and even grey squirrels?'[156] Pete Bundy, Lord Weymouth of Longleat's 'vermin trapper', rued public reluctance to eat greys, given the association with rats. Lord Apsley of Cirencester Park, near Cheltenham, was a connoisseur of squirrel casserole, but a story in the *Gloucestershire Echo* ('Lord Stews Estate Vermin') provoked various reactions from local residents, ranging from 'It's like the Chinese eating dog' to 'it's no worse than the French eating horse.'[157]

A sharpening of the British appetite for squirrel dates from 2006, when Redesdale founded the Red Squirrel Protection Partnership (RSPP) with a £148,000 grant from Defra. RSPP's aim was to kill greys in the exclusion (buffer) zone south of Kielder. But Redesdale wanted carcasses put to good use. A CNN feature showed him frying a grey in the kitchen of his Northumberland cottage, observing that RSPP supplied a lengthening list of 'upscale' butchers.[158] An American documentary in 2009 featured a head chef stung into culinary action by the likes of Redesdale. Marc Sanders, of The Famous Wild Boar Hotel, near Windermere, south Cumbria (now The Wild Boar), was filmed making and serving up grey squirrel pancakes, sourced from the hotel's extensive woodlands.[159]

Redesdale's 'eat to beat' approach was seconded by Inglewood. Consumption struck them (and Saltoun) as the sort of innovative approach a crisis demanded. For, 'unless something radical and imaginative is done' – and here Inglewood resorted to an appropriately culinary Americanism – 'Squirrel Nutkin and his friends and relations are "going to be toast".'[160] He urged celebrity chef Jamie Oliver, who had recently initiated a campaign to revolutionize school dinners, to add grey squirrel meat to the canteen menu as a healthy option. Inglewood – who admitted he had not yet sampled the goods – then invited the Labour government's front bench Defra team to dine on grey squirrel. The venue would be a hotel in the Lake District [The Pheasant Inn, Bassenthwaite Lake] of which he was a director, and the event was to launch the 'eat a grey to save a red' initiative of Save Our Squirrels (SOS).[161]

Lord Bach, Defra's parliamentary under-secretary of state, was disinclined to accept Inglewood's invitation. He recognized greys were edible and that their consumption had generated substantial press interest, but downplayed their appeal to British appetites.[162] Bach's Commons counterpart, Barry Gardiner, was no keener to sample, even though Shadow Minister for Defra Bill Wiggin had informed him that Elvis Presley's favourite dish had been fried (grey) squirrel. Nonetheless, an American reporter identified a boom, in 2009, in fancy London restaurants and rural gastropubs for grey squirrel in various guises: Peking duck-style, Southern fried, stew, curry, confit, pâté, pasty, empanada, burger and canapé.[163]

The Return of Another Native

Other red proponents pinned their hopes on the food preference of a native species. The potential assistance of the pine marten, a small cat-sized predator as beleaguered as the red squirrel but enjoying a modest early 2000s upturn in its fortunes, was not widely discussed in the media until recently. The essential ingredients of the debate were in place, though, seventy years ago. In 1952, a photograph by Frances Pitt in *Country Life* bore the caption: 'Could we have the pine marten back, the ebullience of the grey squirrel might well be curbed.' At the time, on the eve of 'bob a brush', the only breeding populations were in the Scottish Highlands, the Lake District and North Wales. Pitt's photo accompanied an article by an animal ecologist on the importance of predators, whose absence highlighted the futility of trying to protect a 400-hectare (1,000 ac) woodland estate by killing even four hundred greys a year.[164] A few weeks later, *Country Life* editorialized that the pine marten's 'special taste for squirrels' might endear them to those who thought that the fewer squirrels the better.[165] Near-eradication of martens, a journalist reckoned, had made England 'a sort of promised land for the grey squirrel'.[166]

MAFF was less enamoured of its potential as a super-predator because martens mainly lived in conifers, with greys mostly among broadleaves. As such, they reasoned, the red was at greater risk. In North America, where the American red squirrel also dwelled among conifers, the American pine marten (equivalent in size to its European counterpart) took far more reds than greys.[167] A UK government-made film of 1959 observed that the pine marten, 'swift and agile, able to give chase on the ground or in the trees', had been a prime predator of native squirrels before its numbers fell alongside those of its prey, as forest cover receded over the centuries prior to 1900.[168]

Benefiting from full legal protection in the UK since 1988, martens remain critically endangered, though populations are recovering in Scotland and Mid Wales. In these areas of marten revival, also including the island of Ireland, recent studies suggest grey populations are falling.[169] This creates an opportunity to rebuild red populations. Not because martens avoid reds – they eat whatever flesh is available – but because they eat more greys. Co-evolving with martens in the British Isles and Ireland, reds developed avoidance strategies based on chemical detection of the scent clues with which martens communicate and mark their territories. In the Scottish Highlands, the last UK stronghold for red squirrel and pine marten, there is no evidence that martens are detrimental to reds.[170] A landscape of watchful fear for reds, on the other hand, is a landscape of naive unfear for greys without a shared history with martens. At twenty locations in Northern Ireland, confronted with marten scent, reds emitted an unmistakable fear response. Yet greys failed to react, even 'in sniffing distance' of marten faeces solution, feeding on without vigilance. The nimbler red can also more readily evade martens by going out on the thinnest twigs of upper limbs, then leaping into the air, hoping for the best. Unlike the pursued, the pursuer cannot survive such a fall.[171]

The pine marten, though, is no panacea. After all, in 1912, a Board of Agriculture leaflet urged its 'encouragement' as a means of keeping down reds in conifer plantations.[172] So, a nagging fear remains among advocates of martens as biocontrol: what if martens exhaust the supply of the squirrels they are supposed to eat and turn to reds? There are other drawbacks. Nest-confined youngsters, red and grey, are surely equally vulnerable. And being adaptable, greys could adjust to predation sooner rather than later. Besides, research in Northern Ireland suggests that no matter how much marten numbers recover in rural areas – and regardless of the high public acceptability of this form of control compared to lethal methods[173] – they are unlikely to establish urban populations, in Northern Ireland or elsewhere in the UK. Just as greys in densely populated human areas operated effectively beyond reach of trap, gun and poison in the 1950s and '60s, greys inhabiting the 'refugia' of today's cities, towns, suburbs and peri-urban fringe will remain beyond the reach of martens, uncontainable potential reservoirs for spontaneous restocking of the countryside.[174]

Fighting Extinction

In his foreword to a recent compendium on the ecology, conservation and management of red squirrels, broadcaster and wildlife expert Iolo Williams recalls seeing his first red squirrel in Mid Wales at the age of five, circa 1967. Within years, they were all gone, replaced by greys.[175] 'Would you like to explain to a grandchild, or to the spirit of Beatrix Potter, that Squirrel Nutkin is extinct?' This chilling question was directed at Britons in 1997 by the American journalist William Montalbano in a widely syndicated article prompted by the release of the UK Red Squirrel Group's Strategy. After all, in barely a century, the dynamic grey had dislodged reds from the 'perch of treetop privilege ... enjoyed since the ice ages'.[176] Montalbano's vision of Nutkin as endling (the last of a species) harked back to the ice ages to convey just how long (rather than just how short a time) the red squirrel had called Britain home.[177] After the ice ages, or last ice age – circa 10,000 years ago – when lands that are now England, Wales, Scotland and Ireland were connected to what is now mainland Europe, various terrestrial species such as the red squirrel migrated northwestwards into what was then a peninsula of northwest Europe. And around 2,500 years later, rising seas began to create the British Isles and Ireland, which made the final (perhaps tsunami-driven) break from the mainland circa 6,100 BC.[178] According to this temporal watershed, species that had already colonized these future islands, post-ice age, are widely classified as native species; any species that has arrived since geographical separation from the rest of Europe, whether intentionally introduced by people (such as the grey squirrel) or transported inadvertently beyond its historic/natural geographical range, is generally considered non-native.

Montalbano's article focused on Red Alert's campaign to prevent a repetition in Northumberland of what happened in Durham, the county immediately to the south, where the red proportion of the overall squirrel population dropped from 60 per cent to 15 per cent between 1987 and 1997.[179] Northumberland was the scene of operations for two leading opponents of red extinction. Redesdale and his right-hand man, Paul Parker, a pest controller, were cut from the Churchillian cloth that Inglewood admired. They killed remorselessly for the sake of reds. 'To think they have been here since the Ice Age', Parker reflected, 'and we might be seeing the last of them here is something I really don't want to think about.'[180]

As journalist Tim Adams observed, RSPP was 'not a resurgent Tufty Club'.[181] As Redesdale explained: 'we do nothing with red squirrels apart from save them by killing grey squirrels.'[182] Trapped in a wire cage with a hazelnut-baited, spring-loaded, gravity-operated platform, which allowed the release of any reds unharmed, the animals were then shot in the back of the head at close range through a gap in the wire mesh, just behind the ear. There was also cranial despatch: decanted into the corner of a sack, they were bludgeoned over the head.[183] 'We only call ourselves the Red Squirrel Protection Partnership', Redesdale confessed, 'because if we called it the Grey Squirrel Annihilation League people might be a bit less sympathetic.'[184]

Parker told an American reporter in the autumn of 2007 that his personal tally had reached 2,332.[185] A year later, RSPP's volunteer army of nine hundred (Adams summed them up as 'grannies and game wardens, families and farmers') had racked up a total kill count exceeding 19,000.[186] On assignment in Northumberland, Alphonso Van Marsh, of CNN's London bureau, mainly known as a war correspondent, produced *Squirrel Wars in the UK* (3 September 2008), a despatch preceded by a warning: 'This report includes graphic content. Viewer discretion is advised.' The bulletin featured 'squirrel hunter' Redesdale and 'exterminator' Parker shooting greys, as well as RSPP's oldest volunteer, 82-year-old Dorothy Sanderson, who had trapped 29 under her bird table.

Redesdale and Parker also featured in *Squirrel Wars: Red versus Grey*, a 25-minute British-made documentary for Channel 4 which aired on 17 February 2009. Redesdale is mostly filmed in his office in Parliament, trying to raise funds to continue the mission. He referred to threats to biodiversity. He could not do anything, directly, to save rhinos or gorillas, but he could do something to prevent red squirrel extinction. Parker, decked in camouflage, is largely shown at work in wintertime, setting and checking traps, shooting the trapped and tossing them into his van. Queen's 'Another One Bites the Dust' supplies his soundtrack as he drives to the next occupied trap.

Parker has seen plenty of reds over the decades, yet still gets the same thrill as when he saw his first one in the 1970s. And every red reminds him of why he despises greys. Reds are 'mystical' creatures, each with its own character, that 'respect the place they live in'. By contrast, 'a grey squirrel is just a grey squirrel.'[187] And like Saltoun, Parker invokes English social distinctions: reds are 'little gentlemen'.[188]

What makes them worth fighting for, explains a comrade-in-arms, Don Clegg, a retired teacher who handfeeds reds peanuts at his home on the shores of Kielder Lake, is straightforward: 'it's just like they belong here.'[189] For Redesdale, Parker and Clegg, belonging and unbelonging determine the right to exist.

Tom Tew, the senior mammologist for the Joint Nature Conservation Committee (the statutory adviser to the UK government on national and international nature conservation), was a moderate in comparison. 'We have nothing against the American gray squirrel,' he assured an American journalist: 'We'd just rather it remain in America.' Tew only wanted to eliminate unfair competition, favouring 'a balanced enjoyment of grays', not their total annihilation.[190] A stance like Tew's was defeatist for zero-tolerance Parker. Rejecting allegations of 'squirrel racism' in *Squirrel Wars*, he insists that 'everything's got its place in nature.' 'This', he said, holding a dead grey, 'is from America; it shouldn't be here.' 'In the wrong environment', they 'upset the balance' and 'the natural status quo'. In short, 'they just don't fit in.' This reasoning made the grey squirrel killable – and normalized their killing. In an example of Cohen's phenomena of folk devils and moral panics, Parker compares greys to a 'bunch of townies', 'loud and boisterous', disturbing the peace of the countryside. 'What we should have' is reds. 'That's what belongs on this island.'

Parker rejects the view that greys have been here so long that they qualify as members of England's wildlife. They are intruders, plain and simple ('I hate them'). The final scenes of *Squirrel Wars* show Parker at home, declaring that an Englishman's 'true duty' is to fight for his native squirrel. At the end, he paraphrases the accolade of the 'big man' (Winston Churchill) for the plucky airmen who fought off the Luftwaffe during the Battle of Britain: 'never before has so much been done by so few.'

Squirrel Wars reported that RSPP's body count had reached 21,749 (by the time the film was made). But that was less than 1 per cent of the greys' overall British population. Luke Sewell, who produced, directed and filmed *Squirrel Wars*, realized that RSPP's crusade is a lost cause. Whether fighting mice, rats or grey squirrels, there are no permanent victories, just quixotic attempts to control the uncontrollable. Deep down, Parker knows this too; when Defra's grant money that pays his wages runs out, the greys will be back – and laughing at him.

Squirrel Wars was not the only documentary to be released in 2009 that concerned the squirrelly battle for Britain. Across the Atlantic, six months later, a nineteen-minute American-made documentary, shown on PBS, also focused on Northumberland. *Nutkin's Last Stand* featured many of the usual suspects, not least Redesdale and Parker. 'Something is rotten in England' was the strapline, the narrator elaborating that 'a plague of North American gray squirrels threatens the beloved native red squirrel' ('immortalized' by Potter, according to the synopsis). And so 'the English are up in arms, and a band of patriots – including lords, priests, artists and farmers – has come together to fight back.' Parker, quips Redesdale, is 'obviously in love' with reds, and this love, the featured 'patriots' insist, requires that greys die. Or nothing will be left to love.[191]

Nutkin's Last Stand opens with Parker responding to a call about a trapped grey. The scene shifts to the packed interior of a village hall, where grey-haired seniors listen to a talk on the 'Charismatic Pest' by SOS officer Philippa Mitchell. Next on camera is Carri Nicholson, SOS's project manager, who explains that SOS works closely with RSPP by monitoring greys' whereabouts. SOS volunteer Veronica Carnell emphasizes that the Eurasian red squirrel ('its proper name') is Britain's only squirrel. Meanwhile, her pet red flits around her living room, nibbling at potted plants and scampering up and down the curtains.

Nutkin's Last Stand moves to neighbouring Cumbria, where Doreen Hallsworth, a sixty-year-old Church of England minister, recounts how she shot to fame in Carlisle, a town on the northwest rim of Lakeland. In April 2006, Abbey (National Savings Accounts), a building society, featured a squirrel in a television advert. Like the promotional materials of NPI and its successor since 1999, National Provident Life, the advert drew on the squirrel attribute of nut hoarding as a metaphor for the wisdom of saving. The squirrel in Abbey's advert was grey. Livid at the society's promotion of the squirrel responsible for local reds' decline, Hallsworth (an Abbey customer for thirty years), kicked over a 4-foot cardboard version of the advert outside Abbey's Carlisle branch. A spokesperson explained that Abbey had featured a grey because, nationwide, it was the most common squirrel 'now, unfortunately'. Showing clippings from the local press, Hallsworth relates that Abbey, bowing to public pressure, replaced grey with red.[192]

Up next in *Nutkin's Last Stand* is farmer Harrison Martin, who has rigged an aerial crossing up outside his house in order for reds to traverse the road safely. Martin was angry and indignant. Not only did greys fatally infect, fight and kill reds. Seeing them even in Buckingham Palace's grounds was the last straw, something particularly 'terrible'. No corner of England, it seemed, was sacred or impregnable. Parker had a better grasp of London's status as the capital of the grey's empire in England, disclosing (not in *Nutkin's Last Stand*, but to journalist Tim Adams) that what he really wanted was to take the fight to London's parks, hitting the intruder in its 'own backyard'.[193] At his cottage in the upper Rede Valley, Redesdale explained how greys 'scout' for new territory and move 'as an army'. Inspired, perhaps, by manoeuvres at the adjacent Otterburn military training area – Chinooks are regularly seen overhead – he feels like an armchair general directing a quasi-military operation. Parker, the self-styled red's 'guardian', reappears on camera in *Nutkin's Last Stand*, his van's back door displaying the slogan, 'Wanted: 'Red & Alive (in Newcastle and North Tyneside)'. While stroking the tail of a seven-month-old female he has shot in the back of the head, he dismisses greys as 'flying rodents . . . good for nothing . . . apart from shooting'.

The makers of both documentaries, one British, the other American, both knew it was about much more than squirrels. 'Britishness itself', *Squirrel Wars*' Sewell surmised, was at stake. RSPP's cause 'embodied what it meant to be British and proud'. Sewell reckons Parker is anxious about the world changing around him, approaching his anti-grey struggle as his way of pushing back against wider currents of unwanted change. Nicholas Berger, who made *Nutkin's Last Stand*, also appreciates that if you scratch his film's surface, you will find it is about patriotism and national identity, as well as war, invasion and race. He suspected that the red's guardians projected and channelled their feelings about American imperial power 'running roughshod' over the world onto the grey.[194] *Nutkin's Last Stand* was broadcast in Australia as *Battle of Britain* on 21 August 2011 under the strapline 'War is underway in the British countryside ... to kill off the invading grey squirrels in order to save the native red squirrel population.' And yet not one red advocate in *Nutkin's Last Stand* or *Squirrel Wars* mentioned that, as late as the 1920s in England, and until 1946 in Scotland, foresters and gamekeepers killed reds as merrily as Redesdale and Parker were snuffing out greys.

Not surprisingly, *Nutkin's Last Stand* carried a warning: 'includes images of dead squirrels'. An American TV critic reckoned that U.S. viewers 'may well find themselves sad and angry' at footage of the 'stiff little bodies of their compatriots piled in the back of a truck'.[195] A decade earlier, the British zoologist Derek Yalden had argued for the superior rights of native species:

> The sentimental response in some quarters is that any animal has the 'right' to live here. I cannot agree. The native species certainly have the right to be here, more right than we have, but the exotic introductions have no right to be here, and no right to eliminate our native species. The whole point of bio-diversity is that different parts of the world have different faunas and floras.[196]

But Americans seeing the slaughter of American squirrels on screen might feel that the grey had as much 'right' to live in Britain as the native red. Redesdale and Parker insisted that greys were in the wrong environment. From greys' standpoint, though, they were in exactly the right environment. Far from not fitting in, they fit right in. Van Marsh's CNN report characterized a large English oak as a 'McDonald's for squirrels'.[197] Greys felt at home because it was like being back home (more or less), among mast-producing trees. The (white) oak, hickory, beech and walnut back home, eastern North America's dominant deciduous forest ecosystem, was richer in arboreal diversity than Britain's broadleaf cover. But evolving and living there left them pre-adapted to English and Welsh woodlands, not least in that they had developed the capacity to digest tannins-rich acorns.[198] Britain was no strange old world. To use a phrase from Michael Tod's *The Silver Tide* (1994), it was a New America.

Winning Young Hearts and Minds

Nutkin's Last Stand ends with Ernie Gordon, an elderly red squirrel stalwart in Northumberland, a familiar figure in Alnwick's Hulne Park, watching and feeding reds. Gordon had recently published the first volume of *The Adventures of Rusty Redcoat* (2007), and celebrated Rusty, whose home was Hulne Park, as a 'precious part of our heritage in Britain'. To ensure this heritage lived on, you needed to win the next generation's hearts and minds.[199] In 2014, aged 81, Gordon published the second volume. His aim with this series of books was to replicate and refresh the affinity between child and red squirrel that Potter forged with *The Tale of Squirrel Nutkin*.[200] His second book sought to raise youngsters' awareness of the gravity of the 'relentless march north' of 'disease-ridden' greys, not just for Hulne Park's population but for those in Scotland, where SQPV's first fatality was recorded in June 2007.[201] Gordon criticized a string of governments for twiddling their thumbs: 'the nation's children have been badly let down by the complacency of many of their elders for far too long.'[202] He urged pupils at The Duke's Middle School, Alnwick, to write to Prince Charles, patron of RSST (who commended them for 'writing so passionately about red squirrels'), and to Prime Minister David Cameron.[203] Rusty Redcoat's website proclaims the red's 'right to live safely in the woodlands, and gardens, of Great Britain' and beseeches youngsters to 'kick some adult backsides'.[204]

Though warfarin deployment was constrained, the supply of verbal poison was unlimited. Gordon's vitriol against 'dreaded' and 'imported' 'non-indigenous diseased grey squirrels' paled against the vituperative tone of another example of activist squirrel fiction for children: *Stumpy: Hero of the Lakes* (2008). The authors, Lakeland residents Tim and Pat Cook, coordinated the Patterdale and District Red Squirrel Group and donated most of their book's profits to 'helping our native squirrels survive'. The storyline is familiar: colonized natives mount a spirited resistance to turf out foreign overlords. After Stumpy loses a front paw and rear foot in combat with Brownfang, the greys' 'huge, muscular' leader, a human friend (Simon) equips him with artificial limbs, including metal claws. 'Ignorant humans', Stumpy lectures an eleven-year-old girl (Meg), 'say the greys don't attack us, but they do, all the time. They steal the food we've cherished for centuries, so that we, and worse, our babies starve, and they carry an evil disease.'[205] As chief of the Red Defence Force, Stumpy recruits assorted local creatures as 'warriors' to protect their homeland and repossess 'stolen' lands.[206] The red deer stag (Magnus) derides the 'tree-rats' and 'greedy ... ugly brutes' that left their enemies in 'Carolina' and 'don't belong in Europe'. Another ally, the mythical Old Man of Coniston, urges the troops to 'push back this hideous grey tide'. Insectified as a 'swarm of grey woodlice', 'grey vermin' are stripped of their squirrelly credentials.[207]

'We hate grey squirrels. They don't belong here,' announce the ravens who enlist in the 'red squirrel army'. Brownfang is dismissive of the insurrection, especially of Meg: 'She's bound to be one of those sentimental townees [sic] who think we are lovely cuddly creatures and have never seen our prettier red rivals. Pooh ... Might is right, I say, and with our allies we'll soon have the whole of Britain in our grasp.'[208] Stumpy agrees. Deluded animal lovers who think greys are cute and innocuous prevent native wildlife protectors from taking firm action.[209] It 'made my heart sick', Simon discloses, 'to think that people down there [southern England] thought that those foreigners were native English squirrels. They didn't know what a real English squirrel should look like.'[210]

Making Sure the Unthinkable Stays Unthinkable

Two reports in 2011 underscored the urgency, for the red's defenders, of some sort of radical counterstrike. As red populations had declined by more than 50 per cent since 1960 (and by 95 per cent in England), the University of Oxford's Wildlife Conservation Research Unit (WildCRU) reckoned the 'omens' were 'bleak'.[211] Toni Bunnell's survey (a month earlier) for digital channel Eden TV predicted that Britain's sixth most endangered animal species (second only, among mammals, to the Scottish wildcat) might be extinct by 2031.[212] A few months earlier, the *Daily Telegraph* had interviewed Prince Charles. 'We must succeed,' insisted Charles, who had grown up with reds at Sandringham (Norfolk) and Balmoral (Deeside) in the 1950s and '60s. 'We have no choice,' he continued. 'How can we just give up, do nothing and say it is all impossible while witnessing the disappearance of one of this country's most endearing species? It is unthinkable.' Echoing Simon's sentiments in *Stumpy*, Charles wished more Britons appreciated that 'it wasn't so long ago that reds were common all over Britain,' and 'recognised that the grey just doesn't belong here'.[213]

Charles's interviewer, Robin Page, was passionately pro-red too. In his capacity as patron of the Red Squirrel Survival Trust, Charles had recently contributed the foreword to Page's children's book, *Why the Squirrel Hides Its Nuts* (2011). John Paley's illustrations indicate that the nut-hiding squirrels are red. The book's price (£5.50) included a 50 pence contribution to red squirrel conservation groups like RSST. An afterword ('Help!') explains that reds are in deep trouble, having been 'replaced by the larger, more aggressive' grey.[214] The 'tree rats', Page felt, 'would be far prettier if . . . still confined to North America', and he vigorously endorsed RSPP's shoot-to-save policy.[215] He derided 'Professor Acorn' and hailed Shuttleworth as a 'red squirrel hero' for his achievements on Anglesey.[216] In a blogpost, however, Page cautioned that the grey was not to blame for being over here and taking over: landowners brought it home as a 'mobile garden ornament'. Clearly culpable, though, were toothless government departments and quangos more concerned, in his view, about a backlash from animal rights advocates than about the red's welfare.[217]

Given this absence of firm commitment, redoubtable volunteers were left to lead the fight. Page cited the involvement of the Countryside Restoration Trust (which he founded and chaired) in the captive breeding programme of the East Anglian Red Squirrel Group, which was raising reds in Norfolk for a transplant to Mersea Island, Essex. He also praised the British Wildlife Centre, whose Surrey-raised stock had seeded a colony on Tresco, a grey-free Scillies island, in 2017. Charles simultaneously ordered a cull of approximately

4,000 greys on his Duchy of Cornwall estate as a prelude to a reintroduction to Cornwall's extremities, where the last red was recorded in 1984.[218]

Page deplored what he considered the infiltration of conservation by political correctness. As evidence, he quoted Ian Rotherham, an ecological historian who publicized the term 'eco-xenophobia'. For Rotherham, grey squirrel culls 'resonate with ideas growing with the BNP [British National Party]'.[219] Page was known as an advocate of population and immigration controls, once suggesting (as 'a joke', he explained) that contraceptives could be administered to immigrants in raisins.[220] He ran as a Referendum Party candidate for South Cambridgeshire in 1997's general election and again in 2005, this time for the United Kingdom Independence Party (UKIP). Page quit UKIP in 2009, but a prominent UKIP-er, Stuart Agnew, kept the grey in his sights. UKIP spokesperson for agriculture and rural affairs, Agnew was elected MEP (Member of the European Parliament) for East of England in 2009. In 2012 Agnew was the only politician that the ex-Conservative, recent UKIP convert Janice Atkinson-Small could identify who fully shared her anti-'townie' views on greys ('while I welcome visits from our U.S. human cousins, this little export needs to go').[221]

Atkinson-Small, a *Daily Mail* columnist and future MEP for South East England, loved reds as much as she hated greys, whose UK population she inflated to 5 million. Indignant that 'misguided Scottish landowners' once 'persecuted' blameless native squirrels, she reinforced her position through analogy: 'If there was a band of illegal immigrants that cost our economy an estimated £14 million per annum, carried a fatal disease that killed off most of the indigenous population and threatened our wildlife and woodlands too, wouldn't you be keen to go to war with them?' Agnew told her that eradication of greys was part of UKIP's wildlife and forestry policy.[222] His hard line was shared by the ideologically aligned British Democratic Party (British Democrats), which declared on 11 June 2011 that the grey's 'relentless expansion' must end and the species be 'eclipsed from Britain'.[223]

Squexit, the New European Frontier and Scotland's Squirrels

De-greying Britain was unachievable. UKIP's overriding goal was more achievable: pulling Britain out of Europe. Did greys have anything to fear from Leave's victory in the 2016 referendum on EU membership? A skit on the *Russell Howard Hour* ('Why squirrels fear Brexit') raised this question in a send-up of no-deal Brexit preparations.²²⁴ Twenty-eight days before Britain's original exit date of 29 March 2019, stand-up comedian Howard lampooned the more drastic forms of prepping. 'Why squirrels fear Brexit' began with a clip from a recent 'News at Ten' feature. With the newsroom's countdown clock showing 36 days to go until 29 March, presenter Tom Bradby reported that one in twenty Britons was stockpiling food, toiletries and medicine.

Next up for a ribbing from Howard was the even more extreme Mark McLean, a 33-year-old call centre worker from Glasgow who was not just squirrelling away camping equipment, army ration packs and Spam. Woven into Howard's skit was an extract from the *Daily Mail*'s profile of the 'prophecy prepper'. In case of post-Brexit urban unrest – even a Russian invasion – 'Mr McLean has trained himself to hunt squirrels and collect rainwater to drink.'²²⁵ 'Does it get madder than that?', Howard asked. And so 'Squexit' was born. The squirrels McLean planned to subsist on when his supplies ran out were grey by default. As someone commented on a YouTube upload of Howard's sketch (which has had over 900,000 views to date): 'Lord I can't wait for Brexit now! I'm finally going to get revenge on that squirrel.' 'Send the gray squirrels back, this is red squirrel land,' posted another. Others denied that squirrels had anything to fear: 'the country isn't going to go so broke that we need to hunt squirrels.' Greys were plentiful in Glasgow. But where McLean wanted to hole up, the northwest Highlands, squirrels of any kind are thin on the ground. And if they do exist there, they are red only.

Grey squirrels and Brexit resurfaced during the general election campaign of 2019. Liberal Democrat leader Jo Swinson fell victim to a fake news story dubbed the 'Great Squirrel Scandal'.²²⁶ The spoof article (allegedly a *Daily Mirror* report) claimed that a private, eight-year-old Facebook video showed Swinson killing (grey) squirrels in her Dumbartonshire garden with pebbles from a slingshot. 'I don't go for the head because that's too clean a death,' was a fabricated quote. In another made-up quote, she called grey squirrels 'pleb bunnies'.

Screenshots of the so-called article, which claimed animal rights groups had 'blasted' Swinson, were shared initially on Twitter by anonymous pro-Labour accounts. Then, when repeated on Medium, a Brexit Party account with 9,000 followers picked it up (Swinson was a prominent 'Remainer').[227] No matter that the photoshopped image, which had attracted over 20,000 social media interactions, showed Swinson with a grey squirrel. The Brexit Party's Peter Jeffries tweeted: 'Brexiteers . . . we have the truly magnificent Red Squirrel here on the Isle of Wight. When you see one you feel privileged. As a kid in Surrey, I grew up with them declining. Jo Swinson enjoys shooting them and laughing as she does it.' Others who fell for and spread the story also misidentified Swinson's squirrel as red. According to a Lib Dem activist, the party's canvassers were confronted on the doorstep about Swinson and squirrels.[228] Swinson even had to issue a debunking denial on LBC radio, reassuring Ian Dale that she liked squirrels, red and grey.[229]

Another leading Remainer embroiled with squirrels was Conservative grandee and peer Michael Heseltine. Heseltine, who believes Britain's place is squarely within Europe and represents his party's more liberal wing, shared the uncompromising line on greys that the hard, arch-nationalistic right espoused. Thenford, Northamptonshire, home of the former deputy prime minister (1995–7) since 1977, included a 16-hectare (40 ac) arboretum. Writing about their estate in 2016, Heseltine and his wife, Anne, explained the take-no-prisoners approach they adopted after greys gnawed into the entrance holes of thirty nesting boxes. Two gamekeepers shot and trapped more than three hundred over a six-month period. 'These foreign intruders', the Heseltines explained, 'may have a Walt Disney appeal in London parks, but to us they are Public Enemy Number One . . . and are shot without hesitation.'[230] 'They have largely destroyed England's indigenous red squirrels,' they continued, 'but we are not going to let them destroy our trees without a fight.'[231] Just as gung-ho was the European Squirrel Initiative (ESI), a charity consisting of landowners, foresters, scientists and conservationists that the Heseltines keenly supported.[232] ESI unflinchingly depicted dead or targeted greys on the cover of the April 2017 issue of its newsletter, *Squirrel*. A cover image captioned 'Another Good Day in London' showed hundreds of dead greys spread over tarmac. Another cover ('Gotcha!'), of December 2020, showed a grey face staring into the cross-hairs.

When ESI was established in 2002, the UK and Ireland were the only countries where the Eurasian squirrel was in distress nationwide. Recognizing its harrowed status in the British Isles and Ireland, the Bern Convention (on the Conservation of European Wildlife and Natural Habitats) of 1982 proposed to protect the species from being killed, injured or taken. As it was not endangered across most of its range, however, the species was not a beneficiary of habitat protection under the EU Habitats Directive of 1992. IUCN's Red List of 2007 placed it among the species of 'least concern' owing to its 'stable' population trend.[233]

The Red List noted, nonetheless, that reds were under threat in a third European country. As *Stumpy*'s Brownfang boasted, 'we've already got our advance guard in Europe.'[234] In 1948 an Italian diplomat, returning from Washington, DC, brought back two pairs of greys (*scoiattolo grigio* in Italian), releasing them in the woods of Stupinigi, near Turin, in Piedmont. In 1966 the owners of Villa Gropallo at Genoa Nervi, Liguria, turned out five greys that had been introduced from Norfolk, Virginia, at their coastal estate. Nearly thirty years later, in 1994 the city authorities of Trecate, near Milan, Lombardy, released three pairs (procured from an animal importer nearby) in the town park.[235] By 2005 the biggest of northwest Italy's three grey colonies, 10,000 strong, was located southwest of Turin, overlapping with Italy's largest concentration of hazelnut groves (think Nutella).[236] The second largest, at Parco Del Ticino, was 40 kilometres (25 mi.) from the Swiss border.[237]

When populations of the native red (belonging to the subspecies *S. vulgaris fuscoater*) were higher in northwest Italy, foresters treated it as a pest. In the winter of 1979–80, though, Italy's National Wildlife

Institute recorded the last red at Stupinigi.[238] When a trial eradication of greys began in May 1997, the red (*scoiattolo rosso*, or *scoiattolo commune*, in Italian) was protected.[239] The National Wildlife Institute initiated a programme of live trapping, tranquillizing and euthanizing (by gassing) in the mixed deciduous old growth woodlands of Racconigi Park, near Turin. Within a month, following public outcry spearheaded by animal rights activists, a court order halted the cull.[240] The judicial inquiry and appeal against the charges of illegal hunting and cruelty to animals lasted three years. Meanwhile, greys had spread around northwest Italy to the extent that elimination, though still feasible, was no longer politically and financially practicable.[241]

Based on British experience, ESI's Miles Barne expressed concern in 2004 that Italy's greys would expand northwards.[242] In the 'best case' scenario, they would reach the Alps between 2038 and 2048, Switzerland between 2051 and 2066 and France by 2078–83. (There have been no reports of greys travelling to France from England through the Channel Tunnel, since it opened in 1994.) In the 'worst case' simulation, greys arrived in the Alps by 2015, France by 2026–31 and Switzerland by 2031–41.[243] In 2013, when northern Italy's grey population stood at roughly 12,000, and their area of distribution encompassed more than 2,000 square kilometres, the stage of invasion was reckoned to be comparable to England's around the year 1900.[244] That year, 2013, Italy banned their importation, trading and release into the wild. The EU then, in 2016, added greys to the 'species of Union concern' covered by the Invasive Alien Species Regulation. Outlawing importation, keeping in captivity, breeding, movement and release replicated Europe-wide the UK's proscriptions.

The Eurasian squirrel entry in 2007's Red List included an ominous observation on the grey's long-term prospects in mainland Europe: 'it can be expected that it may ultimately spread throughout much of the red squirrel's range [and] in the future [the red's] status in Europe is likely to worsen.' Following the most recent Red List assessment (August 2016), however, the red's conservation status remains of 'least concern'. The population was 'decreasing' but numbers were still substantial and broadly distributed, with widespread threats absent.[245] As ZSL's Peter Cotgreave observed twenty years earlier, the species 'will not go extinct if it vanishes from Britain'.[246] But knowing there are still plenty of reds elsewhere in Europe provides little consolation for Britons who have effectively nationalized the Eurasian squirrel.

Even in Scotland, home to most of the remaining UK reds, anxiety is growing. At northern locales such as Loch an Eilein, in the Spey Valley, the squirrel with what *The Field* described as the distinctive blonde tail and ear tufts of the Scottish strain (7–13 July 1982) reigned supreme in the early 1980s and still does. Across Scotland, though, greys outnumbered reds by 2010 and, year by year, were making themselves more at home.[247] In 2012 the *Washington Post* quoted Prince Charles's statement, from RSST's website, that even Scotland's reds might be 'driven out by the [grey's] relentless Northern march'.[248] Instigated in 2009, the Scottish Wildlife Trust's project, Saving Scotland's Red Squirrels (SSRS), aimed to prevent further northward creep into 'red-only' areas beyond the Highland Boundary Line and to control greys in mixed populations further south.

Without mentioning Nutkin, the Scottish National Party's Roseanna Cunningham, minister for environment in Scotland, announced in 2010 that this 'emblematic, charismatic and iconic' creature had ranked second in a government-led poll of the species 'most important to the Scottish people'. The following year, 37 per cent of 1,055 people interviewed in a Scottish Natural Heritage (SNH) survey were concerned about the red's welfare, making it the species in Scotland of gravest concern.[249] And in 2013, as part of the Year of Natural Scotland, an online poll by SNH and VisitScotland to identify Scotland's favourite wild animal placed the red as runner-up to the golden eagle, attracting 20 per cent of 12,417 votes cast.[250]

This deep attachment to the red as spirit of place furrily embodied explains why, in 2012, campaigners reacted so angrily to a TV advert for Highland Spring in which a grey squirrel, alongside a deer, ladybird, owl, rabbit and mole, promoted its mineral water.[251] The mole – a native species, like the other featured animals bar the grey and the rabbit – actually had the starring role. Nonetheless, Simon Poots, the chairperson of the Perth and Kinross Red Squirrel Group, believed that the grey squirrel's appearance was 'a slap in the face of all those battling tirelessly to save our red squirrels' (the advert's broadcast coincided with a county-wide cull of greys). SSRS's Ken Neil was incredulous: 'Considering they are a Scottish company ... and based in Perthshire ... I cannot understand what they were thinking.' The firm's preference, Highland Spring explained, was for a red. But they could not source a trained one, whereas greys were 'readily available'. Nonetheless, they maintained that a grey more accurately reflected the squirrel profile of the company's Ochil Hills catchment.[252] But reds hung on there. In 2017 the head of Scotland's Heritage Lottery Fund explained why the Fund had given SSRS £2.46 million: 'Many of us' have a 'soft spot' for reds.[253] Yet many Britons also had a soft spot for greys.

7

LEARNING TO LIVE WITH (AND TO LOVE) THE GREY

In a 'country diary' entry published in *The Guardian* on 30 January 1971, William Condry identified various shifts in attitudes to grey and red squirrels since the 1870s. Initially, the grey was 'adored'. Then, as numbers grew and novelty waned, dislike replaced affection, before turning into hatred after a 'propaganda' barrage brought the 'crimes these Americans were committing' to wider attention. Nostalgia for the red and a longing for its return to pre-grey glory paralleled this volte-face. Condry rightly pointed out that the red's current champions were largely unaware that their darling was equally unpopular – and often killed for its own pestiferousness – when and where it was common rather than uncommon. Now it was evident that 'campaigns' to reduce grey numbers had been mostly ineffectual, he wondered whether Britons would start accepting them for what they were worth. Most visitors to the countryside, he wagered, would resign themselves to the grey's victory and recognize the animal as 'a normal and pleasant feature' of rural Britain.

If the future looked grey, why not enjoy it? A MAF official had seen this coming in the early 1940s: few townsfolk 'ever give a thought to the trail of destruction and damage wrought in the countryside by these charming yet mischievous rogues'.[1] 'My name is mud in the village,' a vicar told a reporter in 1947: 'I *like* grey squirrels.' The reporter sympathized. After all, the grey's defects were 'merely the faults of the red squirrel magnified by numbers and over-boldness'.[2] Children's author Cecily Rutley included a grey squirrel among her 'Tales of the Wild Folk' (1949), a series whose other mammals were all native Brits:

mole, harvest mouse, hedgehog, badger and otter. Greykin, his siblings and his parents, their 'soft coats of silver grey' marked with 'reddish-brown', are almost beyond reproach, as they do squirrelly things such as leap from tree to tree, gather nuts in autumn and look for buried nuts in the spring. Greykin savours the inner bark of trees, but Rutley mentions this non-judgmentally. Only on one occasion does Greykin blot his copybook: 'once or twice, alas! he found a bird's nest, and stole an egg.' His only encounter with another species is a close call with a stoat.[3]

In the post-war decades, though, many Britons heavily subscribed to the list of crimes and misdemeanours that had accumulated over the first half of the century. A fresh outrage noted was nibbling through the foil top of doorstep milk bottles to pilfer the cream. And Garth Christian, a nature writer and conservationist, lumped greys with sparrows and tourists as a threat to crocuses. He also tossed them in with people that should 'learn better manners', while the RSPB aligned greys with nest-robbing children.[4] Meanwhile, the allegation that greys attack, kill and eat juvenile reds persisted. A woman in Leicestershire who first saw a grey in her local woods in 1944 claimed there were no reds thirty years later; they had 'eaten the young reds in their nests'.[5] In 1945 two gamekeeper friends of A. W. Boyd, the *Manchester Guardian*'s long-serving country diarist, located three young red bodies underneath a yew in Cheshire; a grey, they reported, had bitten them in the back of the neck, stoat-like.[6]

Sixty years later, there was a fuller appreciation of the nature of the grey's appeal among the red's proponents. MP Alan Beith regarded his constituency (Berwick-upon-Tweed, Northumberland) as a haven for reds, but accepted that greys were well liked and 'quite natural' elsewhere.[7] Fellow Liberal Democrat, Lord Beaumont of Whitley, an Anglican priest, had sharpened his shooting skills against 'tree rats' during a wartime boyhood in Buckinghamshire, where there were still plenty of reds. But he also appreciated that greys now possessed as much value for many people as reds did for him.[8] Baroness Farrington spoke in the same 1998 debate as Beaumont of the bother greys that gained access to attic spaces caused, as recently happened in her own home. She also recognized that many people did not regard the grey as an 'alien invader', not least because they had never seen one of its alleged victims.[9] RSPB's Big Garden Birdwatch survey of 274,000 householders in 2014 confirmed Farrington's observations: only 5 per cent of respondents had seen a red in their garden and just 2.7 per cent saw one on a monthly basis. By contrast, greys were spotted in 90 per cent of gardens.[10] As such, as Beith observed in 2006, eradication was 'neither possible nor popular'.[11]

In his foreword to a state-of-the-art compendium on grey squirrels, John Gurnell, who has studied squirrels since the 1970s, imagines a thought experiment. What if the grey was the native and Britain had been invaded by 'alien red squirrels from continental Europe', who replace the indigenous greys: which squirrel would Britons prefer then? The 'more attractive looking animal' or the native grey?[12] Whether native or non-native, more attractive or less attractive, greys provide a vital connection between people and the natural world – in other words, a valuable cultural ecosystem service. An earlier survey, in 2002, by the Mammal Society indicated that 45 per cent of UK respondents like greys, whereas only 24 per cent dislike them.[13]

'Squirrel Racism'

The shoot-on-sight policy that was advocated by Lord Mackie of Benshie in 1998 to protect the creature he loved so 'dearly' at his home in Angus gave Farrington the shudders. If she endorsed such a policy (she was the government's spokesperson on the environment in the Lords) her mailbox would bulge with letters from outraged parents who 'derive enormous pleasure from taking their children to feed grey squirrels'.[14] The 'final solution sort of way' (her words) in which Lord Rotherwick wanted to deal with greys also worried her.[15]

At Inglewood's seat of Hutton-in-the-Forest, in February 2011, Prince Charles spoke of reds 'returning to the woodlands and gardens where they were once terrorised by greys' and of 'my dream' that reds would again 'thrive throughout the United Kingdom'.[16] In 1989 a pair of sciurologists had speculated on the feasibility of a national eradication campaign. Modelled on the successful elimination of the muskrat in the 1930s and, more recently, of the coypu (another non-native furbearer that had escaped from fur farms), this might take the form of a 'rolling front' fanning outward from the southwest and 'justified to the public by the promised subsequent re-establishment of native British red squirrels'.[17] Dreams of restoration were paralleled, though, by further erosion, since 1989, of the ethos that the only good grey is a dead grey. It was time, contended John Bryant, wildlife officer of the League Against Cruel Sports (LACS), in 1994, to accept it as 'part of our naturalised wildlife'.[18] For broadcaster Cole Moreton, the red squirrel *Conservation Strategy* of 2005 emerged from 'a bloody conflict charged with romance, hatred and a whiff of racism. Reds versus greys. Small versus big. Native versus invaders. The English versus the alien.'[19]

In 2008, shortly before joining the BBC's *Springwatch* as co-presenter, Chris Packham lent his support to greys in a now defunct online article. 'We made a mistake, we shouldn't have brought them in,' he reflected, 'but then we didn't know better back in 1876.' In the New Forest, where he lives, reds were long gone. So, controlling greys there made no sense. But he did not oppose efforts to curb numbers where they constituted a 'real threat to our fluffty tufty Great British Red boys'. For Packham, the desire to purify the British countryside and return to a golden age of native rule by 'ecological cleansing' was an alien and alienating vision. The grey squirrel was over here and here to stay, so live and let live.[20] His description of the anti-grey brigade as 'a small band of lunatics who are insidiously bogged down and blinded by sentimental racism', featured in an Animal Aid 'factfile', also in a critique by journalist Oscar Rickett, who felt that reds symbolized, for the likes of Charles, a 'green and pleasant land' that never was.[21]

Half a century earlier, mammalogists had accepted the grey as a permanent member of Britain's fauna. The Mammal Society of the British Isles' landmark reference work of 1964, *The Handbook of British Mammals*, ranked it among thirteen 'established introductions' – a list which also included the rabbit, black and brown rats, edible dormouse, American mink and deer such as muntjac and sika.[22] 'Professor Acorn' characterized efforts to delegitimize the grey's position among British fauna as 'one small step removed from racism'.[23] Many Britons were hostile to greys, suspected Andrew Tyler, director of Animal Aid, because they were the 'wrong colour'. Rejecting the mentality that 'the only good squirrel is a red squirrel,' he likened anti-grey measures such as 'smashing their dreys and stamping on the young' to a 'pogrom'.[24] On BBC Radio 4's *Today* programme, on 29 January 2001, Tyler claimed that the species was being culled on Anglesey 'for no other reason than that it is grey and foreign'.

'Squirrel racism' allegations extended to Trudy Harrison, MP, whose constituency included *Squirrel Nutkin*'s setting. During a Commons debate in 2019 on potential red extinction, she described the red's decline as a 'national tragedy' and castigated the twentieth-century gamekeepers, foresters and 'country folk' who had killed thousands. After she called for a redoubling of efforts to eliminate greys from areas adjacent to red strongholds, John Woodcock, who represented the south Cumbrian constituency of Barrow and Furness, invited her to respond to those who might accuse her of being 'nothing more than a squirrel racist'. Woodcock assured colleagues that he, too, wanted to secure the red's future. But Harrison's fervency prompted him to play devil's advocate. Because the arrival of Britain's first greys roughly coincided with the creation of Barrow's shipyards in 1871, he compared greys to those who migrated into the area (from elsewhere in Lancashire, but also from Scotland, Ireland and Cornwall) to work in the shipyards. 'I do not imagine', he added, 'that she would suggest herding up Barrovians and removing them from their native Cumbria.'

Harrison reassured Woodcock that 'I of course would not want to see Barrovians rounded up and banished from Barrow.' The under-secretary of state at Defra, Thérèse Coffey, poured scorn on Woodcock's attempts 'to accuse us of being racist about squirrels': 'I have never heard such nonsense.' She recommended that he 'go on an education tour in Cumbria' to appreciate red squirrels' specialness. (Because she grew up in Formby, Coffey – like me – did not realize that not all squirrels were red until she went to live in the southeast, in her case, to attend university in London.)[25] Andrew Hodgkinson of the Penrith and District Red Squirrel Group remonstrated that 'we get called racist because we shoot the greys because they're grey, not red. So it's like we're shooting them because of the colour of their fur, which it really isn't.'[26]

The Grey Squirrel Defence Community

A letter to the editor in 2010 identified a drastic swing in attitudes since the 1950s, when MAF even 'lent out sets of light aluminium poles' to poke the pests out of their dreys, given the widespread feeling that 'the public should help kill them.'[27] Yet even the most extreme of anti-greys had dished out backhanded compliments over the decades. 'There is nothing good to be said for them,' observed a *Guardian* country diarist in 1945, 'except for their rather attractive appearance.' Recalling the 'brown' squirrels sporting in his boyhood garden as one of his strongest youthful memories, William Beach Thomas lamented in 1946 that 'those I write of [now] are grey'. And yet, he conceded that despite 'their alien morals they are most engaging to observe', especially their 'gymnastic' exploits. Moreover, there were few, if any, other animals 'whose ways we can watch daily, even from the windows'.[28] 'So lovely to look at – so hateful to know.' That is how inveterate grey foe James Wentworth Day summed up his mixed feelings in the headline of a *Daily Mail* article on 24 August 1974.

In post-war British fiction, positive portrayals of squirrels extended to greys. Albra Pratten's *Winkie: The Grey Squirrel* (1949) tells the story of a young squirrel who, frightened by a jay, falls out of the tree he is exploring. A small boy picks him up and he begins a new life in a house and garden (he is returned to the woods when old enough to fend for himself). Though mischievous, there is no suggestion that Winkie is troublesome. And in Winkie's world, greys are the norm: there are no reds. But by the early 1950s, a fictive red often encountered greys. Kenneth Grenville Myer's *The Story of Bushy Squirrel and Me* – 'the observations of naturalists' about the 'natural lore of the British Red Squirrel' – handles grey–red tensions delicately. The observation that 'you may be lucky enough to find [one] . . . near you' reminds us that 'this pleasant little [red] creature' is no longer the default squirrel. Myer relates the life of Bushy and family as a series of diary entries. The entry for 2 February reads: 'Saw cousin grey squirrel – we are not friends so did not speak.'[29]

What the 'American squirrel' needed in 1953 – when 'bob a brush' was rolled out – according to the American journalist John Allan May, was 'friends, fast'. May invited a 'Squirl Defense Community' to step forward, so that the 'outlaw' received a fair hearing.[30] But the grey already had British friends. MAF Minister Thomas Dugdale, for instance, regretted the sentimental regard many Britons had for what a journalist dubbed 'natural pets'. The problem (the reporter explained) was that they brought out 'all man's pet-loving instincts'.[31] A prominent member of the 'Squirl Defense Community' was naturalist Richard Mabey. 'I must confess to a thoroughly sentimental affection for urban grey squirrels,' Mabey wrote in 1973. In the countryside, they 'probably need to be controlled'. In urban environments their 'sheer brazen audacity' was hard to resist.[32]

In 1975, in a debate on hare coursing, Lord Aylestone brought up greys as part of a wider discussion of cruelty to animals. He admitted to his own quandary: 'If I have a hate of any animal at all ... it is the grey squirrel.' But, he conceded, 'it is more or less a love–hate relationship.' His garden contained three productive walnut trees, but he never got to taste a single nut. And yet, 'if I were asked to shoot one I could not because the little blighters sit up, using their front paws to hold the walnut in front of them.'[33] Sympathies such as Mabey's and Aylestone's complicated the red defenders' killing mission.

During the Second World War, the secretary of the Humane Education Society – a vegetarian – wrote to the minister of food seeking verification of press reports that his ministry had authorized the killing and eating of greys. 'Surely', he implored, 'there is more than enough slaughter going on already.'[34] MAF, however, saw the problem differently: it was not that people were feeding themselves greys; the difficulty was that they were feeding them. A prime obstacle that the wartime ministry identified was the 'householder' who overlooked its pest status and offered 'encouragement' (that is, gave food).[35] MAF's pamphlet 'You *Versus* Pests: The Grey Squirrel Menace' (April 1953) addressed 'how you can help'. The final injunction: 'Do not feed them.'[36]

In the early 1960s, in the context of rural interests' pressure to reinstate bounty payments, the Conference of Animal Welfare Societies' honorary secretary objected to MAFF's heartless, 'even brutal' killings during the bounty.[37] These humane sensibilities and sensitivities cut across the simple divide of squirrel nationality: that reds are British and belong, whereas greys are American and do not. A minor fracas in London in 1970 – reminiscent of President Eisenhower's squirrel spat in 1955 – provided a telling rejection of this mentality of red or grey.

After a luncheon she hosted, American reporters confronted the wife of the U.S. Ambassador to Great Britain, Leonore Annenberg, about her husband Walter's recent 'hassle' over squirrels.[38] The 6,000 crocus bulbs recently planted at Annenberg's official residence, Winfield House, had attracted the attention of squirrels in adjacent Regent's Park. The gardener shot four. But then 'television crews poured in' and the 'crocus crisis' hit the evening news. Callers jammed the embassy switchboard, some offering advice on killing the bulb snatchers. Most, though, were furious. Rattled, the ambassador halted the cull.[39] According to the *Washington Post*, shooting squirrels in London was 'an atrocity to the animal-loving English and the London press'. A New Yorker (whose tulip beds had been similarly disrupted) agreed that it 'could have been predicted' that shooting 'would arouse the ire of every animal lover in England'.[40] Yet the ambassador's gardener was reportedly following the advice of a Regent's Park official: 'Shoot them, that's the only way. They simply can't be trapped, they're too smart.'[41] The *Post* neglected to mention the offending squirrels' American roots, but British reporters dwelt on them. The embassy's spokesperson pushed back gently, armed with facts: 'it's a little unfair to say the Americans brought them over. It's more likely to have been some returning Britons.'[42]

At Home in a Home Away from Home

The furry colonists themselves could hardly be blamed for prospering in their adopted home. Though galling to squirrel nationalists, the newcomers proved admirably suited to the red's former habitat. In fact, they were better equipped to maximize use of available foods. Their out-of-placeness from red champions' standpoint was snugly fitting in-placeness from greys' perspective. Given their fondness for reds, however, Fitter explained in 1939 that most Britons accepted the *post hoc, ergo propter hoc* argument. The fallacious reasoning behind 'after this, therefore because of this' ran as follows: greys showed up after reds departed, so they must have caused this departure. Instead, greys' appearance 'coincided' with reds' virus-driven disappearance. They occupied a 'vacant ecological niche' across the Home Counties, whose countryside in many ways resembled eastern North America's open deciduous woodlands. The grey, in short, was 'able to step into [the red's] shoes without a fight'.[43] No matter that the replace-rather-than-displace explanation – first associated with Middleton and holding sway among sciurologists for half a century – was superseded in the 1980s by more nuanced explanations of grey advance and red retreat (and of grey prevention of red recolonization).[44] This does not alter the biological reality that while reds belong in Britain in terms of what is considered to be indigenous, greys belong too, and belong even better. Britain was home away from home.

As Scots pine grows comparatively slowly, most trees afforested since 1919 have been fast-growing non-native conifers: Norway and Sitka spruce, Corsican pine, Douglas fir and European and Japanese larches. On maturation, these conifers became the red's British strongholds. Formby squirrel reserve only became a fortress for reds after local landowners tackled dune encroachment and sand blow in the late 1800s and early 1900s by planting Scots, Corsican and Austrian pines.[45] That conifers were the red's 'natural habitat' (as Fitter, among others, argued) is harder to substantiate than the grey's proclivity for hardwoods.[46] When epidemics struck around 1900, the native squirrel largely inhabited – as it had since the last ice age – the deciduous woodland characteristic of England and Wales. Just three conifers – Scots pine, yew and juniper – are native to the British Isles. Yew is not widespread, and juniper more of a shrub than tree. Beyond the Scottish Highlands, Britain had little coniferous cover before afforestation with non-natives began in the late 1700s and early 1800s.[47]

After greys took root and spread, softwood plantations that struck defenders of the English landscape such as Wordsworth as so otherworldly became the backbone of Britain's timber industry and were the only areas that recovering reds were likely to recolonize.[48] Less than a century after Brocklehurst reputedly planted Britain's first greys in Cheshire, the county's reds were effectively confined to Corsican pine-dominated Delamere Forest.[49] At a competitive disadvantage, reds had little choice over where to call home.

On the grey-free Isle of Wight, today, 2,000 reds live among native hardwoods, mixed woodland and conifer plantations. It does not follow, though, that reds on the mainland would have been better off among the broadleaves. In the 1970s and '80s, sciurologists established that deciduous woodland was 'sub-optimal' for a species adapted to conifers across most of its Eurasian range. The only continental European country in which reds inhabit broadleaved woodland is Denmark.[50] Moreover, the British Isles lie at the westernmost limit of *Sciurus vulgaris*'s range, so life for the British branch of the Eurasian red squirrel family has been difficult at the best of times.[51] Nowhere at home was an ideal home. Moreover, the compatibility with conifers that reds acquired after the grey's spread arguably reinforced an existing tendency, given that many English populations are descended from Scandinavian and other resuscitating continental stock.[52]

Britain's reds, unaccustomed to competition, were ill-equipped to cope with a rival congener honed by North America's more competitive squirrel context. And so, Kathleen MacKinnon concluded in 1978, 'over most of Southern Britain the Grey Squirrel is here to stay and the Red gone forever.'[53] Jan Taylor reiterated that the grey did not win the battle for Britain's woods by throwing its weight around. And after interviewing squirrel researcher Jessica Holm, a *Daily Mail* features writer realized that the 'aggressive thug' and 'muscular foreign interloper' was simply better at living in Britain's typical treescapes.[54] Besides, interspecies altercations were no more frequent than intraspecies infighting. 'If you are a poor, persecuted red squirrel,' Holm remarked, 'you are just as likely to lose a toe at the tooth of another red as you are to have it bitten off by a filthy rotten foreigner.'[55]

Don't Be Beastly to the Greys

Conspicuous among those asking fellow Britons not to be 'beastly' to the grey in the 1970s was the journalist and novelist Auberon Waugh.[56] Well known for his fulminations against an 'American takeover', Waugh would support closer European integration in the early 1990s as a mechanism for de-Americanizing Britain and the only realistic alternative to 'political, economic, cultural and intellectual absorption by the United States'.[57] Nonetheless, he was very fond of grey squirrels, and had been since growing up in the Somerset village of Combe Florey in the 1950s.[58] For the editor of a conservative American magazine, Waugh's support for greys was another example of the contrarian and unpredictable views of someone whose hostility to the United States targeted, variously, Hollywood, the Pentagon, McDonald's and the *New York Times*.[59]

Waugh inveighed against '[grey] squirrel-bashers' Lords Bradford and Jersey.[60] Bradford made comments in *The Times* in 1971 that (in Waugh's estimation) were 'of a most violent and unpleasant nature'.[61] Bradford, whose estate produced conifers, urged wholesale killing of the 'alien invader' with no holds barred 'as a matter of extreme urgency' (to protect reds too). Waugh compared Bradford's language to 'the eloquence of an Enoch' – an allusion to former Conservative MP Enoch Powell's 1968 'rivers of blood' speech that warned of the dire consequences of further mass immigration of non-whites from Commonwealth countries.[62] 'In respect of the alien grey immigrant I plead guilty to racial discrimination,' proclaimed a supporter of Bradford's stance.[63]

Waugh did not oppose fox or stag hunting, decrying the 'misty-eyed townspeople', 'nature fanatics' and 'animal sentimentalists' who hated these pursuits.[64] His stance on greys, though, disrupted the divide on wildlife between 'townie' (unrealistic, sentimental) and countryfolk (realistic, unflinching) that many Conservatives identified.[65] Waugh, who still lived in Combe Florey, thought greys had as much right to make their home at Bradford's 405-hectare (1,000 ac) Shropshire estate, Weston Park, as Bradford himself. And so, 'if . . . Bradford does not retire from his campaign, I shall suggest that

Weston Park be taken over as a hostel for immigrant families, on the grounds that immigrants are the more attractive mammals.'[66]

Waugh fired another riposte after a Bradford missive demanding 'total outlawing' of greys. Though the greys on Waugh's property ate 'almost everything', his hardwoods were unscathed. To suggest that these 'delightful animals' destroyed trees was 'fantasy and wild hysteria'.[67] The other 'virulent squirrel-hater' he locked horns with was Lord Jersey. Going beyond Bradford's analogy between greys and the recent outbreak of a new, more potent strain of Dutch elm disease, Jersey (according to Waugh) blamed greys for the disease itself – by claiming they ate the eggs of tits that fed on the elm bark beetle that was the fungus' main vector. Waugh saw this as further scapegoating.[68] Yet in rural areas, more Britons adhered to Jersey's and Bradford's views than to Waugh's.

Championing and Celebrating the Irrepressible Grey

Members of the landed gentry might detest greys, but greys attracted support among members of the Lords with other backgrounds. The Labour peer Lord Stoddart of Swindon (a Welsh miner's son) objected strongly to fellow peers' proposals in the 1990s for reinstating bounty payments on a 'very intelligent and entertaining animal'. That a neighbour might shoot or poison those cavorting in his garden 'horrified' Stoddart.[69] At the time, though, the grey's best friend in the Lords was Baron (Woodrow) Wyatt of Weeford. (Born on 4 July 1918, his parents named him Woodrow after the incumbent American president, Woodrow Wilson.) A former Labour MP, he was appointed a Conservative Party life peer in 1987 after converting to Thatcherism. A decade earlier, he had written two children's books that revived pre-1920s meanings of saucy. The residents of the trees outside the family home in Wiltshire inspired the dapper central character in *The Exploits of Mr Saucy Squirrel* (1976) and *The Further Exploits of Mr Saucy Squirrel* (1977).

Mr Saucy is thoroughly anthropomorphized, with a bank account, fancy waistcoat, bicycle, colour television and antique furniture in his 'maisonette' in a beech tree near Devizes. Still, he prefers 'squirrel food' such as beechmast, pine shoots, birds' eggs and fledglings, also raided plums, strawberries and walnuts (and 'didn't want to stop being a squirrel'). Despite his impeccable Englishness (Union Jack-striped dressing-gown, gentlemanly comportment, cream tea), Mr Saucy Squirrel is a grey squirrel. The first hint is when he explains to his drey-warming party guests that his maisonette would be called a duplex in America. Later, he expresses a preference for Coca-Cola over luncheon wine.[70] His American identity is fully disclosed when he goes to buy a bowler hat. The shop assistant, pointing to the range of hats available, remarks: 'If you were an American squirrel I would pronounce it DERBY not DARBY. But I expect you're a good English squirrel,' to which Mr Saucy replies:

> Dear me . . . I'm always being asked that question. I'm sorry to have to tell you that my family came from America at the end of the nineteenth century. We are not the old kind of English squirrel people call the red squirrel. But we've been here a long time now.

Mr Saucy then takes the opportunity to explode a damaging piece of fake news: that 'we American squirrels' evicted 'English' reds. The truth, Saucy explains, is that they were already 'dying out in many parts of the country when my family arrived. We just took over in various places they'd already left.' He portrays a frictionless British–American coexistence: 'We don't quarrel with English squirrels at all. We eat the same food but we don't fight over it.' Indeed, he would be happy to see more of them around. Wyatt's authorial voice steps in rather clumsily to provide additional details. Greys 'never attacked' reds, but were 'cleverer', with 'a way of getting the best sites and places to find food', which 'probably put the red squirrels off living in the same areas'. Saucy reassures the sales assistant that his patriotic credentials are unimpeachable: 'We came from America so long ago that we are now properly naturalised English. I don't sing "The Star-Spangled Banner" . . . I sing "God Save The Queen".'[71]

In 1996, Wyatt raised eyebrows in the Lords with the claim that Henry VIII's forest destruction was the 'real villain' of the red squirrel's decline, not Saucy's real-life counterpart. (Wyatt's singling out of oak felling for the Tudor navy hardly conformed to the latest research findings, noted the Earl of Lindsay.) According to Wyatt, greys 'happily' and innocently made do with the remaining bits of woodland. Besides, they offered no threat to tree or bird in his garden and were hardly to blame for carrying a virus (SQPV) that they were immune to.[72] The following year, a journalist at *The Guardian*, intrigued by a *Sunday Times* report about a new breed of virus-immune red 'super-squirrel', invited Wyatt to respond, in his capacity as 'the country's leading admirer' of the grey. He was supremely unconcerned unless super-reds started eating greys.[73]

Wyatt's daughter, Petronella, for whom he wrote the *Mr Saucy Squirrel* books, inherited her father's allegiance to greys. In 2001, a few years after his death, as deputy editor of *The Spectator*, she wrote a paean to the furry immigrant as the epitome of immigrant vigour, hustle and bustle. Petronella, whose mother is Hungarian, argued that greys, 'like some of our immigrants, were cleverer than the red'. Moreover, without the hard work of the 'grey-squirrel-like' Poles, Czechs and (Asian) Indians, the British economy would languish. If native squirrels suffered after greys arrived, it was 'partly because they were lazy and lacked initiative'. Most provocative was her claim that Enoch Powell was the 'hero of all Red-squirrelians'.[74]

A fervent believer in the virtues of unrestrained free enterprise took an even harder pro-grey line. In 2006 financial journalist Jonathan Guthrie denounced state intervention to protect 'wimpy red squirrels from competition' by culling greys (bringing to mind a grey's characterization of reds as 'squimps' – 'all soft and gentle' – in Michael Tod's *The Second Wave* (1995)[75]). In the social Darwinian environment of British woodlands, reds were simply paying the price for inadaptability and ineffective use of resources: replacement by more energetic and enterprising newcomers. Comparing the robust Japanese car industry to its sickly British counterpart, Guthrie resorted to a squirrel analogy: 'If a grey squirrel was a car, it would be a Toyota. A red squirrel would be a Rover.'[76]

Lager-Drinking Greys

Toyota did not run adverts featuring grey squirrels. But an advertiser for lager had already tapped into their enviable attributes. The Durham Wildlife Trust's conservation manager, Steve Lowe, addressed grey popularity in 1997 by referring to a 'beer commercial in which a tough-guy gray squirrel completes a derring-do series of commando-like tasks to battle his way to a nut'.[77] Black Label, a brand that the Canadian brewer Carling created in 1927 – probably the first global beer – was exported to Britain in the late 1950s. Available on draught since 1965, Black Label quickly became popular in Australia, New Zealand and South Africa too. A decade before the Black Label advert, in 1979, a British journalist described the fate of British ale using a squirrelly analogy: traditional 'British ale', brown and served at room temperature, was 'rapidly being driven into extinction by a pale and decidedly chilly interloper from Europe and North America. Just like the red squirrel, it never stood a chance.'[78]

Carling's prize-winning TV advert drew on footage from a recent BBC1 documentary presented by squirrel scientist and squirrel book author Jessica Holm: *Daylight Robbery! Portrait of the Grey Squirrel as Thief*, which aired on 29 August 1988. Billed as a 'portrait of the ingenious, intelligent, acrobatic grey squirrel showing how rarely we win in any battle of wits with it', it featured a scenario all too familiar to millions of Britons: the one-sided back garden contest with bird-feeding humans in which undaunted greys deploy versatility and dextrousness to overcome any supposedly squirrel-proof system and 'exploit us'.[79]

Since the late 1970s, Black Label's ad campaign featured various men doing something 'cool, clever or difficult'. The catchphrase: 'I bet he drinks Carling Black Label.' The sequence the ad extracted from *Daylight Robbery!* was the final, all-out attempt to flummox the squirrels with a 'fiendishly difficult' assault course. To access hazelnuts in a bird feeder, a squirrel had to negotiate a pole, seesaw, trapdoor and highwire washing line (upside down), as well as abseil and take tremendous leaps. It took the local (wild) squirrels sixteen days to crack the system. Accompanied by the *Mission Impossible* series soundtrack,

Learning to Live with (and to Love) the Grey

Grey squirrel raiding a bird feeder.

a confident and jaunty 'commando' approaches the nuts. Mission impossible accomplished, one owl says to another: 'I bet he drinks Carling Black Label.'

One of three adverts shortlisted in the Best Performance by an Animal category at the Golden Break Advertising Awards for 1989, Carling's squirrel won. A trained grey, Cyril the Squirrel, attended the awards ceremony as a stand-in for the advert's unknown, unnamed wild squirrel. After being spooked by a dog, Cyril left his perch on the shoulder of host Jonathan Ross, climbed down "Wossy's" back and scampered across the stage at the London Palladium.[80] Cyril had been an eponymous, transatlantic name of endearment for squirrel since at least 1946 and became a general name for a squirrel (regardless of colour) in Britain in the early 2000s.[81] This advert, and the adulation Cyril received, spoke to a generational shift. For those who had never seen a red, nor encountered *Squirrel Nutkin* as a bedtime story, the grey was the only squirrel in town. And Cyril was *the* squirrel, especially after becoming a leading character in the animated series *Maisie* (1999–2000), as well as the hybrid protagonist, neither recognizably red nor grey (brown fur, but bulky and with rounded ears), of Jo Wright's *Cyril the Squirrel* children's book (2010).

Showcasing brazen cheek, fearlessness and breathtaking agility, the Black Label advert won hearts and minds.[82] Peter Wilson, spokesperson for Timber Growers UK, credited the advert with a pernicious influence: 'People should think of the grey squirrel as the North American tree rat ... not as the cuddly animal of TV advertising fame.'[83] Unfortunately for Wilson, *Daylight Robbery!* was such a hit that Holm presented a sequel from the same back garden. *Daylight Robbery 2* (BBC2, 1 April 1991) set an even more daunting series of challenges. Successful completion of a 15-metre (50 ft) 'super assault course', designed by the public, required the ascent of a narrow tube (chimney), negotiating a windmill, holes in spinning discs, a revolving door, and a pole with spinning rollers, then a nearly 2-metre (6 ft) leap. The home stretch involved boarding a red rocket that whizzed through a Perspex tube, then a 2.4-metre (8 ft) jump to access the food. This time, it was no stroll in the park for the local squirrels, who took twice as long (just over a month) to complete the course in one go. This time, the star performer was a female that, once she got the hang of it, repeated the exercise twenty times a day. (Just as well the star of the original *Daylight Robbery!* was male: 'I bet she drinks Carling Black Label' was unimaginable in 1989.) Holm's take-home message? The grey squirrel was a 'natural winner' and 'here to stay'. When *Daylight Robbery!* was repeated, a TV critic remarked that, 'totally devoid of British diffidence', the starring squirrel showed how greys did 'not wait for an introduction or an invitation'.[84] Plaudits were unavoidable.

Early twenty-first-century tales of grey squirrel chutzpah are legion. The British Trust for Ornithology (BTO) reported commandeering of cavities occupied by larger birds and of nest boxes intended for tawny owls, kestrels, stock doves, goosanders and mandarin ducks.[85] Gardening personality Anne Wareham's *Outwitting Squirrels and Other Garden Pests and Nuisances* (2015) contained 101 'cunning stratagems' to deploy against creatures like grey squirrels. Nonetheless, she remained fully aware of the limits of human authority: squirrels were fast learners, not least from other squirrels, and had astonishing memories. The 'persistence, determination, and ingenuity' of the brightest of greys was unbeatable.[86]

Greys' invincibility led some to fall foul of the law. In 2010, 58-year-old window cleaner Raymond Elliot was fined £1,547 (RSPCA's prosecution costs) and given a six-month conditional discharge for unlawfully killing a grey squirrel. After live-trapping a grey that was raiding his garden bird table, he drowned it in a water butt. The RSPCA (a neighbour apparently reported the incident) prosecuted Elliot under the Animal Welfare Act 2006 for causing unnecessary suffering to a 'protected animal' (one caught in a trap and therefore under human protective control) by using a non-instantaneous means of killing. Drowning, according to RSPCA veterinary evidence, could take up to three minutes. The court ruling was the first case brought under the 2006 Act involving a wild creature. The only humane way to kill a grey, the RSPCA maintained, was to take it to a vet and pay for a lethal injection.[87] Deterrents that the RSPB recommended after Elliot's prosecution included a few drops of Tabasco sauce or a dusting of chilli powder to render birdfeed unpalatable (unlike mammals, birds do not have the P-type receptors to detect capsaicin). And to prevent access, domes, baffles, discs and inverted cones could be rigged up. Or Vaseline could be smeared on bird table poles.[88]

Belonging and Becoming British

Someone with no requirement for chilli powder, Vaseline or baffles was 'Professor Acorn'. He also had no time for the distinction between native and non-native (alien) that hinged on whether a species arrived under its own steam (a 'natural' means such as range expansion after the ice age, the Eurasian red's mode of arrival in the British Isles) or depended on human-assisted dispersal.[89] Just because greys did not swim across the Atlantic, nor had proactively stowed away aboard a ship, but arrived because of the likes of Thomas Brocklehurst, was no reason to deny them a legitimate presence in Britain. As many of

Britain's current reds were descended from stock introduced from mainland Europe to augment receding British populations, for 'Professor Acorn' all squirrels in Britain were immigrants. Moreover, all squirrels in Britain were British by being born in Britain.[90] Journalist Michael Leapman, who suspected that 'ingrained anti-American prejudices' fuelled anti-grey feeling, concurred that having been born here, 'they are as British as you and me.'[91]

Britishness, belonging and citizenship shaped the solo-performance *A Kiss from the Last Red Squirrel*. Elyssa Livergant's 'auto-fantasy' (performed at venues from Colchester to Minsk, between February 2009 and February 2011) was a despatch from 'the frontline of the war between red and grey'. Sometimes wearing a grey squirrel headdress, Livergant introduced herself as a Canadian immigrant of Jewish background. Her efforts to secure indefinite leave to remain in the UK inspired the piece. On stage, she adopted the persona of a lecturer representing the European Squirrel Initiative, tub-thumping about grey violence and commanding her audience to destroy the aggressors. Grey versus red, for Livergant, was a springboard to explore 'attempts to hold onto what you love on national, cultural and personal levels'. Her key question (a refutation of the 'where are you "really" from?' question): 'How long does something have to be in a place for it to actually be *from* there?'[92]

Americans were often taken aback by British antipathy towards 'their' squirrels. But they were far from apologetic or contrite. Author Thomas Katheder (who was working on a book about a planter family in colonial Virginia) drew attention to a 'revenge of sorts', comparing the grey's impact on Britain's red to that of the English red fox on the indigenous American grey fox. The colonial gentry that introduced fox hunting to the Chesapeake Bay area of Maryland and Virginia in the 1730s and '40s brought over their quarry as well as their foxhounds. Apparently, the native grey fox ('reclusive and skittish') provided poor sport. 'Clever and wildly audacious', the imported red was the native grey's 'behavioral opposite'. Moreover, just as the grey squirrel 'crowded out' its 'more staid British cousins', Katheder claimed that the English red fox largely outcompeted the grey fox across its range.[93]

Norine Harris and Bill Ortel – galvanized, like Katheder, by a *New York Times* piece on Britain's 'squirrel wars' (21 October 2007) – gave an avifaunal example of a biological invasion in reverse. They saw the grey's exploits as just deserts for the ravages of the 'English' (house) sparrow (introduced to east coast cities in around 1850) and the 'British' (European) starling (1890) that had 'overrun ... our country' and made life awkward for 'indigenous' birds. They perceived a difference, though, in the two nations' responses. Americans 'do not hire thugs to go out and slaughter them all'. 'We thought the Brits loved animals,' they added. Disappointment turned to indignation: 'How dare they kill our squirrels?'[94]

Another American who held up the house sparrow as 'an obnoxious invasive alien from Europe' to even the squirrel score was Mark Blazis. Leafing through a wildlife magazine while visiting England in 2009, the outdoor and environmental affairs columnist was surprised to find an article that 'lamented the introduction of a harmful, invasive alien – our gray squirrel'. 'The Brits', Blazis informed his readers in Massachusetts, 'passionately detest them, complaining they dominate their little native red squirrel.' The latter, he noted, was 'quite different from our red squirrel, having cute ear tufts and gentlemanly behavior'. He was even more astonished that the author 'disparaged our squirrel as "a typical American – obnoxious and aggressive"'. He conceded that greys could be a pest but quoted American ornithologist Roger Tory Peterson's tribute to pesty species. As Peterson told him: 'Any animal that can live with man on man's terms – to not only survive but thrive – has my admiration.'[95] New to the great British squirrel debate, Blazis overestimated the solidity of the anti-grey consensus among Britons, many of whom shared his less censorious views on a squirrel they also considered 'ours'.

Peterson's accolade to pesty species is all about what today's animal historians call non-human agency. With pesty critters like grey squirrels, though, it is not really a question of finding that unacknowledged, overlooked agency and making a persuasive case for it.[96] The agency possessed by greys – or any other 'problem' animal for that matter – is self-evident to the people that live with them. Besides, the problem with pesty, invasive and nuisance creatures, for those that identify them in these ways, is surely that they have *far too much* agency. *Reducing* the amount of their agency is the challenge their human detractors and co-inhabitants face.

Still, the historian ought not be overly generous in their acknowledgement of squirrel agency. Once brought over to Britain, it is easy to assume that greys can largely take the credit for their success. But we should not overlook conventional human agency in Britain's squirrel history. The role of those naive Victorian introducers is worth revisiting. In the early 2000s, wildlife biologists Piran White and Stephen Harris identified 'invasion pressure', the volume, persistency and geographical spread of introductions between 1876 and 1929, as the critical shaper of success.[97] The notion of an all-conquering species, a 'super-squirrel' exuding hybrid vigour forged by extensive interbreeding between separate populations, was challenged by researchers at Imperial College London and ZSL. Initial findings compared twelve DNA markers from 315 squirrels at fourteen sites across four regions, one in Italy (Piedmont) and three in the UK (East Anglia, Northumberland and Northern Ireland). Correlation of genetic diversity with the size of the founding population (Piedmont's original introductions were a tenth of East Anglia's) showed that smaller numbers meant less diversity, which affected a population's invasiveness by reducing the ease and speed of adjustment to a new place.

The implications of the study's UK findings for the Eurasian red's future beyond northern Italy were substantial. Piedmont's greys existed in mostly discrete groups in 2014 but were gradually radiating outward. Lisa Signorile and others warned that the merging of separate introductions could prove disastrous for reds by boosting greys' overall genetic diversity, thereby increasing the likelihood of their northward expansion into and over the Alps.[98] The task for grey squirrel management in Italy was to keep populations apart. Imperial College's press release underlined that the availability of reliable documentation for Piedmont's introductions was the study's cornerstone. Giuseppe Casimiro Simonis (Conte di Vallario) attended various meetings in Washington, DC, in 1947. The diplomat was so captivated by local squirrels' 'extraordinary friendliness' he brought back two pairs, releasing them near his villa in 1948.[99] This introduction's modest size, the smallest the study examined, translated into the lowest genetic diversity among the descendant populations investigated.

The study's full findings drew on a genotype database of nearly 1,500 greys from 59 UK sites as well as locations in Italy and one in North America (West Virginia). The central question, for Britain, was how far genetic diversity enhancement through pooling of populations explained the species' overall success. For pooling resolves the 'genetic paradox' of how a small, 'bottlenecked' population with low diversity, evolutionary potential and reproductive fitness can become an invasive success. After Middleton in 1931 singled out three major expanding centres of population in England, namely, the Midlands, Cheshire and Yorkshire, Shorten in 1954 reported the emergence of a conglomeration through merger of these three nuclei, identifying three further 'expanding nuclei' in Scotland.[100]

The study of biological invasions is deeply historical, often predicated on the reconstruction of a species' likeliest invasion scenario. The starting point of Signorile's team was the sites of release and translocation (internal transfers) extracted from Middleton's records. They also sampled fourteen current populations (specimens culled in 2011), which included those from Woburn, Bedfordshire, as well as individuals taken not far from a site where stock from Woburn was released. Additionally, they worked with DNA samples from eleven squirrels – descendants of the original releases at Woburn – that the British Museum of Natural History collected in 1921–2. The data (from 381 samples in total) indicated that intermixing in Britain had been minimal; genetic diversity among U.S. populations was higher.

'In contrast to the grey squirrels' reputation as a highly invasive species', Signorile's team concluded, 'our data indicate low migration rates, little gene flow, limited mixing of populations after introductions and, most importantly, multiple human-mediated translocations as a key driver for the species' success in the UK'.[101] Imperial College publicized the central finding that it was human actions rather than 'innate ... propensity' to spread and rapidly invade fresh territory 'through the formation of a large, homogenous expansion front' under this headline: 'Don't Blame Grey Squirrels: Their British Invasion Had Much More to Do With Us'.[102] The conclusion that there were human super-spreaders rather than a super-squirrel attracted substantial attention, the takeaway message invariably that of Imperial's press release: we are responsible for the mess, after all. Headlines included: '"Super-Squirrel" a Myth, Reveals New Study of Greys' Anatomy' (*Daily Express*, 27 January 2016); 'Don't Blame Grey Squirrels for Invading the UK! Humans Helped Them Spread' (*Daily Mail*, 26 January 2016); 'Research Says Grey Squirrels Wronged' (*Daily Post*, 27 January 2016); 'Grey Squirrels "Unfairly Blamed" Over Ecological Crimes Made by Humans' (*Western Daily Press*, 27 January 2016); and 'Humans to Blame for Dominance of Grey Squirrels' (*Yorkshire Post*, 26 January 2016).

Ecologist Jason Gilchrist points out what is obvious but often overlooked: they 'did not come over here of their own accord and did not ask to be introduced. Neither do they have any control over the pox that they carry. They do what they do; which is to be grey squirrels.'[103] But being a grey squirrel is not being as powerful as many Britons assume – regardless of how completely individual specimens confound our efforts to rein them in. TV personality Esther Rantzen decried the swaggering, all-powerful grey as a 'brutish, raucous North American import'.[104] But the almost intrinsic expansionist qualities many Britons associate with greys as a collective entity, often merging their traits with those of their human counterparts, is mostly projection.

Having redirected the heat back onto humans, the media's instinct was to name and shame. Throughout the one hundred years' war, grey squirrel foes had excoriated the person held almost single-handedly responsible for Britain's grey squirrel problem: the Duke of Bedford. One of those gunning for him was the Earl of Selborne, a Conservative peer and the RSPB's vice president. The subject of debate in the Lords on that occasion in 2000 was imprisonment as a punishment for wildlife crimes such as harming an endangered species or releasing non-native species into the wild. 'In retrospect', remarked Selborne, 'I should like to have seen the noble Duke . . . have great penalties heaped on him' (short of a custodial sentence).[105] Signorile underscored Bedford's role as peerless super-spreader, gifting greys nationwide.[106]

Human assistance did not end in 1937, when translocations were outlawed.[107] Signorile's database indicates that a more recent translocation explains their presence in Aberdeen, on Scotland's northeast coast. And that is why Aberdeen's colony is more closely related to those in Hampshire's New Forest than to more established groups in areas adjacent to Aberdeen. Greys also suddenly appeared on an island off Scotland's northwest coast: Skye. There were no squirrels of any colour on the island before a grey was captured in October 2010. DNA sampling confirmed this individual's origins, previously anecdotal; it had stowed away under the bonnet of a car driven up from Glasgow. In January 2011 another specimen was apprehended in a garden on Skye, 160 kilometres (100 mi.) from the closest mainland populations.[108]

Simply Being Squirrel

Hitching a ride is something that grey squirrels do. To be a grey squirrel, though, is also to belong to a bigger family of squirrels. Squirrels belong to different species from different continents, but Eurasian red and American grey might share a common ancestor, however distant. In 2003, researchers at Duke University in North Carolina, the heart of eastern grey squirrel country, reported that the earliest material remains of the squirrel family were circa 36 million years old. Found in western North America, they were fossil evidence of the Douglas squirrel (*Douglas-sciurus jeffersoni*), a subspecies of the American red from the coastal forests of California and Oregon.[109] 'We think of "squirrels" as those that invade trash cans on campus,' remarked Duke evolutionary biologist Louise Roth. Yet from that original point of distribution in coastal California and Oregon, squirrels branched out to every continent except Antarctica and Australia, evolving into three major lineages consisting of 273 species within 51 genera (fifty of which she and John Mercer tissue-sampled from museum and live specimens over a ten-year period).[110]

The first fossil record of squirrels in Europe appeared 30 million years ago. The likeliest route into Eurasia was across the land bridge (Beringea) that connected Alaska with Siberia before about 7 million years ago, facilitating intercontinental travel that was eventually cut off by rising seas. Palaeobotanical records suggest Beringea was forested, so able to support tree and flying squirrels.[111] Squirrels were probably heading in the opposite direction to the humans that eventually populated the Americas by heading southeastward across a subsequent land bridge. A newspaper's science editor speculated that 'Squirrel Nutkin may have been an economic migrant.'[112]

Two years after these signs of deep-time squirrel convergence emerged, the Forestry Commission adopted a policy based on disconnection. Michael Tod's novel *The Silver Tide* introduced the notion of 'Ourland' (modelled on Brownsea Island, Dorset) to denote the native squirrel's last strongholds. Actual Ourlands have underpinned the FC's stronghold strategy since 2005. A journalist referred to this defence of defendable insular homelands within a sea of grey as 'squirrel apartheid'.[113] For many red champions, enclaves for reds that were exclusion zones for greys were the only places in which they stood a fighting chance. But for those that objected to killing greys that infiltrated a no-go area, squirrel apartheid smacked of repugnant segregation. Astrid Goldsmith explained that *Squirrel Island* (2016), her award-winning sci-fi animation with puppets, was about 'squirrel apartheid'. The protagonist, Dot, is a grey outlaw trapped on a 'hostile and mysterious' red squirrel island. She and her accomplice, Mr Acorn, then discover sinister reds dressed as soldiers who are busily shipping the island's acorns across the Solent to the English mainland.[114]

Goldsmith got the idea for *Squirrel Island* after visiting the Isle of Wight, where, she emphasized, introducing a grey is punishable by a £5,000 fine or two years in jail. In 1975 the island's MP, Stephen Ross, was consoled by the thought that were the red to be 'driven out of the British Isles generally, at least we could preserve it in one particular quarter'. He related how a grey from the New Forest, nearby on the mainland, had boarded the ferry at Lymington but was spotted when it docked at Yarmouth. The animal then led local RSPCA officials, among others, on a merry dance, running down the ferry's side and even swimming in the harbour. Some of the pursuers apparently got bitten but it was eventually apprehended and returned to the mainland.[115] The island's Red Squirrel Species Action Plan of 2003 includes a Grey Squirrel Contingency Plan that underlines the threat posed by even one grey, which might be pregnant or carrying SQPV.[116] The island's MP in 2005, Andrew Turner, commended the vigilance of his constituents, who report any suspected sighting, which invariably makes front-page news in the local press.[117]

The 'red squirrel propaganda and anti-grey signage' that Goldsmith saw on the Isle of Wight conjured up 'dystopian '70s sci-fi movies and I kept imagining what would happen if a plucky grey squirrel was trapped in that kind of environment'. *Squirrel Island* was her protest against the monstering of the grey and reaction to the red's status as endangered species poster child. 'I suppose it's partly because they are more photogenic than other endangered species (poor old brown trout),' she surmised, 'and partly due to the popularity of anthropomorphised red squirrel characters like Tufty and Squirrel Nutkin.'[118]

Goldsmith captured the potent associative power of Tufty and Nutkin for reds and their defenders. Nonetheless, a countervailing tendency to use Tufty and Nutkin as generic squirrel names is also detectable. Self-confessed squirrel hater, journalist Quentin Letts, traced the rot of 'nursery sweetification' of squirrels in general back to Potter's *Nutkin*.[119] The Nutkin effect worked for greys too, given that city folk rarely saw reds. In the 1970s Amanda Collins, who worked on ITV's nature documentary series *Survival*, regularly shared her suburban home with a female grey named Nutkin, who frequently entered the kitchen (for a hand-fed biscuit) and the living room.[120] 'Rats in Nutkin's Clothing' was a telling description of the masquerading creature that urbanites shared their sandwiches with in parks.[121] Reflecting on the grey's predations in 'our Oxford college garden', Robin Lane Fox, historian of classical antiquity at New College, concurred. Lane Fox, who was Garden Master at New and Exeter colleges, remarked that 'we have ... all been brainwashed by dear Beatrice [*sic*] Potter and her seminal tale of Squirrel Nutkin. How could we possibly punish anything with a brother called something like Twinkle Berry

[sic].'[122] Lane Fox seemed to imply that though Twinkleberry was a red squirrel, like Nutkin, a squirrel of any colour would benefit from connotations of cute furriness. Potter gave Nutkin a flawed personality that embodies reprehensible characteristics ever more widely associated with greys than reds since the early 1900s. Twinkleberry rather than Nutkin is the role model for children, at least in parents' eyes: polite, respectful and industrious. So, this tendency to attach Nutkin's name to greys too is not particularly surprising.

A similarly colour-blind convergence applies to the name Tufty. 'Times are hard for Tufty Fluffytail' was the opening line of an article about the tastiness of squirrel pie made with plump greys that the Forestry Commission had culled in its Rheola plantation near Neath, South Wales. A squirrel not far away, in Singleton Park, Swansea, whose mistreatment a toddler witnessed, was also grey. Nevertheless, the *Daily Mirror* headline ran 'Jo, 2, Sees Tufty Get a Kicking' (1 April 1996).[123]

Grey squirrel motif on the gate of a house ('Grey Squirrels'), Combe Martin, Devon.

We should not read into this blurring and interchangeability a decisive discarding of colour, let alone the dawning of a new era of post-colour squirrel history. Over the coming decades, though, the lives of more Britons are more likely to be enriched by the fleshy counterparts of Wirral the Squirrel than by Squirrel Nutkin's – if simply because there are more real Wirrals than real Nutkins. And switching from emotions to economics, the grey does not really deserve its status as poster villain for invasive non-native species. To increase public support for badger culls, a farmer argued that the badger's image must become 'less like a panda, and more like a grey squirrel'.[124] But in a survey in 2010 of the financial costs that invasive non-natives exact in Britain, the grey squirrel ranked only tenth in the top twenty, its damage infliction calculated at £14 million annually in direct costs. By far the costliest species, at £263 million, was the rabbit, while the rat and house mouse also outranked the grey.[125] Yet these statistics miss the essential point. No matter how much havoc rabbits (Lagomorpha order, not Rodentia) and other rodents wreak and how much we hate them, they do not impinge on the lives of Britain's first squirrel. Also, emotions trump economics (obviously).

A feature in BBC *Wildlife* in 2008 enumerated the points of difference between greys and reds. Among the contrasts in appearance are ears and feet: greys never have tufted ears and their fingers are 'always grey', whereas reds' fingers are 'russet all over'. Variations in habitat, feeding grounds and food were also identified, as well as in athleticism, greys being 'less agile' (a disputable point).[126] And yet, for all the differences of colour, size, weight and provenance – not to mention allegations that the 'American tree rat' is 'not a true squirrel'[127] – traits of shared squirrel-ness are hard to ignore. Arboreal squirrels are similar in overall appearance as well as cranial, dental and other anatomical features.

This common ground extends from morphology to behavioural traits. The squirrel was the first animal featured in H. W. Shepheard-Walwyn's book *The Spirit of the Wild* (1924), a series of British mammal portraits. 'I will not insult my readers,' Shepheard-Walwyn explained in 'Spirit of Pan' – a paean to the 'wee, wild Peter Pans of our sylvan glades' – 'by offering to describe the common red Squirrel indigenous to our lands.' Then he turned to Britain's new squirrel, whose acquaintance 'the veriest guttersnipe can go and make', since the 'grey, or American, form of this fascinating rodent' had readily become 'naturalized in the heart of London'. Given the grey's ubiquity, he thought it enough to note that 'our dear little native species is simply a somewhat smaller edition.'[128]

During the grey's rapid spread in the early 1930s, not all Britons forgot the squirrelly qualities that transcended (or muddied) the colour line. Beneath their brown or grey fur, squirrels, for William Beach Thomas, were characterized by 'half-aimless curiosity and morbid restlessness'.[129] 'Weariness and the squirrel do not consort,' he added. The squirrel's 'master attribute', he continued, 'is restless vitality'. He also underlined the 'singularly similar habits' of red and grey; their nests, for instance, were hard to tell apart.

Red and grey were also accused of the same infamies. Some of the red's 'more sentimental champions', Fitter observed, overlooked that it was 'just as arrant an egg-thief'.[130] There was a crossover, too, in colour terms. Seasonally, Thomas explained, the English brown's fur alternated between summertime's rufous red (chestnut) and winter's brownish-grey and black-lined hue.[131] In fact, when spring moult arrived, the red's coat could be as ash-grey as the 'dowdy old ash-bin fur' of the greys that poet Cecil Day-Lewis saw in Greenwich Park in 1962.[132] In turn, greys' heads and backs turned brownish in summer; even in winter, russet streaks down silver grey bodies, tails and paws.[133] (In fact, this colour crossover reinforced the misbelief in cross-breeding.[134]) In short, greys can be red and reds can be grey.

Other points of convergence abound. Grey and red females have 'similar reproductive potentials'.[135] Neither species hibernates. Both 'scatterhoard' nuts and seeds. And some people like to feed both species. The parkgoer's tradition of feeding greys peanuts (and worse) goes back to Regent's Park in the 1910s. In the mid-1990s many of the 250,000 yearly visitors to the National Trust's Formby Squirrel Reserve fed its reds peanuts (30 to 56 per cent of the squirrels' dietary intake during late winter and spring at the time) and worse, much worse, namely, bacon, burgers, crisps, chips and biscuits.[136] Until recently, the Trust sold food for visitors to feed to the Formby

Grey squirrel, Clydebank, Renfrewshire, Scotland.

squirrels (shades of early 1900s Regent's Park). Feeding by visitors is now 'actively' discouraged and the Trust has also quit so-called supplementary feeding.[137]

Red or grey, peanut-fed or hazelnut, acorn and pine cone-foraging, squirrels are bumptious, nimble charismatics with magnificent tails employed for shade in summer and as blankets in winter. Saucy, chittering livewires (*chuk-chuk-chuk*), they are bundles of nervous energy. Frisky acrobats with liquid-like movements, their activity peaks are early morning and dusk, with a lesser midday burst. Greys can leap up to 2.5 metres (8 ft) – ten times their body length – from tree to tree; and reds are not far behind at 2 metres (6½ ft).[138] (As I try to write this section, in June 2022, four brownish-red young greys fresh from the drey, brimming with *joie de vivre*, take turns to spring from a corrugated garage roof to a nearby tree trunk in a gently parabolic curve, attaching themselves to the tree as if Velcro-ed, then jumping back, and repeating the exercise.) And when, very occasionally, they take a tumble, the tail functions as a parachute to soften the landing.

Using some of their claws as anchors, and with 180-degree outwardly rotatable ankle and wrist joints, they can run downwards head-first at breakneck speed. On the ground, greys can clock up to 29 kilometres per hour (18 mph) in bounds of 1 to 1.5 metres (3 to 5 ft). Reds are no slouches out of the trees either, able to cover 5 metres (16 ft) per second (18 kilometres per hour (10 mph)).[139] Both species are able swimmers too, using their hind feet, doggy-paddle style, and tail as rudder.[140] In his broadside poem 'Squirrel' (1987), Ted Hughes pinpoints arboreal locomotion (if with a sizeable dose of poetic licence). The animal 'with a rocketing rip' has only two gears: 'freeze and top'.[141] Hughes omits his zippy squirrel's colour. But he wrote the poem when living in red-less north Devon. The poem's illustrator, R. J. Lloyd, lived nearby. His painting clears up the ambiguity. Hughes's squirrel is grey. Its fur, though, is tinged with reddish brown.

References

PREFACE

1 The Mammal Society also supplied assessments for 'lower plausible limit' and 'upper plausible limit'. These for reds were 218,000 and 553,000 respectively and, for greys, 1.34 million and 3.79 million. Sub-totals were provided for England, Scotland and Wales, but not for Northern Ireland: Mammal Society, 'Final Red List Version for Mammal Review', www.mammal.org.uk, 30 June 2021; Mammal Society, 'One Quarter of Native Mammals Now at Risk of Extinction in Britain', www.mammal.org.uk, 30 July 2020. Estimates of surviving reds from other sources are considerably lower, widely given as 140,000 – only 15,000 of them in England – whereas the figure for greys is usually only slightly lower, at 2.5 million. See, for example, Toni Bunnell, *Eden Species Report: Top Ten Threatened Animal Species in the UK – Their Distribution and What You Can Do to Conserve Them* (August 2011), p. 3; Helen Keating, 'Red Squirrel Facts', www.woodlandtrust.org.uk/blog, 1 November 2018.

2 'Grey squirrels [House of Lords, hereafter HL]', *Hansard*, 236 (31 January 1962), columns (hereafter, cc.) 1033–5, https://hansard.parliament.uk; Norman Shrapnel, 'NATO Debate without Fire: Less Interesting Than Squirrels', *The Guardian* (1 February 1962).

3 'Text Maniacs', *Daily Star* (10 August 2007). Gilbert is a character in an episode titled 'Irregarding Steve' (December 2006) of the cartoon series *American Dad!* Regarding 'smartness', a study of the comparative problem-solving abilities of greys and reds in Britain, assessed through 'an easy and a difficult food extraction task' designed to measure 'behavioural flexibility', indicated that greys were better at solving the harder assignment – a 91 per cent success rate on the first or a subsequent visit versus 62 per cent for reds. See Pizza Ka Yee Chow, Peter W. W. Lurz and Stephen E. G. Lea, 'A Battle of Wits? Problem-Solving Abilities in Invasive Eastern Grey Squirrels and Native Eurasian Red Squirrels', *Animal Behaviour*, CXXXVII (2018), pp. 11–12, 16, 19; 'Are Alien Grey Squirrels

Cleverer Than Native Red Squirrels in Britain?', https://squirrelweb.co.uk, 21 February 2018; Bethany Minelle, 'Red and Grey Squirrels Go Head to Head in Nut IQ Test', *Sky News UK*, https://news.sky.com, 20 February 2018.

4 'Royals Up For Reds', *Daily Star* (5 June 2010); 'Charles Support for Reds', ibid. (12 March 2011); 'Charles's Cull Order', ibid. (20 October 2014); 'Harry's Nuts on Squirrels', ibid. (26 February 2015).

5 Paul Donnelley, 'Alien Squirrels' Days Numbered', ibid. (8 February 2021); 'Saved: The Red Squirrel', ibid. (9 September 2020); Natasha Wynarczyk, 'Come on You Reds!: Brit Squirrels in Tree-Mendous Revival', ibid. (12 September 2020).

6 Patrick Williams, 'Stevie's Backing the Reds', ibid. (18 September 2016). The red's conservation status is 'endangered' according to the Mammal Society's 'Red List for Britain's Mammals', the first official Red List compiled specifically for the UK: Mammal Society, 'One Quarter of Native Mammals Now at Risk of Extinction in Britain' (2020).

7 Jessica Holm, *Squirrels* (London, 1987), p. 37.

8 John Harrison, 'Girl of 2 Is Savaged by Squirrel in Knutsford', *Daily Star* (7 November 2002); Anthony Walton, 'Nutty Squirrel: Bobby Is Attacked by "Beast"', ibid. (27 October 2003); 'Pool Goes Nuts', ibid. (17 May 2014).

9 Monica Charsley, 'Violent Squirrel Leaves Town "Afraid to Leave Their Homes" After Attacking 18 People', *Daily Star* (28 December 2021); Barney Davis, '"Bloodthirsty" Squirrel Attacks 18 People in Small Welsh Village in Two-Day Christmas Rampage', *Evening Standard* (29 December 2021). Under the 2019 law, an injured or sick grey, or an orphaned kit, that an animal rescue centre has rehabilitated also cannot lawfully be set free: Nada Farhoud, 'Orphaned Baby Squirrels Rescued Not Allowed Back in the Wild Due to Strict Laws', *Daily Mirror*, www.mirror.co.uk, 1 May 2022.

10 'Swindon Squirrels Blamed for Damage to Car Brakes', *BBC News*, www.bbc.co.uk, 15 October 2010.

11 'Squirrel Hell Hits Workers', *Daily Star* (1 August 2016). Ticks, lice and the other parasitic vectors of Lyme disease have been found in UK greys: Caroline Millins et al., 'An Invasive Mammal (the Gray Squirrel, *Sciurus carolinensis*) Commonly Hosts Diverse and Atypical Genotypes of the Zoonotic Pathogen *Borrelia burgdorferi Sensu Lato*', *Applied and Environmental Microbiology*, LXXXI/13 (2015), pp. 4236–45.

12 Ross 'Bushy Tail' Kaniuk and James 'Nutter' Wickham, 'Sven Would Be Nuts Not to Take Me!', *Daily Star* (21 April 2006); 'Drives Fans Nutty', ibid. (26 April 2006); 'Height of Cheek', ibid. (1 October 2012). Sid became an animated squirrel character in the Disney series *101 Dalmatian Street* (2019) and a lead figure in Jane Whitehouse's children's book *My Furry Friend Sid the Squirrel* (2022).

13 Sophie Bateman, 'Animals Rescued from Very Strange Places', *Daily Star* (30 December 2019); 'Squirrel Kill Rap is "Nuts"', ibid. (20 July 2010).

14 'Come Over Here, Take Our Nuts . . .', *Daily Star* (22 March 2020).

15 'Nuts on Squirrel', *Daily Star* (28 October 2011); 'Dish of the Grey',

ibid. (10 February 2015); Alexander Brown, 'Furballs are Dish of the Day', ibid. (4 February 2019).
16 Liam McInerney, 'Punters Go Nuts for Fried Squirrel', *Daily Star* (16 October 2019).
17 'Missing Lynx', *Daily Star* (12 July 2013).
18 Joe Faretra, 'Scientists to Put Squirrels on the Pill to Stop Them Causing Havoc to Countryside', *Daily Star Online*, www.dailystar.co.uk, 11 July 2022.
19 Chris Riches, 'Why Squirrels are Just Like Humans', *Express Online*, www.express.co.uk, 11 September 2021; Jaclyn R. Aliperti et al., 'Bridging Animal Personality with Space Use and Resource Use in a Free-Ranging Population of an Asocial Ground Squirrel', *Animal Behaviour*, CLXXX (October 2021), pp. 291–306.
20 Though it is technically in Freshfield, to avoid confusion I refer to the Formby reserve and to Formby's squirrels.
21 In 2015, Cairngorm Brewery, Aviemore, brewed a seasonal ruby red ale, 'Autumn Nuts'. A portion of proceeds went to the Scottish Wildlife Trust's Red Squirrel Project: 'Sounds Nuts – Craft Beer to Preserve Squirrels', British Guild of Beer Writers, news, www.beerguild.co.uk, 27 August 2015.
22 'Little Grey Brother': 'Home-Made Toys for Baby's Christmas Stocking', *Daily Mail* (17 November 1922); 'The Art of Stockings: Onion Colour and Squirrel Grey', *Manchester Guardian* (14 June 1922).
23 M. R., 'Destructive Squirrels' (letter to the editor, hereafter LTTE), *Country Life*, XXVII/702 (18 June 1910), p. 926.
24 F. W. Rich, 'Damage by Squirrels', *Country Life*, XXXI/801 (11 May 1912), p. 703.
25 'A Plea for the Squirrel', *Country Life*, XXXI/798 (20 April 1912), p. 562.
26 W. Beach Thomas, 'Nuts or Squirrels', *The Spectator* (24 December 1937), p. 1147.
27 Dan Russell, 'Grey Squirrels', *Saturday Review of Politics, Literature, Science and Art*, 4,244 (6 February 1937), p. 101.
28 Frank Finn, 'The Sins of Alien Animals', *The Graphic*, XCVI/2487 (28 July 1917), p. 120.
29 'Grey Squirrels: Not So Bad as They Are Painted', *Daily Mail* (31 March 1922).
30 H. N. Southern, review of Frederick S. Barkalow and Monica Shorten, *The World of the Gray Squirrel* [1973], in *Animal Ecology*, XLV/2 (1976), p. 605.
31 Harry V. Thompson and T. R. Peace, 'The Grey Squirrel Problem' (1960), MAF 131/53, The National Archives, Kew (TNA). MAF is the abbreviation for the records of the Ministry of Agriculture and Fisheries and the Ministry of Agriculture, Fisheries and Food.
32 Polly Pullar, *A Scurry of Squirrels: Nurturing the Wild* (Edinburgh, 2021), foreword, pp. 1–2, 4, 7.
33 Gregory Blaxland [pseudonym], 'In the Country: The Apologists', *Punch*, CCXLII/6354 (20 June 1962), p. 949.
34 Richard Fitter, 'Cards for Country-Lovers', *The Observer* (1 December 1963).
35 Brian C. R. Bertram and David-Petter Moltu, *The Reintroduction of Red Squirrels into Regent's Park* (London, 1987), pp. 1–2.
36 Jamie Allen, 'Consider the Squirrel', *Oxford American: A Magazine of the*

South, 85 (Summer 2014), at www.oxfordamerican.org.
37 R.S.R. Fitter, 'The Grey Squirrel', *The Observer* (31 May 1959).
38 Stanley Cohen, *Folk Devils and Moral Panics: The Creation of the Mods and Rockers* [1972], 3rd edn (2002) (Abingdon, Oxon, 2011), p. 1. On the media's development of a 'problem frame' that identifies specific threats, then bigged up to generate and spread fear, see David Altheide, *Creating Fear: News and the Construction of Crisis* (New York, 2002), pp. 84, 89.
39 Cohen, 'Introduction to the Third Edition' (2002), *Folk Devils*, pp. xxiii–xxv.
40 Ibid., p. xii.
41 Joseph Langland, 'Sacrifice of a Red Squirrel', *New Yorker*, 38 (26 January 1963), p. 32.
42 H. G. Lloyd, 'Past and Present Distribution of Red and Grey Squirrels', *Mammal Review*, XIII/2–3 (1983), p. 80.
43 'Grey Neighbour from Hell', *Forestry and British Timber* (1 November 1998), p. 8.
44 William D. Montalbano, 'British Wage War on U.S. "Invaders"; Imported Gray Squirrels, Reviled as "Tree Rats", Are Pushing the Native Red Variety toward Extinction', *Los Angeles Times* (2 January 1997).
45 P.W.W. Lurz, 'Changing "Red to Grey": Alien Species Introductions to Britain and the Displacement and Loss of Native Wildlife from Our Landscapes', in *Displaced Heritage: Responses to Disaster, Trauma and Loss*, ed. Ian Convery, Gerard Corsane and Peter Davis (Woodbridge, Suffolk, 2014), pp. 269–70.
46 J. Bryce, S. Cartmel and C. P. Quine, 'Habitat Use by Red and Grey Squirrels: Results of Two Recent Studies and Implications for Management' (Forestry Commission Information Note 76, Edinburgh, October 2005), p. 9.
47 MAF, Minute sheet, 12 March 1942, MAF 44/45.
48 Patrick Barkham, '"Kill Them, Kill Them, Kill Them": The Volunteer Army Plotting to Wipe Out Britain's Grey Squirrels', *The Guardian*, www.theguardian.com, 2 June 2017.
49 Mike Dunn et al., 'Public Attitudes towards "Pest" Management: Perceptions on Squirrel Management Strategies in the UK', *Biological Conservation*, 222 (2018), pp. 55, 60. An earlier, if smaller, study of two places, the Sheffield area and Cumbria, anticipated the finding that urban dwellers view greys more favourably. In total, 13 per cent of respondents from Sheffield and 36 per cent from Cumbria felt negatively towards greys: Ian D. Rotherham and Stuart Boardman, 'Who Says the Public Only Love Red Squirrels?', *ECOS*, XXVII/1 (2006), pp. 30–33.
50 Craig Shuttleworth, 'Grey Squirrels Are Bad for the British Countryside – Full Stop', *The Conversation*, 6 April 2017, https://theconversation.com.
51 Patrick Barkham, 'Reader Poll: Help Choose Our National Icon', *BBC Wildlife*, XXXI/6 (June 2013), p. 72.

1 A TALE OF TWO SQUIRRELS

1 Graham Johnson, 'Rooney and the Vice Girl: His Note after Bathroom Sex', *Sunday Mirror* (25 July 2004); Graham Johnson, 'Rooneygate: The

Vice Video: Caught!', *Sunday Mirror* (22 August 2004).

2 'Rooneygate: The Vice Video: Roo: I Confess', *Sunday Mirror* (22 August 2004); Lara Gould, 'Sex Pest Rooney: Wonder Boy's Lost the Plot', *Sunday Mirror* (5 June 2005); Dan Evans, 'Wayne's Girl Lays Down the Law', *Sunday Mirror* (29 August 2004).

3 'Rooney and Coleen Split: She Slaps Him and Leaves over Hooker', *The Sun* (26 July 2004); 'Coleen Chucks Ring Away: £25K Rooney Gem Hurled to Squirrels', *The Sun* (27 July 2004).

4 Chris Millar, 'Coleen Throws Wayne's Ring to Squirrels', *Evening Standard*, www.standard.co.uk, 26 July 2004.

5 'Rooney's Ring Peril for Squirrels', *Western Mail* (3 August 2004); 'Don't Disturb the Squirrels', *Birmingham Post* (28 July 2004); Rachel Cooke, 'Year of Their Wives', *The Guardian* (30 December 2006); Peter Conrad, 'The Making of the Girl Next Door', *The Observer*, www.theguardian.com, 3 December 2006; Carole Cadwalladr, 'Meet the Real Footballers' Wives', *The Observer*, www.theguardian.com, 4 June 2006; Lucy Mangan, 'Coleen and Wayne: The Unanswered Questions', *The Guardian*, www.theguardian.com, 20 February 2007.

6 Fiona Cummins and Brendon Williams, 'My Rock: Coleen with Ring She "Dumped"', *Daily Mirror* (28 July 2004); 'Coleen: My Story', *Daily Mirror* (19 February 2007); Annabel Crabb, 'True Brits', *Sunday Age* (Australia) (1 August 2004); 'Britons Beseeched Not to Look for $60,000 Diamond Ring Tossed to Uncommon Squirrels', *National Post* (Canada) (28 July 2004); Hugh Adami, 'Soccer Stars Do Scandals Right', *Ottawa Citizen* (1 August 2004); Shaun Phillips, 'Rooney's Riches: Man U Pays $68 Million for England Wonderkid', *Herald Sun* (Australia) (2 September 2004); 'Squirrels Stand in the Way of Wayne Rooney Engagement Ring', *Agence France Presse* (27 July 2004).

7 Adami, 'Soccer Stars'; David Thomas, 'Worra Story!', *Daily Mail* (10 February 2006); Nigel Pauley, 'The Sign that Says "I Love Roo"', *Daily Star* (27 August 2004).

8 David Thomas, 'A Looney Rooney Love Life', *Daily Mail* (31 July 2004).

9 '"Good Job I'm an Actor": Hidden Art: Stars Versus the Squirrel on Paper', *Birmingham Post* (15 September 2000); 'Stars Sketch to Save Squirrels', BBC News, http://news.bbc.co.uk, 15 September 2000; John Ingham, 'Stars Go Nuts to Help Save the Red Squirrel', *Daily Express* (15 September 2000).

10 'Macca's Squirrel in the Pink', *Manchester Evening News* (15 September 2000).

11 'Full Text: Clegg Reform Speech', http://news.bbc.co.uk, 19 May 2010.

12 'Nick Clegg Calls for Ideas on Laws to Be Repealed', BBC News, www.bbc.co.uk, 1 July 2010.

13 'Grey Squirrels and the Law', *Country Life*, LXXXII/2118 (21 August 1937), p. 184. For currency conversions, I have used the National Archives' 'Currency Converter: 1270–2017', www.nationalarchives.gov.uk/currency-converter.

14 Drumlanrig, 'Ginger Jibe a Nutty Mistake for Harman', *Scotland on Sunday* (31 October 2010).

15 Patrick Kingsley, 'G2: 2010 Review: The Gaffes', *The Guardian*, www.theguardian.com, 30 December 2010;

'Danny's Tufty Decision', *Mail on Sunday* (14 November 2010).

16 Emily Ashton, 'Clegg Accuses Harman of "Outrageous Discrimination"', *Press Association News* (10 November 2010).

17 Ben Borland, 'Harman Forced to Take Back Her Anti-Ginger Jibes', *Express on Sunday* (31 October 2010); 'Trading Places: Cross Bencher', *Express on Sunday* (7 November 2010).

18 Andy Philip and Scott Macnab, 'SNP and Labour Clash in Economy Debate', *Press Association News* (4 November 2010).

19 'Deregulation Bill, Schedule 11: Other Measures Relating to Animals, Food and the Environment', Public Bill Committee, House of Commons [HC], 11 March 2014, cc.341–2, https:// publications.parliament.uk; Tamara Cohen, 'Grey Squirrel Law Scrapped', *Mail Online*, www.dailymail.co.uk, 19 March 2014; Law Commission, *Wildlife Law*, vol. 1: *Report, 50 Years* (9 November 2015), p. 329.

20 Sarah L. Crowley, Steve Hinchliffe and Robbie A. McDonald, 'Killing Squirrels: Exploring Motivations and Practices of Lethal Wildlife Management', *Environment and Planning E: Nature and Space*, 1/1–2 (2018), p. 137.

21 'Deregulation Bill, Schedule 11', column (hereafter, c.) 345; Kashmira Gander, 'Ministers Defeated in Grey Squirrel Battle', *The Independent*, www.independent.co.uk, 18 March 2014.

22 Cohen, 'Grey Squirrel Law Scrapped'.

23 Jonathan Walker, 'Ministers Raise White Flag to Grey Squirrels', *The Journal* (Newcastle) (18 March 2014); Jonathan Walker, 'Defeat in Red Squirrel Fight', *Newcastle Evening Chronicle* (18 March 2014).

24 'Red Squirrel Population [HC]', *Hansard*, 448 (28 June 2006), c.103; 'Red Squirrels [HC]', ibid., 272 (21 February 1996), c.331, https:// hansard.parliament.uk.

25 Ingrid H. Tague, 'The History of Emotional Attachment to Animals', in *The Routledge Companion to Animal–Human History*, ed. Hilda Kean and Philip Howell (London, 2018), pp. 356–8.

26 Ian Niall, 'Country Life', *The Spectator* (12 March 1954), p. 287; W. G. Teagle, 'Mammals in the London Area 1961', *London Naturalist*, 43 (July 1964), pp. 127–8.

27 J.H.L., 'A Country Diary', *Manchester Guardian* (30 May 1953).

28 F. R. Cann, 'Grey Squirrel Damage in 1960 in the South East Region' (memo), 23 November 1960, MAF 131/53; C. Marston to W. A. Williams, 'Damage by Grey Squirrels in Private Houses', 7 May 1964, MAF 131/175.

29 'Forestry [HL]', *Hansard*, 417 (23 February 1981), c.946, https:// hansard.parliament.uk/Lords.

30 Alison Blunt and Robyn Dowling, *Home* (Abingdon, Oxon, 2006), pp. 117–21.

31 Robert C. Stauffer, 'Haeckel, Darwin and Ecology', *Quarterly Review of Biology*, XXXII/2 (1957), pp. 138, 140.

32 Xia Mingfang, 'The Ecology of Home', *RCC Perspectives/ Transformations in Environment and Society*, 3 (2017), p. 27.

33 Verity Elson and Rosemary Shirley, 'Creating the Countryside', in *Creating the Countryside: The Rural Idyll Past and Present*, ed. Verity Elson and Rosemary Shirley (London, 2017), pp. 27, 29; Alice Carey, 'This Land', in *Creating the Countryside*, ed. Elson and Shirley, pp. 80–81; Patrick

Deer, *Culture in Camouflage: War, Empire, and Modern British Literature* (Oxford, 2009), pp. 175–6.
34 'Forestry Bill [HL]', *Hansard*, 35 (14 July 1919), c.540.
35 'Grey Squirrels', *Manchester Guardian* (29 January 1946).
36 Monica Shorten, 'Introduced Menace: American Grey Squirrel Poses Threat to British Woodlands', *Natural History*, LXXIII/10 (1964), pp. 43–4.
37 David Stapleford, *An Affair with Red Squirrels: Forty Years of Squirrelmania* (Dereham, Norfolk, 2003), pp. 4, 7.
38 For 'occupation', see, for example, *The Handbook of British Mammals*, ed. H. N. Southern (Oxford, 1964), p. 29.
39 Franklin Ginn, 'Extension, Subversion, Containment: Eco-Nationalism and Post(colonial) Nature in Aotearoa New Zealand', *Transactions of the Institute of British Geographers*, XXXIII/3 (2008), pp. 335–53.
40 On Deep England, see Deer, *Culture in Camouflage*, pp. 173–6, 60, 180, 106–7, 134–6; Patrick Wright, *On Living in an Old Country: The National Past in Contemporary Britain* (Oxford, 2009), pp. 77–83. Berkshire, Buckinghamshire, Essex, Hertfordshire, Kent, Middlesex and Surrey are usually identified as the Home Counties, though the term can be stretched to encompass Bedfordshire, Cambridgeshire, Hampshire, Oxfordshire and Sussex. On the routine (and wilful) conflation of the English landscape with British identity by Englanders, see David Lowenthal, 'British National Identity and the English Landscape', *Rural History*, II/2 (1991), pp. 209–10.
41 David Matless, Paul Merchant and Charles Watkins, 'Animal Landscapes: Otters and Wildfowl in England, 1945–1970', *Transactions of the Institute of British Geographers*, 30 (2005), pp. 191–205; Matless, 'Versions of Animal–Human: Broadland, c. 1945–1970', in *Animal Spaces, Beastly Places: New Geographies of Human–Animal Relations*, ed. Chris Philo and Chris Wilbert (London, 2000), pp. 128–36.
42 Paul Readman, *Storied Ground: Landscape and the Shaping of English National Identity* (Cambridge, 2018), pp. 140–41.
43 Hilda Kean, 'Save "Our" Red Squirrel: Kill the American Tree Rat: An Exploration of the Role of the Red and Grey Squirrel in Constructing Ideas of Englishness', in *Seeing History: Public History in Britain Now*, ed. Hilda Kean et al. (London, 2000), p. 56.
44 'Friends and Enemies of the Farmer: Rat and Rabbit Pests', *The Times* (23 May 1944).
45 Kean, 'Save "Our" Red Squirrel', p. 52; Kean, 'Imagining Rabbits and Squirrels in the English Countryside', *Society and Animals*, IX/2 (2001), p. 164.
46 Henry David Thoreau, *Walden and Civil Disobedience* [1854] (Harmondsworth, 1983), p. 272.
47 Philip Howell, *At Home and Astray: The Domestic Dog in Victorian Britain* (Charlottesville, VA, 2015), p. 20.
48 Peter Coates, '"Unusually Cunning, Vicious and Treacherous": The Extermination of the Wolf in United States History', in *The Massacre in History*, ed. Mark Levene and Penny Roberts (New York, 1999), p. 173.

49 Susan Nance, 'Animal History: The Final Frontier?', *American Historian*, 6 (November 2015), p. 29.
50 Anthropologists have pioneered this approach. See Miriam Kahn, 'Your Place and Mine: Sharing Emotional Landscapes in Wamira, Papua New Guinea', in *Sense of Place*, ed. Z. Feld and K. H. Basso (Santa Fe, NM, 1996), pp. 167–96; Kay Milton, *Loving Nature: Towards an Ecology of Emotion* (Abingdon, Oxon, 2002). See also (if squirrel-less) Andrew Flack and Dolly Jørgensen, 'Feelings for Nature: Emotions in Environmental History', in *The Routledge History of Emotions in the Modern World*, ed. Katie Barclay and Peter N. Stearns (London, 2023), pp. 235–51; Dolly Jørgensen, *Recovering Lost Species in the Modern Age: Histories of Longing and Belonging* (Cambridge, MA, 2019).
51 'Wildlife and Countryside Bill [HL]', *Hansard*, 416 (3 February 1981), c.1167.
52 Sara Ahmed, 'Affective Economies', *Social Text*, XXII/2 (2004), p. 119.
53 Tague, 'History of Emotional Attachment to Animals', pp. 345–66.
54 Jean Comaroff and John Comaroff, 'Naturing the Nation: Aliens, Apocalypse and the Postcolonial State', *Journal of Southern African Studies*, XXVII/3 (2001), pp. 628, 631, 637.
55 A. D. Middleton, *The Grey Squirrel: The Introduction and Spread of the American Grey Squirrel in the British Isles* (London, 1931), pp. 1–2.
56 Eugene Kinkead, 'Department of Amplification', *New Yorker* (1 September 1975), p. 55.
57 Monica Shorten, *Squirrels* (London, 1954), pp. x, 185.
58 On constructing 'problem' species, specifically 'rats with wings', see Colin Jerolmack, 'How Pigeons Became Rats: The Cultural-Spatial Logic of Problem Animals', *Social Problems*, LV/1 (2008), pp. 72–94.
59 Antoon De Vos, Richard H. Manville and Richard G. Van Gelder, 'Introduced Mammals and Their Influence on Native Biota', *Zoologica*, 41 (1956), p. 168.
60 John Kelly, 'Brits Rally to Save Red Squirrels from Onslaught of Invasive Grays', *Washington Post* (11 April 2012).
61 William Condry, 'A Country Diary', *The Guardian* (22 April 1972).
62 'Squirrels: Pest or Pet?', *The Guardian*, www.theguardian.com, 17 June 2010.
63 'U.S. Squirrels Lose Battle in Britain', *New York Times* (4 March 1956).
64 'Squirrels: Pest or Pet?', *The Guardian*.
65 Jeffrey Stinson, 'U.S. Varmints Pushing British Relatives Out: England and Scotland Are at War with an Ugly American Intruder – the Gray Squirrel', *USA Today*, 8 December 2008, www.global-factiva-com.
66 Mike Ellison, 'Novel Way to Become a Successful Writer', *The Guardian* (20 January 1994).
67 Fred Patten, *Furry Tales: A Review of Essential Anthropomorphic Fiction* (Jefferson, NC, 2019), p. 143.
68 Michael Tod, *The Silver Tide* (London, 1994), pp. 37, 1, 4, 124.
69 Ibid., pp. 1–2, 4, 20, 79.
70 Ibid., pp. 47, 19, 18.
71 Ibid., pp. 34, 84, 21–2, 19.
72 Ibid., pp. 106, 218.
73 Philip Sherwood, 'The Testing Invasion', *Forum for the Discussion of New Trends in Education*, XX/3 (1978), pp. 72–4; Peter Regent, 'Putting Possession in Context' (LTTE), *The Guardian* (14 November 1983);

Richard Heller, 'Britain on the Griddle', *The Guardian* (21 March 1995); Richard Donkin, 'It's People We Are Talking About', *Human Resources* (March 2005), p. 16; George Trefgarne, 'Is It Time for Tea?', *The Spectator* (20 January 2007), p. 65; Stephen Bloomfield, *Venture Capital Funding: A Practical Guide to Raising Finance*, vol. II (London, 2008), p. 231; Ruth Campbell, 'Let's Hear It for the Fairy Cakes', *Northern Echo* (26 November 2009).

74 Middleton, *Grey Squirrel*, p. 22.

75 MAF, 'Grey Squirrels', Advisory Leaflet 58 (1943), MAF 44/45; MAF, 'The Grey Squirrel' (February 1953), MAF 131/83; Forestry Commission, 'Hints on Controlling Grey Squirrels' (1955), MAF 131/53.

76 Patrick Deer, 'The Dogs of War: Myths of British Anti-Americanism', in *Anti-Americanism*, ed. Andrew Ross and Kristin Ross (New York, 2004), pp. 158–9.

77 William J. Lederer and Eugene Burdick, *The Ugly American* (New York, 1958); Alain Frachon, 'America Unloved', *Le Monde*, 24 November 2001; Simon Schama, 'The Unloved American: Two Centuries of Alienating Europe', *New Yorker*, www.newyorker.com, 10 March 2003.

78 Janice Atkinson-Small, 'Why It's Time We Went to War with the Grey Squirrel', *Mail Online*, http://atkinsonsmallblog.dailymail.co.uk, 5 October 2012.

79 Peter Pringle, 'Diary; Oh to Be in England, with Its Self-Congratulatory Press, Squirrel-Slaughtering Residents and Siberian Railway Stations', *New Statesman*, CXXV/4305 (11 October 1996), p. 6.

80 Paul Johnson, 'Sir Mulberry Hawke Is the Latest Beneficiary of Moral Relativism', *The Spectator* (3 February 2001), p. 24.

81 Paul Johnson, 'Tit for Tat and the Fury of a Squirrel with an Einstein Brain', *The Spectator* (23 February 2002), p. 29.

82 Paul Johnson, 'The Decline and Fall of Anti-Americanism in Britain', *The Spectator* (1 November 1997), p. 30.

83 William D. Montalbano, 'British Wage War on U.S. "Invaders"; Imported Gray Squirrels, Reviled as "Tree Rats", Are Pushing the Native Red Variety toward Extinction', *Los Angeles Times* (2 January 1997).

84 Irene Littlejohns, 'Terrorist Squirrels', *Northern Echo* (8 September 2005).

85 C. H. Pulford, 'Kick Off', *Daily Mail* (17 August 1954); Charles Anthony, 'Taxation', *The Guardian* (30 August 1975); 'Eton Adopts the Old School Jeans', *Daily Mail* (11 June 1975); Lord Paget, 'The Queen's Speech [HL]', *Hansard*, 436 (9 November 1982), c.180; Robert Low, 'An Englishman Abroad', *The Observer* (9 September 1984); Sandra Barwick, 'A Tale of Old County Folk', *The Spectator* (1 February 1992), p. 18; 'Smallweed', *The Guardian* (25 June 1994), p. 22; Anthony Steen, 'Genetically Modified Crops [HC]', *Hansard*, 317 (30 July 1998), c.612.

86 Michelle Hanson, 'That Awkward Age', *The Guardian* (11 June 1996).

87 Julian Champion, 'At Home with the Real Flintstones', *Daily Mail* (25 November 2000); 'Diary', *Computer Weekly* (26 April 2005), p. 72.

88 'Which Invasive Species Are You Most Concerned About?', *Horticulture Week* (22 May 2009), p. 17.

89 Claire D. Stevenson-Holt and William Sinclair, 'Assessing the Geographic Origin of the Invasive Grey Squirrel Using DNA Sequencing: Implications for Management Strategies', *Global Ecology and Conservation*, 3 (2015), p. 22.

90 Sandra Swart, 'The Other Citizens: Nationalism and Animals', in *Routledge Companion to Animal–Human History*, pp. 32, 37.

91 Patrick Barkham, 'Hedgehog Wins UK Natural Emblem Poll', *The Guardian*, www.theguardian.com, 31 July 2013.

92 Patrick Barkham, 'A Wildlife Icon for Britain', *BBC Wildlife*, XXXI/6 (June 2013), pp. 69–73; Ben Hoare, 'Revealed: Your Favourite British Species', *BBC Wildlife*, XXXI/9 (August 2013), pp. 60–62. Building on the poll's momentum, the National Hedgehog Preservation Society wants its prickly darling enthroned as official national animal: Charlotte Dear, 'Should the Hedgehog Become Britain's National Symbol?', *Country Living*, www.countryliving.com/uk/news, 12 November 2015.

93 'What's Your Favourite British Mammal?', *BBC Wildlife*, XXVI/6 (June 2008), p. 47.

94 Sophie Stafford, 'What's Your Favourite Animal?', *BBC Wildlife*, XXVI/6 (June 2008), p. 5; Mark Stratton, 'The Rise of the Otter', *BBC Wildlife*, XXVI/10 (September 2008), pp. 58–9.

95 'How Prince Charles Plans to Sterilise the Nation's Squirrels – With Nutella', *The Guardian*, www.theguardian.com, 26 February 2017.

96 Philip Howell and Hilda Kean, 'The Dogs That Didn't Bark in the Blitz: Transpecies and Transpersonal Emotional Geographies on the British Home Front', *Journal of Historical Geography*, LXI (2018), pp. 44, 52.

97 Chris Pearson, '"Four-legged *Poilus*": French Army Dogs, Emotional Practices, and the Creation of Militarized Human–Dog Bonds, 1871–1918', *Journal of Social History*, LII/3 (2019), p. 733.

98 Monique Scheer, 'Are Emotions a Kind of Practice (and Is That What Makes Them Have a History)? A Bourdieuian Approach to Understanding Emotion', *History and Theory*, LI/2 (2012), pp. 209, 212.

99 Thomas Nagel, 'What Is It Like to Be a Bat?', *Philosophical Review*, LXXXIII/4 (1974), pp. 435–50. On animal emotions, see Marc Bekoff, *Emotional Lives of Animals: A Leading Scientist Explores Animal Joy, Sorrow, and Empathy, and Why They Matter* (Novato, CA, 2007).

100 'Squirrels Brown and Grey', *The Spectator* (13 November 1909), p. 779.

101 Frank Finn, 'Public Pets in Regent's Park', *The Graphic*, LXXXII/2134 (22 October 1910), p. 638.

102 'Forestry Bill [HL]', c.541.

103 Nature Conservancy letter to Colonel Boyle (Fauna Preservation Society), 13 March 1951, FT1/24 (Records Created by and Inherited by the Nature Conservancy, Squirrels, 1950–1968), TNA; 'The Red Squirrel', *Oryx*, I/3 (1951), pp. 147–9.

104 J. E. Bowlby, 'Squirrels Destroying Young Pheasants', *The Field*, LXVIII/1974 (25 October 1890), p. 632; H.G.H., 'Squirrels Eating Young Pheasants', *Country Life*, XXXI/801 (11 May 1912), p. 701. On reds' flesh consumption as norm rather than exception, see E. Adrian

Woodruffe-Peacock, 'The Squirrel's Food' (LTTE), *Country Life*, L/1283 (6 August 1921), p. 176.

105 Robert Darnton, 'Workers Revolt: The Great Cat Massacre of the Rue Saint-Séverin', in *The Great Cat Massacre and Other Episodes in French Cultural History* (New York, 1984), pp. 75–104; Michael G. Vann, 'Of Rats, Rice, and Race: The Great Hanoi Rat Massacre, an Episode in French Colonial History', *French Colonial History*, IV (2003), pp. 191–203; Nancy Jacobs, 'The Great Bophuthatswana Donkey Massacre: Discourse on the Ass and Politics of Class and Grass', *American Historical Review*, CVI/2 (2001), pp. 485–507.

106 Matthew Holmes, 'The Perfect Pest: Natural History and the Red Squirrel in Nineteenth-Century Scotland', *Archives of Natural History*, XLII/1 (2015), pp. 114–18, 119–20, 122.

107 L.C.R. Cameron, *The Wild Foods of Great Britain: Where to Find Them and How to Cook Them* (London, 1917), p. 21.

108 'Grey Squirrels [HL]', *Hansard*, 236 (31 January 1962), c.1035; 'Wildlife and Countryside Bill [HL]', *Hansard*, 416 (3 February 1981), c.1076.

109 In 1942, allegedly, the owner of two pairs bought from a local pet shop, finding it difficult to procure peanuts, released his pets into the pinewoods at Formby Point. Apparently, these four spawned a colony that spread north and south along the Sefton coast: John Lucas, 'Come In, You Reds', *Daily Telegraph* (25 July 1992). For an account of the earlier origins of Formby's reds – probably more reliable – see Chapter Two.

110 E. M. Barratt et al., 'Genetic Structure of Fragmented Population of Red Squirrel (*Sciurus vulgaris*) in the UK', *Molecular Ecology*, VIII (1999), pp. 55–63; M. L. Hale, P.W.W. Lurz and K. Wolff, 'Patterns of Genetic Diversity in the Red Squirrel (*Sciurus vulgaris* L.): Footprints of Biogeographic History and Artificial Introductions', *Conservation Genetics*, 5 (2004), pp. 167–79.

111 'British and Foreign', *Cornhill Magazine*, VII/37 (July 1886), pp. 95–8.

112 Natasha Wynarczyk, 'Come on You Reds!: Brit Squirrels in Tree-Mendous Revival', *Daily Star* (12 September 2020). The Eurasian red squirrel is in fact Denmark's national *mammal*. The mute swan, often cited as the Danish national animal, appears to be the national bird: 'Nationalplanter og -dyr', https://naturstyrelsen.dk, accessed 15 August 2022.

113 Stephen Harris, Carl D. Soulsbury and Graziella Iossa, 'Is the Culling of Grey Squirrels a Viable Tactic to Conserve Red Squirrel Populations?' (University of Bristol/Advocates for Animals, 2006), p. 2.

114 Stephen Harris, 'The Red Has Lost – So Accept the Grey', *BBC Wildlife*, XXIV/9 (September 2006), pp. 36–9.

115 'The Zoological Society: Grey Squirrels', *The Field*, CXIX/3083 (27 January 1912), p. 187.

116 W.S.B., 'Nature Notes: The Grey Squirrel', *Country Life*, XLVII/1205 (7 February 1920), p. lviii.

117 For 'domiciled alien', see Alexander Porter Morse, *The Citizen in Relation to the State: A Paper Read Before the American Bar Association at the Seventh Annual Meeting* (Washington, DC, 1884), pp. 9–10.

118 James Fair, 'Wildlife to Work: The Results', *BBC Wildlife*, XXIX/11 (November 2011), pp. 70–71; Steve

Harris, 'Our Findings about Mammals', *BBC Wildlife*, XXIX/11 (November 2011), p. 72.
119 Harry Wallop, 'Squirrel Fans Fume at Radio Gardeners', *Daily Telegraph* (17 June 2010); Wallop, 'Threats to Radio 4 Experts After Tips on Killing Squirrels', *Daily Telegraph* (17 June 2010).
120 'Squirrels: Pest or Pet?', *The Guardian*.
121 Erik Stokstad, 'The Bloody Battle to Save the Red Squirrel', *Science*, www.sciencemag.org/news, 9 June 2016.
122 Arnold Arluke and Clinton R. Sanders, *Regarding Animals* (Philadelphia, PA, 1996), p. 178.
123 Jonathan L. Clark, 'Uncharismatic Invasives', *Environmental Humanities*, VI (2015), p. 45.
124 Eva Hayward, 'Sensational Jellyfish: Aquarium Affects and the Matter of Immersion', *Differences*, XXIII/3 (2012), p. 177.
125 Steve Nash, 'Desperately Seeking Charisma: Improving the Status of Invertebrates', *BioScience*, LIV/6 (2004), pp. 487–94. Octopuses, on the other hand, have risen considerably up some people's sociozoological scale of late, spurred by the revelatory film *My Octopus Teacher* (2020).
126 Thom van Dooren, 'Invasive Species in Penguin Worlds: An Ethical Taxonomy of Killing for Conservation', *Conservation and Society*, IX/4 (2011), p. 290.
127 Donna J. Haraway, *When Species Meet* (Minneapolis, MN, 2008), pp. 85, 80.
128 Swart, 'Other Citizens', pp. 44–5.

2 RED BEFORE GREY

1 Archibald Thorburn, *British Mammals*, vol. I (London, 1920), p. 80.
2 W. H. Marston, 'Damage by Squirrels', *Journal of the Forestry Commission*, XIX (1948), p. 101.
3 Judith J. Rowe, *Grey Squirrel Control* (London, 1973), p. 1.
4 Gerald E. H. Barrett-Hamilton and M.A.C. Hinton, *A History of British Mammals*, vol. IIC: *Order Rodentia, Part XX* [1918] (London, 1910–21), p. 719; A. D. Middleton, 'The Ecology of the American Grey Squirrel (*Sciurus carolinensis* Gmelin) in the British Isles', *Proceedings of the Zoological Society of London*, C/3 (1930), p. 831.
5 'Squirrel Colonists in Scotland', *Pall Mall Gazette* (29 July 1886), p. 4.
6 J. E. Panton, 'Highways and Byeways, *English Illustrated Magazine*, XIX (April 1885), p. 470.
7 Auceps, 'The Squirrel Carnivorous', *The Field*, XII/294 (14 August 1858), p. 135; G. R., 'Squirrels Carnivorous', *The Field*, XII/295 (21 August 1858), p. 156; 'Occasionally Carnivorous', *The Field*, XIII/337 (11 June 1859), p. 474; 'Chickens Eaten by Squirrels', *The Field*, LIV/1386 (19 July 1879), p. 89.
8 J. G. Wood and Theodore Wood, 'The Squirrel Tribe', *Practical Teacher*, I/1 (March 1881), p. 6.
9 Edmund Selous, *Tommy Smith's Animals* (London, 1899), pp. 173–4.
10 T. C. Bridges, *Scud: The Life Story of a Squirrel* (London, 1907), p. 160.
11 Ibid., p. 11. 'Scamper' and 'dash' are synonyms for 'scud'.
12 MAF, 'Grey Squirrels' (minute sheet), 4 February 1943, MAF 44/45.
13 J. A. Harvie-Brown, *The History of the Squirrel in Great Britain* (Edinburgh, 1881), pp. 31, 86.
14 Ibid., p. 16; Harvie-Brown, 'Early Chapters in the History of the Squirrel in Great Britain', *Proceedings*

of the *Royal Physical Society of Edinburgh*, 6 (1880–1881), pp. 33, 36–42.

5 E. M. Barratt et al., 'Genetic Structure of Fragmented Population of Red Squirrel (*Sciurus vulgaris*) in the UK', *Molecular Ecology*, VIII/SI (1999), p. 56.

6 J. A. Harvie-Brown, 'The Great Spotted Woodpecker (*Picus major*, L.) in Scotland', *Annals of Scottish Natural History*, I/1 (1892), p. 14.

7 James Ritchie, *The Influence of Man on Animal Life in Scotland: A Study in Faunal Evolution* (Cambridge, 1920), pp. 353–4; Harvie-Brown, *History of Squirrel*, pp. 13, 59. Christopher Smout, Scottish woodland history expert, thinks the role of pinewood deforestation has been over-emphasized, arguing for disease as the likelier operative consideration. See T. C. Smout, 'Birds and Squirrels as History', in *Local Places, Global Processes: Histories of Environmental Change in Britain and Beyond*, ed. Peter Coates, David Moon and Paul Warde (Oxford, 2016), p. 59.

8 Harvie-Brown, 'Early Chapters', pp. 44–5.

9 'North-Country Naturalists', *Edinburgh Review*, CXLVI/299 (July 1877), p. 127.

10 Veronica Heath, 'A Country Diary', *The Guardian* (11 March 1983); Barratt et al., 'Genetic Structure of Fragmented Population', p. 56; Barrett-Hamilton and Hinton, *History of British Mammals*, vol. IIC, p. 692.

11 Harvie-Brown, 'Late Chapters in the History of the Squirrel in Great Britain', *Proceedings of the Royal Physical Society of Edinburgh*, 6 (1881), pp. 115–17. For details of reintroductions – not all instigated by aristocrats and some later ones were internal transplants – see Matthew Holmes, 'The Perfect Pest: Natural History and the Red Squirrel in Nineteenth-Century Scotland', *Archives of Natural History*, XLII/1 (2015), pp. 116–17.

22 Harvie-Brown, 'Late Chapters', pp. 142–3.

23 Ibid., p. 148.

24 Ibid., p. 118.

25 Barrett-Hamilton and Hinton, *History of British Mammals*, vol. IIC, pp. 710–11; Ritchie, *Influence of Man*, p. 295.

26 Ritchie, *Influence of Man*, p. 354.

27 'Agriculture (Miscellaneous Provisions) Bill [HL]', *Hansard*, 326 (30 November 1971), c.183.

28 Graham Anderson, 'Rooks and Farmers', *Murray's Magazine: A Home and Colonial Periodical for the General Reader*, VI/36 (December 1889), p. 830.

29 R.S.R. Fitter, *The Ark in Our Midst: The Story of the Introduced Animals of Britain* (London, 1959), pp. 129–30; V.P.W. Lowe and A. S. Gardiner, 'Is the British Squirrel (*Sciurus vulgaris leucourus* Kerr) British?', *Mammal Review*, XIII (1983), p. 58.

30 Jerzy Sidorowicz, 'Problems of Subspecific Taxonomy of Squirrel (*Sciurus vulgaris* L.) in Palaearctic', *Zoologischer Anzeiger*, CLXXXVII/3–4 (1971), pp. 139–40.

31 J. R. Ellerman and T.C.S. Morrison-Scott, *Checklist of Palaearctic and Indian Mammals, 1758–1946* (London, 1966), pp. 472–5.

32 Robert Kerr, *The Animal Kingdom or Zoological System, of the Celebrated Sir Charles Linnaeus. Class I. Mammalia* (Edinburgh, 1792), p. 256.

33 'Squirrel film' (transcript) (January 1959), pp. 1–2, MAF 131/53.

34 *The Handbook of British Mammals*, ed. H. N. Southern (Oxford, 1964), p. 268.
35 Kerr, *Animal Kingdom*, p. 255.
36 Barrett-Hamilton and Hinton, *History of British Mammals*, vol. IIC, pp. 697, 700; Sidorowicz, 'Problems of Subspecific Taxonomy', p. 127.
37 Oldfield Thomas, 'The Seasonal Changes in the Common Squirrel', *The Zoologist*, XX (1896), pp. 401–2; G. S. Miller, *Catalogue of the Mammals of Western Europe in the Collection of the British Museum* (London, 1912), p. 908; Peter W. W. Lurz, John Gurnell and Louise Magris, 'Sciurus Vulgaris', *Mammalian Species*, 769 (15 July 2005), p. 2.
38 A subspecies originally linked to northern Germany's Harz Mountains, *S. v. fuscoater* has since been identified from Alsace-Lorraine to Ukraine.
39 W. Segar, personal communication, in C. Simms, 'Terrestrial Mammals in the Dunes of South West Lancashire, 1961–64', *Annual Report No. 53, Lancashire and Cheshire Fauna Society* (1968), p. 18. See also M. Garbett (pers. comm.), in John Gurnell and Harry Pepper, 'A Critical Look at Conserving the British Red Squirrel *Sciurus vulgaris*', *Mammal Review*, XXIII/3–4 (1993), p. 128.
40 Audrey Insch, 'A Country Diary', *The Guardian* (13 February 1999).
41 Monica Shorten, *Squirrels* (London, 1954), p. 15.
42 F. J. Johnston, 'The Grey Squirrel in Epping Forest', in *The London Naturalist for the Year 1937* (London, 1938), pp. 99, 94.
43 F. J. Stubbs, 'Remarks on the Squirrels of Epping Forest', Appendix C of H. B. Watt, 'On the American Grey Squirrel in the British Isles', *Essex Naturalist*, XX/3–4 (1923), p. 205; Stephen Harris, 'The History and Distribution of Squirrels in Essex', *Essex Naturalist*, XXXIII/2 (1973/4), p. 65.
44 R.S.R. Fitter, 'The Distribution of the Grey Squirrel in the London Area', in *The London Naturalist for the Year 1938* (London, 1939), p. 16.
45 Ibid.
46 R. C. Preece, 'Faunal Remains from Radiocarbon-Dated Soils within Landslip Debris from the Undercliff Isle of Wight, Southern England', *Journal of Archaeological Science*, XIII/2 (1986), pp. 194, 197.
47 Lowe and Gardiner, 'Is the British Squirrel . . . British?', p. 66.
48 Marie L. Hale and Peter W. W. Lurz, 'Morphological Changes in a British Mammal as a Result of Introduction and Changes in Landscape Management: The Red Squirrel (*Sciurus vulgaris*)', *Journal of Zoology*, CCLX/2 (2003), pp. 159, 164.
49 'Red Squirrels [HL]', *Hansard*, 661 (27 May 2004), c.1440.
50 Lowe and Gardiner, 'Is the British Squirrel . . . British?', pp. 58, 60–61.
51 Barratt et al. 'Genetic Structure of Fragmented Population', p. 61.
52 Hale and Lurz, 'Morphological Changes', pp. 160–63.
53 Marie L. Hale et al., 'Impacts of Land Management on the Genetic Structure of Red Squirrel Populations', *Science*, CCXCIII/5538 (21 September 2001), pp. 2246–8; Hale and Lurz, 'Morphological Changes', pp. 163–5.
54 Marie L. Hale, Peter W. W. Lurz and Kirsten Wolff, 'Patterns of Genetic Diversity in the Red Squirrel (*Sciurus*

vulgaris L.): Footprints of Biogeographic History and Artificial Introductions', *Conservation Genetics*, v/2 (2004), pp. 176, 167, 173, 175.

5 David Derbyshire, 'Red Squirrels Face Extinction', *Daily Telegraph* (7 April 2004).

6 Maurice Burton, 'The Chequered Story of Red Squirrels', *Illustrated London News*, CCXVIII/5842 (7 April 1951), p. 546.

7 Gilbert White, *The Natural History and Antiquities of Selborne* [1788] (London, 1837), p. 461. This information was originally published as 'The Naturalist: Squirrels', *The Mirror of Literature, Amusement, and Instruction*, 596 (23 March 1833), p. 183.

8 'The American Sportsman', *The Field*, X/239 (25 July 1857), p. 65.

9 Harvie-Brown, *History of Squirrel*, p. 101.

0 'The Ethics of Dining', *London Society*, XLIV/263 (November 1883), p. 560.

1 'Squirrel Pie', *Daily Telegraph* (30 September 1943); Rice Gaither, 'Squirrels Across the Sea', *New York Times* (28 November 1943).

2 Kathleen Walker-Meikle, *Medieval Pets* (Woodbridge, Suffolk, 2012), p. 94.

3 James J. Rorimer, 'The Unicorn Tapestries Were Made for Anne of Brittany', *Metropolitan Museum of Art Bulletin*, I/1 (1942), pp. 10, 13–14, 16, 18.

4 'Houses of Benedictine Monks: Priory of St Swithun, Winchester', in *A History of the County of Hampshire*, vol. II, ed. H. Arthur Doubleday and William Page (London, 1903), pp. 108–15.

5 Walker-Meikle, *Medieval Pets*, pp. 5, 14, 50–51, 48–9, 64.

66 Edward Topsell, 'Of the Squirrel', in *The History of Four-Footed Beasts and Serpents* [1607–8] (London, 1658), pp. 509–10.

67 Harvie-Brown, *History of Squirrel*, p. 12. See also Jacob Larwood and John Camden Hotten, *The History of Signboards: From the Earliest Times to the Present Day* (London, 1908), p. 163.

68 'Squirrels at the Zoo', *The Spectator* (1 September 1894), p. 267.

69 'Pets and Petters', *Peter Parley's Annual: A Christmas and New Year's Present, for Young People* (London, 1865), pp. 101, 103. On pet-keeping in American cities during the first half of the nineteenth century, see Etienne Benson, 'The Urbanization of the Eastern Gray Squirrel in the United States', *Journal of American History*, C/3 (2013), p. 693.

70 On squirrel pet-keeping as an elite practice before 1800, see Ingrid H. Tague, *Animal Companions: Pets and Social Change in Eighteenth-Century Britain* (University Park, PA, 2015), pp. 41–2, 44–5, 92.

71 White, *Natural History of Selborne*, p. 461.

72 'Something about Squirrels', *Peter Parley's Annual* (1863), p. 168.

73 'Natural History and Management of Pets', *Peter Parley's Annual* (1861), p. 295.

74 'A Tame Squirrel', *Penny Magazine of the Society for the Diffusion of Useful Knowledge* (14 May 1842), pp. 187–8.

75 John Ruskin, 'The Sloth and the Squirrel', *The Leisure Hour* (February 1883), p. 128.

76 Ruskin, 'The Squirrel Cage: English Servitude' (6 June 1874), in *The Genius of John Ruskin: Selections from His Writings*, ed. John D. Rosenberg (Charlottesville, VA, 1998), p. 401.

77 Jay Fellows, *Ruskin's Maze: Mastery and Madness in His Art* (Princeton, NJ, 1981), p. 96; 'The Parliamentary Treadmill', *Punch*, XXI/525 (2 August 1851), p. 63.

78 Phil Robinson, 'On a Kentish Heath', *Contemporary Review*, XLVIII (October 1885), p. 567.

79 Vesper, 'The Squirrel', *The Field*, XI/276 (10 April 1858), p. 307; Douglas English, *A Book of Nimble Beasts* [1910] (London, 1925), pp. 78–80.

80 'Mr. Stillman and His Squirrels' [*Billy and Hans*], *Review of Reviews* (May 1897), p. 496. One authority explained that gamekeepers pre-emptively shot squirrels to 'prevent the recurrence of ... fights between the keepers and the gangs of hooligans': W. J. Stokoe, *The Observer's Book of Wild Animals of the British Isles* [1938] (London, 1958), p. 153.

81 'Rare Birds in London', *Aberdeen Weekly Journal* (21 May 1885), p. 2; 'Cat and Squirrels', in White, *Natural History of Selborne*, pp. 460–61.

82 Aeolus, 'Tame Squirrels with Tin Collars – "A Wrinkle"', *Kidd's Own Journal*, II/27 (3 July 1852), p. 13.

83 William Kidd, 'Domestic Pets – the Squirrel', *National Magazine*, I/4 (February 1857), p. 254; 'The Squirrel', *Kidd's London Journal*, I/11 (18 March 1852), p. 169; 'A Squirrel's Cage', ibid., p. 170.

84 Kidd, 'Domestic Pets – the Squirrel', p. 255.

85 Skuggy was based on *skugg* (also *skug* or *scrug*), a word apparently of Scandinavian origin that was an English synonym for 'squirrel', especially in London.

86 'Domestic Pets – the Squirrel – No. 1', *Kidd's London Journal*, I/8 (21 February 1852), pp. 113–14.

87 Jerome K. Jerome, 'Novel Notes', *The Idler*, 2 (January 1893), p. 301; 'Varieties', *Preston Chronicle* (29 November 1845); 'Miscellaneous', *Bristol Mercury* (29 November 1845); 'Gleanings', *Preston Chronicle* (27 July 1850); Reginald Riley, 'Cat Suckling Young Squirrels', *The Field*, XCIII/2410 (6 May 1899), p. 623; Hannah Velten, *Beastly London: A History of Animals in the City* (London, 2013), pp. 183–4.

88 Bridges, *Scud*, p. 39.

89 The release or escape of pet squirrels, many of them imported, contributed further to the hybridity of Britain's subspecies: Lowe and Gardiner, 'Is the British Squirrel . . . British?', p. 58.

90 Potter to Mrs Wicksteed (17 January 1903), Beatrix Potter Collection, Victoria and Albert Museum, London, quoted in Linda Lear, *Beatrix Potter: A Life in Nature* (London, 2007), p. 160; Potter to Angela, Denis and Clare Mckail (1 January 1903), quoted in Margot Strickland, *Angela Thirkell: Portrait of a Lady Novelist* (London, 1977), pp. 19–20. Keeping pet squirrels was common elsewhere in Europe. In *Perri: The Youth of a Squirrel* (London 1938) by Austro-Hungarian novelist Felix Salten (best known for *Bambi: A Life in the Woods* (1923)), an adult squirrel, Flame-Red, tells Perri, a young female red, about his capture, imprisonment in a 'turning cage' with a wheel and eventual escape (pp. 119–28, 134–5).

91 Kate Fowler-Reeves, *With Extreme Prejudice: The Culling of British Wildlife* (Tonbridge, Kent, 2007), p. 46; Kate Fowler, 'Saving Greys: Don't Demonise Successful Squirrels', *The Guardian*, www.theguardian.com, 24 October 2011.

92 'Game in Season in America', *Sheffield and Rotherham Independent* (25 September 1880).

93 C. J. Cornish, *Wild England of To-Day and the Wild Life in It* (New York, 1895), p. 204. On cruel boys, see also 'Going a-Nutting', *Peter Parley's Annual* (1870), p. 165.

94 Bridges, *Scud*, p. 22.

95 'The Pleasures of the Chase', *Saturday Review*, LVII/1481 (15 March 1884), p. 333.

96 Stokoe, *Observer's Book of Wild Animals of the British Isles*, p. 153.

97 'Mr Stillman and His Squirrels', *Review of Reviews*, p. 496.

98 Bridges, *Scud*, pp. 174, 182, 185, 188.

99 Harvie-Brown, *History of Squirrel*, pp. 104–8.

100 Harvie-Brown, 'Great Spotted Woodpecker', p. 15.

101 *Journal of Beatrix Potter from 1881 to 1897*, transcribed by Leslie Linder (London, 1966), p. 250.

102 G. E. Lodge, 'Poachers Furred and Feathered', *English Illustrated Magazine*, 75 (December 1889), pp. 221–2.

103 'Country Jottings', *Chambers's Journal*, III/137 (14 August 1886), p. 521.

104 'The Destructive Squirrel', *The Field*, XCVIII/2547 (19 October 1901), p. 646.

105 Middleton, 'Ecology of American Grey', p. 835.

106 Fitter, 'Distribution of Grey', p. 14.

107 Middleton, 'Ecology of American Grey', pp. 835–6.

108 'Howling Mobs of Squirrel Hunters', *The Field*, XCIII/2403 (14 January 1898), p. 61; Gaither, 'Squirrels Across the Sea'.

109 A.E.H., 'Squirrels and Their Ways', *The Field*, XCVII/2519 (6 April 1901), p. 475.

110 C. F. Gordon-Cumming, 'The Lowlands of Moray, in the Fourteenth and Nineteenth Centuries', *National Review*, IV/23 (January 1885), p. 647.

111 Barked patches also provide sites for scent-marking territory: Jessica Holm, 'The Real Squirrel Nutkin', BBC *Wildlife*, III/7 (July 1985), p. 339.

112 Keith Laidler, *Squirrels in Britain* (Newton Abbot, Devon, 1980), p. 121.

113 'Squirrel Colonists in Scotland', p. 648.

114 Gordon-Cumming, 'Lowlands of Moray', pp. 647–9.

115 J.D.B., 'Squirrels and Forest Trees', *The Field*, LII/1354 (30 November 1878), p. 713; Harvie-Brown, 'Late Chapters', p. 158.

116 'Squirrels and Fir Trees', *The Field*, XLVIII/1227 (1 July 1878), p. 21.

117 Harvie-Brown, 'Late Chapters', p. 176. Italics in original.

118 Ibid., p. 144.

119 Ibid., pp. 174–5; Gordon-Cumming, 'Lowlands of Moray', p. 649.

120 M. G. Watkins, 'My Four-Footed Friends', *Belgravia: A London Magazine*, XXXV/138 (April 1878), pp. 220–21.

121 Harvie-Brown, 'Late Chapters', pp. 171–2.

122 Ritchie, *Influence of Man*, p. 300.

123 Mark Louden Anderson, *A History of Scottish Forestry*, vol. II, ed. C. J. Taylor (Edinburgh, 1967), p. 652.

124 Ritchie, *Influence of Man*, pp. 300, x, 181, 290.

125 T.A.C., 'A Country Diary', *Manchester Guardian* (25 September 1931). See also A. D. Middleton, *The Grey Squirrel* (London, 1931), pp. 79–80.

126 Ritchie, *Influence of Man*, p. 181. Originally the Ross-shire Squirrel

Club, the name was changed to Highland Squirrel Club in 1907: Annual Statements for 1903 and 1906, HCA/D1693, Highland Archive Centre, Inverness, www.ambaile.org.uk.
127 Highland Squirrel Club, Annual Statements (1935, 1936, 1937, 1938, 1944 and 1946), HCA/D1693, Highland Archive Centre.
128 Lady Rolleston, 'The Destruction of Squirrels: Evidence for the Defence' (LTTE), *Manchester Guardian* (15 September 1936).
129 'What Are Vermin?', *Saturday Review*, LXXIV/1923 (3 September 1892), p. 275; Discipulus, 'Vermin in England', *National Review*, XV/88 (June 1890), p. 477.
130 'Billy and Hans: Mr Stillman's Squirrels', *Daily News* (10 June 1897).
131 W. J. Stillman, *Billy and Hans, My Squirrel Friends: A True History* [1897] (Portland, ME, 1914), pp. xiii–xx, 25.
132 A.E.H., 'Squirrels and Their Ways'.
133 'Country Notes', *Country Life*, VII/155 (23 December 1899), p. 796.
134 K., 'Damage Done by Squirrels' (LTTE), *Country Life*, XXXII/822 (5 October 1912), p. 472.
135 Shorten, *Squirrels*, pp. 5, 38.
136 Monica Shorten [Mrs Vizoso], 'Damage Caused by Squirrels in Forestry Commission Areas in 1954 and 1955', pp. 10, vi, MAF (Infestation Control Division), MAF 131/103.
137 Shorten, *Squirrels*, p. 38; A.W.B., 'Squirrels', *Manchester Guardian* (3 December 1954).
138 'Peter Quennell Reviews the New Books', *Daily Mail* (12 November 1954); K. C., 'Squirrels', *The Observer* (21 November 1954).
139 Middleton, *Grey Squirrel*, p. 72; Middleton, 'Ecology of American Grey', pp. 833, 839–40.
140 Middleton, *Grey Squirrel*, p. 72.
141 Middleton, 'The Decrease in Red Squirrels', *Manchester Guardian* (17 January 1929).
142 I. F. Keymer, 'Diseases of Squirrels in Britain', *Mammal Review*, XIII/2–4 (1983), pp. 155–6.
143 Middleton, 'Ecology of American Grey', p. 835; Middleton, 'The Grey Squirrel (*Sciurus carolinensis*) in the British Isles, 1930–32', *Journal of Animal Ecology*, I/2 (1932), p. 166; Jessica Holm, *Squirrels* (London, 1987), p. 39.
144 Middleton, 'Decrease in Red Squirrels'; Middleton, 'The Grey Squirrel (*Sciurus carolinensis*) in the British Isles', p. 166.
145 John Bonner, 'Virus Blamed on Invading Squirrels', *New Scientist*, 1999 (14 October 1995), www.newscientist.com.
146 S. J. Ball, P. Daszak, A. W. Sainsbury and K. R. Snow, 'Coccidian Parasites of Red Squirrels (*Sciurus vulgaris*) and Grey Squirrels (*Sciurus carolinensis*) in England', *Journal of Natural History*, XLVIII/19–20 (2014), p. 1225.
147 Keymer, 'Diseases of Squirrels', p. 156.
148 Fitter, 'Distribution of Grey Squirrel', p. 14. The 'teaser trailer' for filmmaker Terry Abraham's documentary, *Cumbrian Red*, forthcoming in 2023, depicts a red in a coniferous plantation: Jacob Colley, 'WATCH: New Cumbrian Documentary to Shine Light on Plight of Red Squirrels', *cumbriacrack*, https://cumbriacrack.com, 6 April 2022.
149 'Open Spaces in Towns', *British Architect*, VII/20 (18 May 1877), p. 304.
150 'Birds in London', *Once a Week*, IX/210 (2 September 1878), p. 79.
151 Mrs Marshall, 'The Story of a Day', in *The Sunday at Home: A Family

Magazine for Sabbath Reading, 1520 (16 June 1883), p. 376.
152 Wood and Wood, 'Squirrel Tribe', p. 3.
153 Sidney Colvin, 'East Suffolk Memories', *Magazine of Art*, 8 (January 1885), p. 271.
154 A. H. Palmer, 'James Clarke Hook, R.A.', *The Portfolio: An Artistic Periodical*, 19 (January 1888), p. 42.
155 W. S. Rockstro, 'Winks', *Belgravia*, XLVIII/191 (September 1882), pp. 287, 290.
156 Hugh Boyd Watt, 'The American Grey Squirrel', *Glasgow Naturalist*, V (1913), p. 43.
157 David Bruce, 'A Nest Hunt Among the Grampians', *Good Words*, 28 (December 1887), p. 306; F. H. Trench, 'The Nutter', *New Review*, XV/90 (November 1896), p. 611; Wood and Wood, 'Squirrel Tribe', p. 3.
158 'The Royal Gardens, Kew', *The Field*, CI/2632 (6 June 1903), p. 953; Ella Courtney, 'Here and There in West Hertfordshire', *Temple Bar*, CXXIX/521 (April 1904), p. 485; Netta Syrett, 'The Disenchanted Squirrel', *Longman's Magazine*, XLIII/253 (November 1903), p. 84; 'The Squirrel', *School Music Review*, XXII/258 (November 1913), p. 121; Emily Huntley, 'The Story of Squirrel Brownie', *Quiver*, XLV/3 (January 1910), pp. 319–20.
159 'Bristol Institution', *Bristol Mercury* (13 February 1841); 'Yorkshire Naturalists' Club', *York Herald and General Advertiser* (7 February 1852); 'Pets from Across the Sea', *Leeds Mercury* (13 December 1884); Richard Jeffries, *Wood Magic: A Fable* (London, 1882), Chapter Six, 'The Squirrel'.
160 H. W. 'The Dream-City', *Saturday Review*, CXXXI/3417 (23 April 1921), p. 340; 'A Naturalist's Notes on the Squirrel', *The Leisure Hour: A Family Journal of Instruction and Recreation*, 275 (2 April 1857), p. 219.
161 W. J. Loftie, 'Windsor. X', *The Portfolio*, 16 (January 1885), p. 196.
162 Philip Stewart Robinson, *In Garden, Orchard and Spinney* (London, 1897), pp. 255, 260.
163 Edmund Selous, *Tommy Smith's Animals* (London, 1899), p. 167.
164 'Squirrels at the Zoo', pp. 267–8.
165 Rockstro, 'Winks', pp. 287, 290.
166 William H. Dallinger, 'The English Squirrel: A Survival of Remote Ancestral Knowledge', *Wesleyan-Methodist Magazine*, 122 (May 1899), pp. 368, 372.
167 Barrett-Hamilton and Hinton, *History of British Mammals*, vol. IIC, p. 688.
168 Charlotte Elizabeth Bowen, *Aunt Louisa's London Toy Books: Frisky the Squirrel* (London, 1869), p. 1; Charlotte Elizabeth Bowen, *Frisky the Squirrel* (New York, 1873); Bridges, *Scud*, p. 119; Arthur Scott Bailey, *The Tale of Frisky Squirrel* (New York, 1915), p. 9.
169 Lady Farren, *Frisky Tales: True Nature Stories* (London, 1928), p. 1.
170 *Journal of Beatrix Potter*, p. 417.
171 Phil Robinson, 'British Wild Beasts', *Belgravia*, LII/205 (November 1883), p. 41; J. W., 'Nuts', *The Graphic* (17 September 1887).
172 'The Bullfinch', *Arts and Letters: An Illustrated Review*, 3 (July 1889), p. 5.
173 'A Saucy Squirrel', *Reynolds's Miscellany of Romance, General Literature, Science, and Art*, XV/378 (6 October 1855), p. 170; Josiah D. Canning, 'To a Red Squirrel, Barking at Me While Passing Through a Wood', in *The Harp and Plow by the Peasant Bard* (Greenfield, MA, 1852), pp. 104–5.

174 Beatrix Potter to Norah Moore, 25 September 1901, Hobbs and Whalley Collection/Linder Bequest, National Art Library, V&A, https://collections.vam.ac.uk.
175 Lear, *Potter*, pp. 156, 160–61.
176 Polly Pullar, *A Scurry of Squirrels: Nurturing the Wild* (Edinburgh, 2021), p. 14.
177 Timothy M. Williams, Kim and Gareth Williams, 'Excessive Impertinence or a Missed Diagnosis?', *British Medical Journal*, CCCXI/7021 (23–30 December 1995), pp. 1700–1701. The first author is the son of the third author, a professor of medicine at Royal Liverpool University Hospital. The second author, Kim, is a dog.
178 Potter, *The Tale of Squirrel Nutkin* (London, 1903), pp. 9–11, 14, 18, 33, 41, 56.
179 Judy Taylor et al., *Beatrix Potter, 1866–1943: The Artist and Her World* (London, 1987), p. 111; M. Daphne Kutzer, *Beatrix Potter: Writing in Code* (New York, 2003), pp. 26–7, 29.
180 For the sociopolitical context that shaped Potter's tales, see Kutzer, *Beatrix Potter*, pp. 27–31.
181 Katherine R. Chandler, 'Thoroughly Post-Victorian, Pre-Modern Beatrix', *Children's Literature Association Quarterly*, XXXII/4 (Winter 2007), pp. 301–2; Alison Lurie, *Don't Tell the Grown-Ups: Subversive Children's Literature* (Boston, MA, 1990), p. 95. An American reporter recently identified *Squirrel Nutkin* as a 'cautionary tale of impertinence (and maybe working-class rebellion)': Karin Brulliard, 'To Save Their Beloved Red Squirrels, Brits Will Monitor and Kill Invasive American Ones', *Washington Post*, www.washingtonpost.com, 6 March 2017.
182 Faber, 'Squirrels Destroying Chickens', *The Field*, LX/1545 (5 August 1882), p. 227.
183 C.F.T.Y., 'My Friend "Jack!" or the Life of a Tame Squirrel', *Kidd's Own Journal*, V/10 (1854), pp. 146–8.
184 Lear, *Potter*, pp. 164, 179.
185 Camilla Hallinan, *The Ultimate Peter Rabbit: A Visual Guide to the World of Beatrix Potter* (London, 2002), p. 43.
186 'Picture Books', *The Academy and Literature*, 1648 (5 December 1903), p. 651; 'Children's Books', *New York Times* (5 December 1903).

3 GREY AND RED

1 *Reports of the Council and Auditors of the Zoological Society of London for the Year 1905* (London, 1906), p. 46.
2 Judy Taylor et al., *Beatrix Potter, 1866–1943: The Artist and Her World* (London, 1987), p. 144; Camilla Hallinan, *The Ultimate Peter Rabbit: A Visual Guide to the World of Beatrix Potter* (London, 2002), p. 82.
3 Potter to Angela, Denis and Clare Mckail, 1 January 1903: quoted in Margot Strickland, *Angela Thirkell: Portrait of a Lady Novelist* (London, 1977), p. 20.
4 Jonathan Leake, 'First They Cloned Dolly the Sheep. Now They're Targeting Grey Squirrels', *Sunday Times* (5 January 2020).
5 Taylor et al., *Potter*, p. 144; Linda Lear, *Beatrix Potter: A Life in Nature* (London, 2007), p. 237; Hallinan, *Ultimate Peter Rabbit*, p. 82.
6 Taylor et al., *Potter*, p. 145.
7 One reviewer considered it 'as direct and charming as ever', with no take-home moral, which the 'modern child likes': 'Tales of the Untamed',

The Bookman, XLI/243 (December 1911), pp. 137–8.
8 Lear, *Potter*, p. 237.
9 'Red Squirrel Population [HC]', *Hansard*, 448 (28 June 2006), c.112.
10 Hallinan, *Ultimate Peter Rabbit*, p. 83.
11 Christopher Andreae, 'Silly Squirrel Acrobatics', *Christian Science Monitor* (11 April 1995).
12 Two species of flying (more accurately, gliding) squirrel are endemic to North America. Members of its wider *Sciuridae* family include chipmunks, marmots and prairie dogs.
13 John Gurnell, *The Natural History of Squirrels* (London, 1987), p. 11.
14 A. D. Middleton, 'The Ecology of the American Grey Squirrel (*Sciurus carolinensis* Gmelin) in the British Isles', *Proceedings of the Zoological Society of London*, C/3 (1930), p. 823; Gurnell, *Natural History of Squirrels*, p. 9.
15 Nancy Cox and Karin Dannehl, 'Flemish Eights – Flying Grey', in *Dictionary of Traded Goods and Commodities, 1550–1820* (Wolverhampton, 2007), available at www.british-history.ac.uk.
16 John Lawson, *A New Voyage to Carolina* (London, 1709), p. 124; Samuel Clarke, *A True, and Faithful Account of the Four Chiefest Plantations of the English in America. To Wit, of Virginia, New-England, Bermudus, Barbados* (London, 1670), p. 35.
17 Rice Gaither, 'Squirrels Across the Sea', *New York Times* (28 November 1943).
18 Ralph C. Jackson, 'Migration of Gray Squirrels', *Science*, LXXXII/2136 (6 December 1935), pp. 549–50; 'Two Squirrel Tribes in Grim Conflict', *New York Times* (6 December 1935); 'Nuts: Red Squirrels Make War Upon Grays', *Daily Boston Globe* (13 March 1937); Hans G. Uhlig, *The Gray Squirrel in West Virginia* (Charleston, WV, 1956), p. 33. Greys also moved in search of food, to escape disease and to relieve overcrowding.
19 Ernest Ingersoll, 'A Study of Squirrels', *The Field*, LXXXVIII/2293 (5 December 1896), p. 905.
20 William T. Hornaday, *The American Natural History: A Foundation of Useful Knowledge of the Higher Animals of North America* (New York, 1904), pp. 69, 71.
21 Robert T. Hatt, 'The Red Squirrel; Its Life History and Habits, with Special Reference to the Adirondacks of New York and the Harvard Forest', *Roosevelt Wild Life Annals*, II/1 (1929), pp. 12, 42, 130.
22 Middleton, 'Ecology of American Grey', p. 839.
23 Ernest Thompson Seton, *Bannertail: The Story of a Gray Squirrel* (New York, 1922), pp. 65, 67–9, 105–8, 167, 265.
24 W. Beach Thomas, 'Red or Grey?', *The Spectator* (26 January 1945), p. 82; 'Long John Silver', *New York Times* (19 November 1944).
25 Hatt, 'Red Squirrel', pp. 135–6; Elisha Jarrett Lewis, *The American Sportsman: Containing Hints to Sportsmen, Notes on Shooting, and the Habits of the Game Birds and Wild Fowl of America* (Philadelphia, PA, 1857), pp. 358–9; John D. Godman, *American Natural History*, vol. II: *Mastology* (Philadelphia, PA, 1826), p. 131.
26 'America and West Indies: June 1728, 6–10', in *Calendar of State Papers Colonial, America and West Indies*, vol. XXXVI: *1728–1729*, ed. Cecil Headlam and Arthur Percival Newton (London, 1937), pp. 113–28.

27 Ingersoll, 'Study of Squirrels', p. 965; Rev. W. Bingley, *Bingley's Natural History* (Cincinnati, OH, 1871), p. 297.
28 Hatt, 'Red Squirrel', p. 139.
29 'Squirrel Hunting in America', *Edinburgh Literary Review* (21 February 1829), pp. 207–8.
30 William Priest, *Travels in the United States of America: Commencing in the Year 1793, and Ending in 1797* (London, 1802), p. 91.
31 John L. Koprowski, Karen E. Munroe and Andrew J. Edelman, 'Gray Not Grey: Ecology of *Sciurus carolinensis* in their Native Range in North America', in *The Grey Squirrel: Ecology and Management of an Invasive Species in Europe*, ed. Craig M. Shuttleworth, Peter W. W. Lurz and John Gurnell (Stoneleigh Park, Warks, 2016), p. 11.
32 Ibid., p. 2.
33 Greys were quickly eliminated from the Adelaide area and last seen in Melbourne in the late 1940s. A colony stemming from a later introduction in Victoria (Ballarat) was gone by the early 1970s: J. H. Seebeck, 'The Eastern Grey Squirrel, *Sciurus carolinensis*, in Victoria', *The Victorian Naturalist*, CI/2 (1984), pp. 60–66. Greys fared better in South Africa, where Cecil Rhodes turned them out at Rondebosch on Table Mountain's slopes. According to one account, Rhodes wanted to fill what he regarded as an empty space in non-native oak forests: A. Vos, R. H. Manville and R. G. Van Gelder, *Introduced Mammals and Their Influence on Native Biota* (New York, 1956), p. 181.
34 Craig Shuttleworth, 'Grey Squirrels in Western Canada', *Squirrel*, 35 (November 2017), p. 9; Karl Larson and Craig Shuttleworth, 'Nutkin Ventured, Nutkin Gained', *Squirrel*, 34 (April 2017), p. 9.
35 Mark Downey, 'Getting Squirrelly', *Great Falls Tribune* (11 July 1999); Deborah Richie, 'Forest Sentinels', *Montana Outdoors*, XXIX/6 (1998), p. 10.
36 Karl W. Larsen, 'Grey Squirrel Invasions in Western Canada: History Repeats Itself (Again, and Again)', in *The Grey Squirrel*, ed. Shuttleworth et al., pp. 27–9.
37 Richie, 'Forest Sentinels', p. 10; Michelle McNiel, 'Squirrels: Locals Love 'Em, But the State Doesn't', *Wenatchee World* (26 June 2010).
38 Benjamin Franklin (BF) to Deborah Franklin (DF), 28 January 1772, in *The Papers of Benjamin Franklin*, vol. XIX: *January 1 through December 31, 1772*, ed. William B. Willcox (New Haven, CT, 1975), pp. 42–5.
39 Georgiana Shipley (GS) to BF, 22 September 1772, ibid., p. 300.
40 DF to BF, 15 May 1772, ibid., pp. 140–42.
41 GS to BF, 22 September 1772, ibid., p. 300; BF to DF, 14 February 1773, in *The Papers of Benjamin Franklin*, vol. XX: *January 1 through December 31, 1773*, ed. William B. Willcox (New Haven, CT, 1976), pp. 58–9.
42 John Kelly, 'The Amazing True Story of a Pennsylvania Squirrel in King Arthur's Court', *Washington Post*, www.washingtonpost.com, 11 April 2017; Kelly, 'Benjamin Franklin's Squirrel Died in England in 1772 and Was Buried Here', ibid., 12 April 2017.
43 BF to GS, 6 September 1772, in *Papers of Benjamin Franklin*, vol. XIX, pp. 301–3.
44 DF to BF, 29 October 1773, in *Papers of Benjamin Franklin*, vol. XX,

pp. 449–50. In addition to *squerel*, British vernacular terms for squirrel included *scorel*, *skuggie*, *skoog* and *skarale*.

45 GS to BF, 1 May 1779, in *The Papers of Benjamin Franklin*, vol. XXIX: *March 1 through June 30, 1779*, ed. Barbara B. Oberg (New Haven, CT, 1992), pp. 407–9.

46 GS to BF, 22 December 1774, in *The Papers of Benjamin Franklin*, vol. XXI: *January 1, 1774, through March 22, 1775*, ed. William B. Willcox (New Haven, CT, 1978), pp. 396–7.

47 GS to BF, 1 May 1779, in *Papers of Benjamin Franklin*, vol. XXIX, pp. 407–9.

48 'Zoological Gardens', *Liverpool Mercury* (13 September 1833).

49 In 1868 a grey squirrel pair 'of Canada' was donated to the museum of the Leeds Philosophical and Literary Society: *Leeds Mercury* (21 October 1868).

50 'The Grey Squirrel' (LTTE), *Cambrian Quarterly Magazine*, 2 (1830), pp. 351–2; George Marshall, 'Red and Grey Squirrels', *Country Life*, LXV/1679 (23 March 1929), p. 420; 'Squirrels', *The Spectator* (28 November 1896), pp. 761–2.

51 Phillippa Francklyn, 'Red and Grey Squirrels' (LTTE), *Country Life*, LXV/1681 (6 April 1929), p. 497.

52 A. E. Reveirs-Hopkins, 'The Grey Squirrel' (LTTE), *The Observer* (13 April 1930).

53 Hugh Boyd Watt, 'Red and Grey Squirrels' (LTTE), *Country Life*, LXV/1683 (20 April 1929), p. 569.

54 Hugh Boyd Watt, 'On the American Grey Squirrel (*Sciurus carolinensis*) in the British Isles', *Essex Naturalist*, 20 (1923), p. 191.

55 Middleton, 'Ecology of American Grey', p. 812; Dorothy Bentley-Smith, 'Were Macclesfield Gentry to Blame for Demise of Red Squirrels?', *Warrington Guardian*, www.warringtonguardian.co.uk, 16 January 2007; Caroline Renshaw, 'Who Killed Squirrel Nutkin?', *Cheshire Life* (November 1996), pp. 92–3.

56 D. T. Max, 'Squirrel Wars', *New York Times* (7 October 2007). The widespread assertion in the secondary literature on faunal introductions and biological invasions that this was the first recorded introduction is rarely, if ever, corroborated. The original source, to the best of my knowledge, is Middleton, 'The Ecology of the American Grey Squirrel (*Sciurus carolinensis* Gmelin) in the British Isles' (1930). Middleton quotes as follows (p. 812) from informant R. E. (Richard Ernest) Knowles: 'Mr. T. U. Brocklehurst brought two back from America, kept them in a large cage on the wall of the house, and exhibited them to his friends until, tiring of them, they were liberated into the adjacent woods (in 1876). Mr. T. U. Brocklehurst died in 1886, being succeeded by his son, Mr. W. W. Brocklehurst.' 'It appears that the grey squirrels increased steadily until his death in 1918, and before the death of Mr. W. W. Brocklehurst they became so troublesome that he gave orders to kill them down,' Middleton interjects. Then he quotes from Knowles again: 'which order his son Mr. W. A. Brocklehurst in turn more heartily backed up. They had however got a very strong hold.'

57 Gaither, 'Squirrels Across the Sea'.

58 Etienne Benson, 'The Urbanization of the Eastern Gray Squirrel in the

United States', *Journal of American History*, c/3 (2013), pp. 696–7; Sadie Stein, 'Alien Squirrel', *New York Magazine*, https://nymag.com, 3 February 2014.
59 Middleton, 'Ecology of American Grey', pp. 812, 840.
60 Middleton, *The Grey Squirrel* (London, 1931), p. 22.
61 Eric Mendelsohn, 'The Squirrel Wars' (LTTE), *New York Times* (21 October 2007); Max, 'Squirrel Wars', ibid.
62 W.A.W., 'The Pigeons at Our Window' (LTTE), *The Spectator* (9 February 1907), p. 207.
63 'The American Grey Squirrel: Spread in the British Isles', *Manchester Guardian* (21 April 1930).
64 Monica Shorten, 'Introduced Menace: American Grey Squirrel Poses Threat to British Woodlands', *Natural History*, LXXIII/10 (December 1964), p. 43.
65 Hugh Boyd Watt, 'The American Grey Squirrel in Britain', *The Field*, CXXV/3259 (12 June 1915), p. 1044; Middleton, 'Ecology of American Grey', p. 812.
66 Koprowski et al., 'Gray Not Grey', pp. 1–2.
67 'Notes: Cheshire Bibliography IV', *Cheshire Notes and Queries*, VII (13 August 1887), p. 175.
68 'A Wild Beast Warehouse', *The Leisure Hour* (December 1887), p. 820.
69 'This Morning's News', *Daily News* (4 January 1892).
70 'Possible Pets', *The Spectator* (2 January 1892), p. 13; C. J. Cornish, *Life at the Zoo: Notes and Traditions of the Regent's Park Gardens* (London, 1895), p. 282.
71 'Squirrels at the Zoo', *The Spectator* (1 September 1894), pp. 267–8.
72 Middleton, *Grey Squirrel*, p. 18.
73 'Grey Squirrels', *Manchester Guardian* (29 January 1946); John Allan May, 'Come What May: Mutiny of the (Squirrel Bounty)', *Christian Science Monitor* (17 March 1953).
74 For a table of introductions (1876–1929), see Middleton, *Grey Squirrel*, p. 22.
75 John Paterson, 'Grey Squirrel on Loch Longside and Loch Lomondside', *Glasgow Naturalist*, 4 (1911–12), p. 136; Hugh Boyd Watt, 'The American Grey Squirrel', *Glasgow Naturalist*, 5–6 (1912–14), p. 42.
76 R.S.R. Fitter, 'The Grey Squirrel', *The Observer* (31 May 1959).
77 '"Zoo" Experiment: Grey Squirrels at Large in Regent's Park', *Daily Mail* (26 September 1906); *Reports of the Council and Auditors of the Zoological Society of London for the Year 1906* (London, 1907), p. 59.
78 Watt, 'American Grey Squirrel in Britain', *The Field* (1915), p. 1044.
79 Herbert Walter Macklin, *Bedfordshire and Huntingdonshire* (London, 1917), p. 11.
80 Middleton, 'Ecology of American Grey', p. 812.
81 'Children's Column', *Leeds Mercury* (30 January 1892).
82 T. C. Bridges, *Scud: The Life Story of a Squirrel* (London, 1907), p. 50.
83 Ibid., pp. 124–5, 166.
84 Gerald E. H. Barrett-Hamilton and M.A.C. Hinton, *A History of British Mammals*, vol. IIC: *Order Rodentia, Part XX* [1918] (London, 1910–21), pp. 695, 690, 717; Bridges, *Scud*, p. 18.
85 Barrett-Hamilton and Hinton, *History of British Mammals*, vol. IIC, pp. 690–91.
86 Ibid., pp. 718–19.

87 Watt, 'American Grey Squirrel', *Glasgow Naturalist* (1912–14), p. 43.
88 Cecil Brown, 'The Grey Squirrel: More Evidence for the Prosecution' (LTTE), *The Observer* (4 March 1917).
89 Vernon Watney, 'A Crusade Against the Grey Squirrel' (LTTE), *Country Life*, LXIV/1643 (12 July 1928), p. 65. See also 'Walking Through the Woods', *Saturday Review of Politics, Literature, Science, and Art*, CXXXIII/3466 (1 April 1922), p. 333; E. Marlin Duncan, 'Wild Life in an English Forest', *Quiver* (March 1923), p. 458; J.E.C., 'A Country Diary', *Manchester Guardian* (24 June 1931).
90 C.W.R. Knight, *Wild Life in the Tree Tops* (London, 1921), pp. 45, 49–50.
91 Peter Atkins, 'Introduction', in *Animal Cities: Beastly Urban Histories*, ed. Atkins (Farnham, 2012), pp. 2–3.
92 W. J. Stillman, *Billy and Hans, My Squirrel Friends: A True History* (Portland, ME, 1914), pp. xx–xxi, 12.
93 Frank Finn, 'The Sins of Alien Animals', *The Graphic*, XCVI/2487 (28 July 1917), p. 120.
94 Benson, 'Urbanization of Eastern Gray', p. 694.
95 R.S.R. Fitter, 'The Distribution of the Grey Squirrel in the London Area', in *The London Naturalist for the Year 1938* (London, 1939), p. 8.
96 Ibid., p. 12.
97 *Report of the Council of the Zoological Society of London for the Year 1889* (London, 1890), p. 52. A Small Cats' House was converted into a Squirrels' House in 1885 and among donations in 1887–8 were an Arizona squirrel (*Sciurus arizonensis*), a red-and-white flying squirrel (from Szechuen, China), three American flying squirrels and three common (red) squirrels: *Report of the Council of the Zoological Society of London for the Year 1885* (London, 1886), p. 21; *Report of the Council of the Zoological Society of London for the Year 1887* (1888), p. 36; *Report of the Council of the Zoological Society of London for the Year 1888* (1889), p. 47.
98 'Report of the Zoological Society of London for the Year 1887', *Quarterly Review*, CLXIX/338 (October 1889), p. 542.
99 '"Zoo" Experiment'; 'American Grey Squirrel: Britain's Latest Peril', *The Press* [South Island, New Zealand]', LIII/15929 (16 June 1917), p. 14, https://paperspast.natlib.govt.nz/newspapers.
100 'Snow at the Zoo', *The Spectator* (6 March 1909), p. 370.
101 J. C. Warburg, 'Grey Squirrels in the London Parks', *Country Life*, XLIV/1125 (27 July 1918), pp. 78–80.
102 '"Zoo" Experiment'; 'Birds and Beasts for Reintroduction', *The Spectator* (15 August 1908), p. 224.
103 L.G.M. [Leslie G. Mainland], 'New Pets at the Zoo: Squirrels with a Taste for Chocolate', *Daily Mail* (16 November 1908).
104 L.G.M., 'Odd Tastes at the Zoo: Bun Diet that Killed a Seal', *Daily Mail* (30 December 1919).
105 '"Zoo" Experiment'.
106 Benson, 'Urbanization of Eastern Gray', pp. 695, 698, 700–704, 708. Peanuts are not particularly good for squirrels. High phosphorus content interferes with calcium absorption, causing, in extreme cases, rickets in youngsters as growing bones are calcium deficient: Jessica Holm, *Squirrels* (London, 1987), p. 89.
107 Benson, 'Urbanization of Eastern Gray', pp. 703–4.
108 L.G.M., 'New Pets'.

109 M. Mostyn Bird, 'Wild Life in the London Parks', *Pall Mall Magazine*, XLV/206 (June 1910), pp. 1002–3.
110 L.G.M., 'New Pets'. Among Mainland's books were *Secrets of the Zoo* (1923), *The Hidden Zoo* (1925) and *True Zoo Stories* (1931).
111 Warburg, 'Grey Squirrels', pp. 78–9.
112 L.G.M., 'New Pets'.
113 Warburg, 'Grey Squirrels', pp. 79–80.
114 '"Zoo" Experiment'; 'Birds and Beasts for Reintroduction'.
115 F. Anstey, 'An Acclimatised Colonial', *Punch*, CXL/3666 (22 March 1911), pp. 200, 203.
116 H.M.S., 'Kew in Winter', *Manchester Guardian* (13 January 1913); H.M.S., 'Kew Gardens in Whit-Week', *The Observer* (11 June 1916).
117 Fitter, 'Distribution of Grey', p. 15.
118 Brown, 'Grey Squirrel: More Evidence for Prosecution'.
119 Bunny appears to have been a popular name for a squirrel in the United States. See W.O.C., 'Our Squirrel' and 'More About Bunny the Squirrel', *The Nursery: A Monthly Magazine For Youngest Readers*, V (Boston, MA, 1869), pp. 116, 159.
120 Warburg, 'Grey Squirrels', p. 78; 'Last Free Day at Kew: Children's Good-Bye to "Jimmy", the Squirrel', *Daily Mail* (17 January 1916).
121 'Last Free Day at Kew'.
122 L.G.-S, 'The Squirrels in Kensington Gardens' (LTTE), *Country Life*, XLII/1074 (4 August 1917), p. 118.
123 Ibid.
124 Leon Kelsey, '"Zoo" Squirrel in South Kensington' (LTTE), *Daily Mail* (16 October 1906).
125 W. M., 'The Progress of "Zoo" Squirrels' (LTTE), *Daily Mail* (17 October 1906).
126 L.G.M., 'Grey American Squirrel: Migration from Regent's Park to North Middlesex', *Daily Mail* (4 January 1909).
127 '"Zoo" Experiment'.
128 W. Beach Thomas, 'A Wonderful Home: How a Great Naturalist Lives', *Daily Mail* (29 November 1909).
129 W. B. Thomas, 'A Country Notebook', *Daily Mail* (3 September 1909).
130 L.G.M., 'Grey American Squirrel'.
131 Ibid.
132 P. Chalmers Mitchell, *Official Guide to the Gardens of the Zoological Society of London*, 9th edn (London, 1911), p. 74.
133 'Pets at the Zoo', *The Spectator* (5 October 1912), p. 503.
134 Constance Innes Pocock, *Highways and Byways of the Zoological Gardens* (London, 1913), pp. 189–91.
135 F. W. Frohawk, 'Scarcity of Squirrels', *The Field*, CXIX/3084 (3 February 1912), p. 242; C.H.C., 'Scarcity of Squirrels in West Norfolk', ibid., CXIX/3083 (27 January 1912), p. 188.
136 John Grubb, 'Scarcity of Squirrels', ibid., CXIX/3086 (17 February 1912), p. 304; W.S.H., 'Scarcity of Squirrels', ibid., CXIX/3089 (9 March 1912), p. 502; J. O. Coussmaker, 'Scarcity of Squirrels', ibid., CXIX/3087 (24 February 1912), p. 400.
137 'North Midland' and Editor, 'Scarcity of Squirrels', ibid., CXIX/3091 (23 March 1912), p. 597.
138 'Grey Squirrel in England', ibid., CXXIV/3218 (29 August 1914), p. 429.
139 A.K.R., 'The Grey Squirrel' (LTTE), *Country Life*, XXXVI/925 (19 September 1914), p. 406.
140 M. C., 'The American Grey Squirrel', *Country Life*, XXXVI/927 (10 October 1914), p. 502.

141 W. H. St. Q, 'The Faults of the Grey Squirrel' (LTTE), *Country Life*, XXXVI/928 (17 October 1914), p. 532.
142 Ibid.
143 G.H.F., 'The American Grey Squirrel' (LTTE), *Country Life*, XXXVI/390 (31 October 1914), p. 597.
144 H. B. Watt, 'The American Grey Squirrel' (LTTE), *Country Life*, XXXVI/931 (7 November 1914), p. 630; Albert K. Rollit, 'The American Grey Squirrel', *The Field*, CXXVI/3265 (24 July 1915), p. 180. For Watt's preliminary findings, see Watt, 'American Grey Squirrel in Britain', *The Field* (1915), p. 1044.
145 E. G. Lawton, 'The Grey Squirrel in Kew Gardens', *The Field*, CXXIX/3366 (30 June 1917), p. 920.
146 Geoffrey Beven, 'The Distribution of the Grey Squirrel in the London Area (1953–6)', in *The London Naturalist No. 36 for the Year 1956* (June 1957), p. 6.
147 'The Grey Squirrel: Sir Frederick Treves on a Coming Plague' (LTTE), *The Observer* (25 February 1917).
148 'American Grey Squirrel: Britain's Latest Peril', *The Press*.
149 Arnold H. Mathew, 'The Grey Squirrel: More Evidence for the Prosecution', *The Observer* (4 March 1917).
150 'American Grey Squirrel: Britain's Latest Peril'.
151 'The Regent's Park View', *The Observer* (4 March 1917).
152 'American Grey Squirrel: Britain's Latest Peril'.
153 'A Plea for the Squirrel', *Country Life*, XXXI/798 (20 April 1912), p. 562.
154 J. De Vitre, 'A Bellicose Squirrel' (LTTE), *Country Life*, XXXIV/867 (16 August 1913), p. 244.
155 R.M.R., 'Squirrels and Birds' Eggs' (LTTE), *Country Life*, XXXIV/880 (15 November 1913), p. 690.
156 W. H. Hudson, *Bird Notes and News* (London, 1917), p. 85.
157 Hugh Boyd Watt, 'The American Grey Squirrel in Britain', *The Field*, CXXX/3381 (13 October 1917), p. 536.
158 E. G. Lawton, 'The American Grey Squirrel' (LTTE), *The Field*, CXXX/3385 (10 November 1917), p. 686.
159 Oldfield Thomas, 'The Grey Squirrel in England' (LTTE), *The Field*, CXXIX/3357 (28 April 1917), p. 625.
160 Ibid.
161 Archibald Thorburn, *British Mammals*, vol. I (London, 1920), p. v.
162 David Attenborough, 'Introduction', in *Thorburn's Mammals* (London, 1974), p. 8.
163 Thorburn, *British Mammals*, vol. I, pp. 79–80; 'Some British Beasts', *The Observer* (16 January 1921).
164 Y. Y., 'Wild Life in London', *New Statesman*, XX/497 (21 October 1922), p. 70.
165 Maurice Burton, 'The Grey Squirrel in Britain', *Illustrated London News*, 5626 (15 February 1947), p. 211.
166 'Pretty, but Vermin', *Daily Mail* (6 February 1931).
167 B.C.C., 'The American Grey Squirrel Near London' (LTTE), *The Field*, CXXXIV/3478 (23 August 1919), p. 280.
168 'Regent's Park (Dogs) [HC]', *Hansard*, 12 (4 November 1909), cc.2150–51, https://api.parliament.uk/historic-hansard/written-answers.
169 'Notes of the Week' ('Naturam Expellas'), *Saturday Review of Politics, Literature, Science, and Art*, CLV/4038 (18 March 1933), p. 254.

4 AMERICAN HUSTLE, c. 1919-39

1. 'A Charming Pest: Farmers' War on Grey Squirrel', *Daily Mail* (1 February 1924); 'Grey Squirrels and the Law', *Country Life*, LXXXII/2118 (21 August 1937), p. 184.
2. 'Forester', 'The Grey Squirrel: More Evidence for the Prosecution' (LTTE), *The Observer* (4 March 1917).
3. 'Squirrels (Depredations) [HC]' [written answers], *Hansard*, 94 (7 June 1917), c.370.
4. 'Forestry Bill [HL]', *Hansard*, 35 (14 July 1919), cc.525, 537–40, 545–7. *The Field*'s report did not distinguish between red and grey: 'The Squirrel in Parliament', *The Field*, CXXXIV/3475 (2 August 1919), p. 166.
5. J. F. Peters, 'Squirrels and Trees' (LTTE), *The Field*, CXXXV/3505 (28 February 1920), p. 286; E. Adrian Woodruffe-Peacock, 'Jays and Squirrels' (LTTE), ibid., CXXXV/3509 (27 March 1920), p. 462.
6. M. G. Meugens, 'The Grey Squirrel in Richmond Park', *Country Life*, LI/1307 (21 January 1922), p. 71.
7. Paul Bewsher, 'Love in Autumn', *Daily Mail* (3 October 1921).
8. B. T., 'Rival Squirrels: Foreign or Friend?', *The Observer* (27 November 1921).
9. H. Howard-Vyse, 'Grey Squirrels' (LTTE), *The Field*, XXXVII/3599 (17 December 1921), p. 776.
10. H. J. Massingham, 'Countrifying London', *New Statesman*, XIX/469 (8 April 1922), p. 11; Massingham, *Sanctuaries for Birds and How to Make Them* (London, 1924), pp. 32, 37–8.
11. 'The King's Haven for Wild Birds: Buckingham Palace Sanctuary', *Daily Mail* (8 November 1922); 'Destruction of Grey Squirrels', 1916–38, Office of Works and Successors, Records of the Royal Parks Division, WORK 16/574, TNA.
12. 'The Grey Raiders', *Daily Mail* (7 May 1924).
13. Massingham, *Sanctuaries for Birds*, pp. 39, 38.
14. Ibid., pp. 38–9.
15. 'Squirrel War Over: Comparative Safety for Them in the Parks', *Daily Mail* (16 December 1925).
16. 'A Plague of Grey Squirrels', *Manchester Guardian* (22 September 1923); E. T., 'The Plague of Grey Squirrels' (LTTE), *Country Life*, LXII/1593 (30 July 1927), p. 174.
17. 'A Charming Pest: Farmers' War on Grey Squirrel', *Daily Mail*.
18. E.M.N., 'The Squirrel War', *The Spectator* (31 July 1925), pp. 191–2.
19. 'Spread of the Grey Squirrel', *The Field*, CXLV/3777 (14 May 1925), p. 747; 'The Depredations of the Grey Squirrel' (LTTE), *The Spectator* (22 December 1928), p. 961; William Beach Thomas, 'Squirrel Census', ibid., 8 December 1928, p. 852.
20. M. P., 'The Grey Squirrel' (LTTE), *Country Life*, LXVIII/1765 (15 November 1930), p. 629.
21. William Beach Thomas, *How England Becomes Prairie* (London, 1927), pp. 9, 49, 52, 55, 82, 123; William Beach Thomas, *Why the Land Dies* (London, 1931), pp. 5, 9.
22. William Beach Thomas, 'The Vanishing Labourer', *The Spectator* (11 February 1938), p. 225.
23. On the countryside as national heritage and its preservation as patriotic, see Patrick Abercrombie, *The Preservation of Rural England: The Control of Development by Means of Rural Planning* (Liverpool, 1926);

References

Clough Williams-Ellis, *England and the Octopus* (London, 1928); Clough Williams-Ellis, ed., *Britain and the Beast* (London, 1937); Michiel Dehaene, 'A Conservative Framework for Regional Development: Patrick Abercrombie's Interwar Experiments in Regional Planning', *Journal of Planning Education and Research*, XXV (2005), pp. 132–4; Martin Weiner, *English Culture and the Decline of the Industrial Spirit, 1850–1980* (Cambridge, 1982), p. 73; Malcolm Chase, 'This Is No Claptrap; This Is Our Heritage', in *The Imagined Past: History and Nostalgia*, ed. Christopher Shaw and Malcolm Chase (Manchester, 1989), pp. 128–46; David Matless, 'Definitions of England, 1828–89: Preservation, Modernism and the Nature of the Nation', *Built Environment*, XVI/3 (1990), pp. 179–91; John Stevenson, 'The Countryside, Planning and Civil Society in Britain, 1926–1947', in *Civil Society in British History: Ideas, Identities, Institutions*, ed. Jose Harris (Oxford, 2003), p. 192; Paul Ward, *Britishness since 1870* (London, 2004), p. 54; Roy Strong, *Visions of England* (London, 2011), p. 146; Paul Readman, *Storied Ground: Landscape and the Shaping of English National Identity* (Cambridge, 2018), pp. 14, 19.
24 Francis Younghusband, *The Heart of Nature; or, The Quest for Natural Beauty* (London, 1921), pp. 121–2.
25 William Wordsworth, *A Description of the Scenery of the Lakes in the North of England*, 3rd edn (London, 1822), pp. 86, 90, 72, iv; Andrew Hazucha, 'Neither Deep nor Shallow but National: Eco-Nationalism in Wordsworth's *Guide to the Lakes*', *Interdisciplinary Studies in Literature and Environment*, IX/2 (2002), pp. 65–6.
26 Wordsworth, *Description of Scenery*, p. 116.
27 Younghusband, *Heart of Nature*, p. 121; Wordsworth, *Description of Scenery*, pp. 111–30.
28 William Beach Thomas, *The English Landscape* (London, 1938), p. 108.
29 B. T., 'A Wandering Squirrel: Mammal Migrants', *The Observer* (1 March 1925).
30 Vernon Watney, 'A Crusade Against the Grey Squirrel' (LTTE), *Country Life*, LXIV/1643 (14 July 1928), p. 65.
31 L. Gardiner, 'A Crusade Against the Grey Squirrel' (LTTE), *Country Life*, LXIV/1644 (21 July 1928), p. 105.
32 M. Portal, 'A Crusade Against the Grey Squirrel', ibid.
33 A. G. Herbert and R. Neil Chrystal, 'A Crusade Against the Grey Squirrel', ibid.
34 P. Chalmers Mitchell, 'A Crusade Against the Grey Squirrel', ibid.
35 'The Grey Squirrel', *Western Gazette* (26 December 1930).
36 W. Beach Thomas, 'Grey Squirrels Yet Again', *The Spectator* (13 September 1930), p. 337.
37 W. Beach Thomas, 'Country Life', *The Spectator* (15 November 1930), p. 722.
38 T. A. C., 'Squirrels: Red and Grey', *Manchester Guardian* (27 April 1931).
39 'Topics of the Times: Gray Squirrels Painted Black', *New York Times* (2 September 1930).
40 Humbert Wolfe, *Kensington Gardens* (London, 1924), p. 41.
41 A. H. Chaytor, 'Grey and Red Squirrels', *The Times* (7 February 1929).
42 W.R.C., review of *Near Neighbours*, in *The Bookman*, LXXIX/472 (1931), p. 251.

43 For Elton, the grey squirrel's rapid growth in numbers in Britain was a prime example of an 'ecological explosion': Charles S. Elton, *The Ecology of Invasions by Animals and Plants* (London, 1958), pp. 15, 73.

44 A. D. Middleton, 'Red and Grey Squirrels' (LTTE), *Country Life*, LXV/1678 (16 March 1929), p. 376; 'Red and Grey Squirrels' (LTTE), *The Times* (16 March 1929).

45 Middleton, 'Grey Squirrels', *Daily Mail* (16 January 1930); 'Grey Squirrels', *Manchester Guardian* (24 January 1930); 'The Grey Squirrel', *The Observer* (6 April 1930).

46 Middleton, 'Red and Grey Squirrels', *The Times*; Middleton, 'The Decrease in Red Squirrels', *Manchester Guardian* (17 January 1929); Middleton, *The Grey Squirrel* (London, 1931), p. 76.

47 Robert T. Hatt, 'The Red Squirrel; Its Life History and Habits, with Special Reference to the Adirondacks of New York and the Harvard Forest', *Roosevelt Wild Life Annals*, II/1 (1929), p. 131; Middleton, *Grey Squirrel*, pp. 9, 69–70.

48 Richard Dawkins, 'The Digital River', in *Evolution, Literature, and Film: A Reader*, ed. Brian Boyd, Joseph Carroll and Jonathan Gottschall (New York, 2010), p. 59. By contrast, the closely related grey and fox squirrels can interbreed in the wild in North America. The so-called black (melanic) squirrels sometimes seen in parts of southeast England such as Bedfordshire, Cambridgeshire and Hertfordshire (originally escapees from private zoos, since the 1910s), that currently number up to 25,000, according to some estimates, are grey squirrels with a mutated pigment gene passed on through interbreeding with fox squirrels in North America, where melanic squirrels are widespread, not least in regions with cold winters; black fur may well provide a thermal advantage: Patrick Barkham, 'Black Squirrel "Super" Species? No, Just a Darker Shade of Grey', *The Guardian*, www.theguardian.com, 13 August 2019.

49 Middleton, *Grey Squirrel*, pp. 7–8.

50 A. D. Middleton, 'The Ecology of the American Grey Squirrel (*Sciurus carolinensis* Gmelin) in the British Isles', *Proceedings of the Zoological Society of London*, C/3 (1930), pp. 809–43.

51 Middleton, *Grey Squirrel*, pp. 2, 75–6, 78, 100.

52 Middleton, 'Ecology of American Grey', p. 826; Middleton, *Grey Squirrel*, pp. 45–6, 49, 50, 51–4, 49, 91.

53 Middleton, 'Ecology of American Grey', pp. 826, 827–8; Middleton, *Grey Squirrel*, p. 53.

54 Middleton, 'The Grey Squirrel in the British Isles', *Journal of the Ministry of Agriculture*, XXXVII/11 (February 1931), pp. 1075, 1078.

55 Middleton, *Grey Squirrel*, pp. 67, 72; 'Ecology of American Grey', pp. 832, 831; Middleton, 'Grey Squirrel in British Isles', p. 1077. If greys truly did drive out reds, this was 'but a sorry exchange for the forester', remarked James Ritchie, Regius Professor of Natural History at Aberdeen University: Ritchie, *Beasts and Birds as Farm Pests* (Edinburgh, 1931), p. 50.

56 'Country Notes', *Country Life*, LXVIII/1763 (1 November 1930), p. 537.

57 'The American Grey Squirrel: Spread in the British Isles', *Manchester Guardian* (21 April 1930).

58 'Gray Squirrel in England Gains Faster Than in American Home', *New York Times* (3 August 1930).
59 'Topics of the Times: Gray Squirrels Painted Black'.
60 Middleton, 'Ecology of American Grey', pp. 832, 839.
61 Middleton, 'Decrease in Red Squirrels'.
62 T.A.C., 'The Grey Colonist: Menace of the American Squirrel', *Manchester Guardian* (20 July 1931).
63 'American Grey Squirrel', *Manchester Guardian*.
64 E. M. Nicholson, 'Meddling with Nature', *New Statesman*, xxv/643 (22 August 1925), p. 524.
65 Y. Y., 'This Freedom', *New Statesman*, xxxv/895 (21 June 1930), pp. 329–30. Concern regarding the aspidistra most likely related to viruses the pot plant harboured or harmful microbes in its soil; there is no suggestion that the plant itself is invasive.
66 B. T., 'Rival Squirrels'.
67 B. T., 'The Open Air: The Fashion in Aliens', *The Observer* (15 November 1931).
68 'War on the Grey Squirrel', *Manchester Guardian* (16 February 1931).
69 Middleton, *Grey Squirrel*, pp. 99–100; Percy W. D. Izzard, 'Radio to Help in Attack on Pest: Grey Squirrel Menace', *Daily Mail* (14 February 1936).
70 'Grey Squirrels Again', *Gloucester Citizen* (5 February 1935); 'War on Grey Squirrel', *Manchester Guardian*.
71 F. J. Johnston, 'The Grey Squirrel in Epping Forest', in *The London Naturalist for the Year 1937* (London, 1938), p. 96.
72 L. W. Swainson, 'Grey Squirrels: The National Campaign to Destroy Them' (LTTE), *Manchester Guardian* (20 September 1937).
73 'A Villain with a Dainty Coat: War of Extermination', *The Observer* (8 November 1931).
74 'War on Grey Squirrel', *Manchester Guardian*.
75 'Grey Squirrel [HC]', *Hansard*, 251 (20 April 1931), cc.598–9; 'Grey Squirrel [HC]', *Hansard*, 252 (14 May 1931), cc.1367–8; 'Grey Squirrels [HC]', *Hansard*, 252 (21 May 1931), cc.2192–3; 'Grey Squirrels [HC]', *Hansard*, 253 (15 June 1931), c.1424.
76 According to Middleton, grey density in Burnham Beeches averaged three per acre. Still, this was ten times higher than red squirrel density in the Scottish Highlands in the early 1900s: Middleton, *Grey Squirrel*, p. 30.
77 'England Declares War on Squirrels: U.S. Pest Now Ravaging the Countryside. Enormous Damage', *Daily Mail* (19 June 1931).
78 The *Daily Paper*, quoted in *Punch*, CLXXXI/4747 (25 November 1931), p. 581.
79 'Board of Trade [HC]', *Hansard*, 309 (27 February 1936), cc.951–2.
80 'News From the Bird Sanctuaries', *Manchester Guardian* (16 July 1931); 'No More Grey Squirrels in London Parks: Their "Criminal" Habits', *Times of India* (23 November 1931).
81 'Bird News from the London Parks', *Manchester Guardian* (4 July 1932).
82 'Grey Squirrels [HC]', *Hansard*, 252 (21 May 1931), cc.2192–3.
83 R.S.R. Fitter, *London's Natural History* (London, 1945), p. 215; 'War on Grey Squirrels', *Daily Mail* (9 November 1931).
84 Geoffrey Beven, 'The Distribution of the Grey Squirrel in the London Area (1953–6)', in *The London*

84 *Naturalist No. 36 for the Year 1956* (London, 1957), p. 7; Edwin Bayliss [chairman, London County Council Parks Committee], 'Squirrels at Bay' (LTTE), *Manchester Guardian* (9 March 1956).
85 'Topics of the Times: A War on Gray Squirrels', *New York Times* (25 November 1931).
86 'Pet or Plague?', *Daily Mail* (8 September 1932).
87 L. R. Brightwell, 'The Squirrel Does Not Mean to Starve: Ready for Years of Rainy Days', *Daily Mail* (19 September 1930).
88 R.S.R. Fitter, review of Middleton, *The Grey Squirrel*, in *London Naturalist*, 10 (1931), p. 7.
89 'Clansman', 'The Grey Squirrel' (LTTE), *The Observer* (13 December 1931).
90 'Bolsheviks in Fur and Feather', *Saturday Review of Politics, Literature, Science, and Art*, CLIV/4001 (2 July 1932), p. 5.
91 'Grey Squirrels (Prohibition of Importation and Keeping) Order, 1937 [HL]', *Hansard*, 105 (29 June 1937), c.863; AM (Records of the Office of the Parliamentary Counsel) V/215, TNA.
92 MAF, Minute sheet (Mr Glenower), 16 July 1943, MAF 44/45.
93 'Grey squirrels [HC]', *Hansard*, 301 (2 May 1935), cc.563–4; Izzard, 'Radio to Help in Attack'.
94 Untitled, *Manchester Guardian* (10 November 1937).
95 B. H. King, 'Grey Squirrel Damage', *Journal of the Forestry Commission*, XXI (1950), p. 106.
96 Judith J. Rowe, *Grey Squirrel Control* (Forestry Commission Leaflet 56) (London, 1973), p. 16–17; FC, Committee on Grey Squirrels (CGS), Item No. 8, 'Claim for Award in Respect of Alleged Invention of Squirrel Pole', Minutes, Meeting, 5 May 1954, MAF 130/81.
97 Forestry Commission, 'War on the Grey Squirrel: New Bonus Scheme for Tails', February 1953, MAF 131/88; FC, *Hints on Controlling Grey Squirrels* (Leaflet 58), 1953, MAF 217/33.
98 'Country Notes', *Country Life*, LXVIII/1763 (1 November 1930), p. 536; Middleton, *Grey Squirrel*, p. 30.
99 M. P., 'Grey Squirrel', *Country Life*.
100 Ibid.
101 H.B.C.P., 'The Grey Squirrel', *Country Life*, LXIX/1792 (23 May 1931), p. xlviii.
102 Christopher Addison, 'Grey Squirrels [HC]', *Hansard*, 249 (2 March 1931), c.20.
103 Monica Shorten, *Squirrels* (London, 1954), pp. 38–9.
104 A. D. Middleton, 'Grey Squirrels: Appeal for Assistance in a Survey', *The Observer* (9 April 1937); The Grey Squirrel (Prohibition of Importation and Keeping) Order, 1937, 6 May 1937, MAF 44/22; 'The Grey Squirrel Outlawed: Placed in Same Category as Musk Rat', *Gloucester Citizen* (30 June 1937).
105 'Destructive Imported Animals Act, 1932 [HC]', *Hansard*, 324 (27 May 1937), c.557.
106 Ibid., c.559.
107 Ibid., c.556.
108 'Grey Squirrels (Prohibition of Importation and Keeping) Order, 1937 [HL]', *Hansard*, 105 (29 June 1937) c.863.
109 Frances Pitt, 'Miss Phyllis Kelway' (obituary), *Nature*, 155 (19 May 1945), p. 600.
110 Phyllis Kelway, *Swift Movement in the Trees and at Their Roots* (London, 1937), pp. 21, 25.

111 Middleton tabulated 32 introduction points between 1876 and 1929, the majority of them internal redistributions from centres such as Woburn: Middleton, *Grey Squirrel*, pp. 21–2.

112 'Grey Squirrels (Prohibition of Importation and Keeping) Order, 1937 [HL]', *Hansard*, c.871; 'Grey squirrels (Destruction) [HC]', *Hansard*, 342 (28 November 1938), c.47; 'Grey Squirrels and the Law', *Country Life*, LXXXII/2118 (21 August 1937), p. 184.

113 Anthony Brunner, 'Grey Squirrels' (LTTE), *Country Life*, LXXXII/2118 (21 August 1937), p. 206; 'Grey Squirrels and Law', ibid., p. 184.

114 'Britain Starts War on Grey Squirrels', *Baltimore Sun* (23 October 1938).

115 J. W. Parry, 'The Grey Squirrel's Depredations' (LTTE), *Manchester Guardian* (14 July 1937).

116 'Grey Squirrels (Prohibition of Importation and Keeping) Order, 1937 [HL]', *Hansard*, c.869.

117 Ibid., c.867.

118 'Grey Squirrel [HC]', *Hansard*, 326 (13 July 1937), cc.1083–4.

119 R.S.R. Fitter, 'The Distribution of the Grey Squirrel in the London Area', in *The London Naturalist for the Year 1938* (London, 1939), p. 6.

120 Helen Vaughan Williams, *Squirrel War; or, The Fight for the Dolls' House* (London, 1936), pp. 28–9, 36, 73, 40, 76, 78.

121 Ibid., pp. 84–9.

122 Ibid., pp. 53–5.

123 Ibid., pp. 57–9, 61, 62–3.

124 Ibid., pp. 63, 109.

125 Winifred Stamp, *'Doctor Himself': An Unorthodox Biography of Harry Roberts, 1871–1916* (London, 1949), p. 69.

126 Harry Roberts, 'Jew and Blackshirt in the East End', *New Statesman and Nation*, XII/298 (7 November 1936), p. 698.

127 A.W.B., 'Reply to Correspondent', *Manchester Guardian* (25 September 1936).

128 Lady Rolleston, 'The Destruction of Squirrels: Evidence for the Defence' (LTTE), *Manchester Guardian* (15 September 1936).

129 Basil Viney, 'The Destruction of Squirrels: London Parks the Poorer' (LTTE), *Manchester Guardian* (29 September 1936).

130 Rolleston, 'Grey Squirrels: Harmless and Friendly Creatures', *Manchester Guardian* (10 July 1937).

131 Viney, 'Grey Squirrels', *Manchester Guardian* (14 October 1937).

132 J. W. Parry, 'The Grey Squirrel's Depredations: A Destructive Pest' (LTTE), *Manchester Guardian* (14 July 1937).

133 Harper Cory, 'The Grey Squirrel: A Naturalist's Comment on the Destruction Order', *Manchester Guardian* (16 July 1937).

134 F. Howard Lancum, 'Grey Squirrels', *Manchester Guardian* (15 September 1937).

135 Alfred Bulley, 'The Grey Squirrel', *Manchester Guardian* (9 October 1937).

136 'The Grey Squirrel: A Reluctant Verdict of Guilty', *Manchester Guardian* (1 October 1937).

137 W.S.B., 'Nature Notes: The Grey Squirrel', *Country Life*, XLVII/1205 (7 February 1920), p. lviii.

138 Cang Hui et al., 'Defining Invasiveness and Invasibility in Ecological Networks', *Biological Invasions*, 18 (2016), p. 972.

139 'Defining Invasiveness and Invasibility in Ecological Networks', news release, DSI-NRF Centre of Excellence in Invasion Biology, Stellenbosch University, South Africa, April 2016, http://academic.sun.ac.za.

140 Dan Russell, 'Grey Squirrels', *Saturday Review of Politics, Literature, Science and Art*, CLIX/4244 (6 February 1937), p. 101.

141 'Conservation of Wild Creatures and Wild Plants (Amendment) Bill [HL]', *Hansard*, 388 (26 January 1978), c.500; 'Red Squirrels [HL]', *Hansard*, 587 (25 March 1998), c.1321.

142 W. Beach Thomas, 'Astraea Redux', *The Spectator* (28 July 1944), p. 79. On the absence of evidence of 'interference competition', that is, direct aggression, disruption of mating chases or attempts to mate with oestrus red females, see John Gurnell, Peter W. W. Lurz and Lucas A. Wauters, 'Years of Interactions and Conflict in Europe: Competition Between Eurasian Red Squirrels and North American Grey Squirrels', in *Red Squirrels: Ecology, Conservation and Management in Europe*, ed. Craig M. Shuttleworth, Peter Lurz and Matthew W. Hayward (Stoneleigh Park, Warks, 2015), pp. 22–3.

143 Middleton, 'Decrease in Red Squirrels'; Gurnell et al., 'Years of Interactions and Conflict', pp. 19, 28. On direct and indirect competition, see A. Okubo, P. K. Main, M. H. Williamson and J. D. Murray, 'On Spatial Spread of the Grey Squirrel in Britain', *Proceedings of the Royal Society of London, Series B, Biological Sciences*, 238 (1989), pp. 113–25.

144 'The Zoological Society: Grey Squirrels', *The Field*, CXIX/3083 (27 January 1912), p. 187.

145 Ibid.

146 H. W. Shepheard-Walwyn, *The Spirit of the Wild* (London, 1924), pp. 4, 5–6, 8–9.

147 Kelway, *Swift Movement*, pp. 25, 26–7, 33–4, 42, 29, 30, 42.

148 C. Somerville Watson, 'How to Manage Pets in the Winter', *The Boy's Own Paper* (6 October 1900), p. 12; Gerald E. H. Barrett-Hamilton and M.A.C. Hinton, *A History of British Mammals*, vol. IIC: *Order Rodentia, Part XX* [1918] (London, 1910–21), p. 688.

149 William Bliss, 'The Grey Squirrel' (LTTE), *The Observer* (6 December 1931).

150 G. McB, 'A Country Diary', *Manchester Guardian* (13 October 1949); John Timson, 'Britain's Surviving Red Squirrels Feel the Pressure', *New Scientist*, CXXIV/1689 (4 November 1989), p. 35; Andrew Wilson, 'Tufty's Last Stand', *Daily Mail* (5 December 1998).

151 Michael Viney, 'Why Our Red Squirrels Are Out on a Limb', *Irish Times* (18 November 2000).

152 John Dumbrell, *A Special Relationship: Anglo-American Relations in the Cold War and After* (New York, 2001), p. 14.

153 Tim Radford, 'Red Squirrel Has U.S. Family Tree', *The Guardian* (21 February 2003).

154 'British War on U.S. Squirrels: Pets Forty Years Ago Become Pests. Country Overrun', *Daily Mail* (21 January 1931).

155 Mary Gordon, 'The Grey Squirrel' (LTTE), *The Observer* (29 November 1931).

156 T. A. Coward, 'Correspondence: Squirrel: Red v. Grey', *Manchester Guardian* (18 June 1923).

157 John Still, 'Secluded Races', *The Observer* (20 November 1932).

158 Frances Pitt, *Wild Animals in Britain* (London, 1938), pp. 66–8.
159 E.M.N., 'Squirrel War'.
160 MAF Pest Officer, leaflet, MAF 217/33.
161 Leonard Robert Brightwell, 'The Grey Raiders', in *The Tiger in Town* (London, 1930), pp. 143–5.
162 Ibid., pp. 145–7, 148–9.
163 Ibid., pp. 150, 152, 157, 158, 160–66.
164 Hui et al., 'Defining Invasiveness and Invasibility', p. 971.
165 B. T., 'Rival Squirrels'.
166 T.A.C., 'Grey Colonist'.
167 W. Beach Thomas, 'The Squirrel and the Sanctuary', *The Spectator* (5 December 1931), p. 767.
168 W. Beach Thomas, 'A New Squirrel Crime', *The Spectator* (12 July 1930), p. 50.
169 C. Denton Hornby, 'The Grey Squirrel', *Journal of the Ministry of Agriculture*, XLIX–L (1942–4), p. 235.
170 Pitt, *Wild Animals*, pp. 66–8.
171 'Gray Squirrel in England Gains Faster'.
172 Kathleen MacKinnon, 'Competition Between Red and Grey Squirrels', *Mammal Review*, VIII/4 (1978), p. 187.
173 R. M. Keane and M. J. Crawley, 'Exotic Plant Invasions and the Enemy Release Hypothesis', TRENDS *in Ecology and Evolution*, XVII/4 (2002), pp. 164–70.
174 T.A.C., 'Squirrels: Red and Grey'. For 'promised land', see also 'Forestry Commission Joins in War on Grey Squirrel', *Manchester Guardian* (8 January 1954).
175 A.W.B., 'The Little Owl: Witness for the Defence and Prosecution', *Manchester Guardian* (17 January 1936).
176 Wordsworth, *Description of Scenery*, pp. 88, 90, 71, 92.

5 WAGING WAR ON THE 'GREY PERIL', *c.* 1939–73

1 E.M.N., 'The Squirrel War', *The Spectator* (1 August 1925), p. 191.
2 Julian Huxley, 'Is War Instinctive – and Inevitable? A Scientist Answers a Question Which Is Fundamental to the Problem of World Peace', *New York Times* (10 February 1946).
3 MAF (Infestation Control Division), 'You *Versus* Pests: The Grey Squirrel Menace' (April 1953), MAF 131/83.
4 Forestry Commission (FC), 'War on the Grey Squirrel: New Bonus Scheme for Tails' (February 1953); FC, 'The Grey Squirrel: A Woodland Pest', leaflet no. 31 (1960), MAF 149/116.
5 MAF, 'Grey Squirrels', Advisory Leaflet 58 (1943), MAF 44/45.
6 'Grey Squirrels and the Law', *Country Life*, LXXXII/2118 (21 August 1937), p. 184.
7 'Squirrel War Over: Comparative Safety for Them in the Parks', *Daily Mail* (16 December 1925).
8 'Squirrel Pie "a Real Delicacy"', *Daily Telegraph* (26 September 1984).
9 Richard Church, *A Squirrel Called Rufus* (London, 1941), p. 53.
10 Frances Bird, *New Statesman and Nation*, XXII/564 (13 December 1941), p. 498.
11 Church, *Squirrel Called Rufus*, pp. 11, 18, 24, 31.
12 Ibid., pp. 49–50.
13 Ibid., pp. 79–80, 118, 141, 149.
14 'Lancashire Farmers' Campaign Against Pests: Rats and Moles Head the "Black List"', *Manchester Guardian* (2 February 1946).
15 'Pest Control [HC]', *Hansard*, 389 (27 May 1943), c.1775; 'Pest Destruction (Cartridges) [HC]',

Hansard, 392 (21 October 1943), cc.1518–19; 'Pest (Destruction) [HC]', *Hansard*, 428 (28 October 1946), c.281; MAF, Minute Sheet, 10 September 1943, MAF 44/45.

16 D. E. Vandepeer (Second Secretary, MAF), 'The Grey Squirrels Order, 1943' (15 September 1943), MAF 44/45; 'The Grey Squirrel' (February 1953), MAF 131/83.

17 W. Beach Thomas, 'Traps and the W.A.E.C.s', *The Spectator* (28 December 1945), p. 622; 'Rat and Sparrow Club', *Manchester Guardian* (10 April 1956).

18 Secretary to the Committee, Agricultural Executive Committee, County of Surrey, letter to MAF ('Destruction of Grey Squirrels'), 14 October 1942, MAF 44/45; A.W.B., 'A Country Diary', *Manchester Guardian* (20 March 1945).

19 'Emergency Powers (Defence): Killing of Grey Squirrels (Kent) Order', 10 February 1942; Memorandum to Executive Officers of County War Agricultural Executive Committees in England and Wales, Destruction of Grey Squirrels: Grey Squirrels Order, 1943, MAF 44/45.

20 Elspeth M. Veale, *The English Fur Trade in the Later Middle Ages* (London, 2003), pp. 22–33.

21 MAF, press notice, 4 January 1943, MAF 44/45; Charles L. Coles, 'Grey Squirrel Pelts' (Item 3/4), FC, Committee on Grey Squirrels (CGS), Minutes, Meeting, 22 July 1953, p. 2, MAF 131/53.

22 Timms (Ministry of Food) to F. Winch (MAF), 4 January 1943, MAF 44/45.

23 A. D. Middleton, *The Grey Squirrel: The Introduction and Spread of the American Grey Squirrel in the British Isles* (London, 1931), p. 79.

24 Board of Education, *Good Fare in War-Time: Food Education Memo No. 3* (London, February 1941).

25 A.P.F. Grant (Ministry of Food, Meat and Livestock Branch) to Timms (Ministry of Food), 4 December 1942, MAF 44/45.

26 'A Game Dinner in America', *The Field*, LII/1354 (7 December 1878), p. 743; H. P., 'Duck Shooting in Western Marshes', *The Field*, XCVII/2508 (19 January 1901), p. 90.

27 'Corrigeen' (Joseph Adams), 'A Squirrel Hunt in an American Forest', *The Field*, LXXVI/1996 (28 March 1891), p. 439.

28 William T. Hornaday, *The American Natural History: A Foundation of Useful Knowledge of the Higher Animals of North America* (New York, 1904), p. 69.

29 Thomas Unett Brocklehurst, *Mexico To-day: A Country with a Great Future* (London, 1883), p. 2.

30 Gerald E. H. Barrett-Hamilton and M.A.C. Hinton, *A History of British Mammals*, vol. IIC: *Order Rodentia, Part XX* [1918] (London, 1910–21), p. 718.

31 L.C.R. Cameron, *The Wild Foods of Great Britain: Where to Find Them and How to Cook Them* (London, 1917), p. 21. See also Hugh Boyd Watt, 'The American Grey Squirrel in Britain', *The Field*, CXXX/3381 (13 October 1917), p. 536.

32 Roberta Meikle, 'The Grey Squirrel: More Evidence for the Prosecution: Why Not Eat Him?', *The Observer* (4 March 1917).

33 H.B.C.P., 'Grey Squirrel Control', *Country Life*, LXX/1822 (19 December 1931), p. xxvi.

34 H. Noble Hall, 'Uses of the Squirrel' (LTTE), *The Times* (29 August 1936).
35 'The Game-Book's "Various"', *Manchester Guardian* (16 September 1941).
36 W. Beach Thomas, 'Harmful Blessings', *The Spectator* (3 November 1939), p. 619. A recipe for squirrel, roasted or braised, served on a bed of polenta, featured in the game section of a landmark American cookbook first published (privately) in 1931 and in print continuously since 1936. The author preferred greys over reds (less 'gamy'): Irma S. Rombauer and Marion Rombauer Becker, *Joy of Cooking* [1936] (Indianapolis, IN, 1975), p. 515.
37 'Lucio', 'Squirrel en Casserole', *Manchester Guardian* (20 January 1941).
38 Ibid.
39 'Absence of Rabbits in the Fields', *The Times* (31 May 1945).
40 'Lucio', 'Squirrel en Casserole'.
41 Vicomte de Mauduit, *They Can't Ration These* (London, 1940), pp. 54–5.
42 W. Beach Thomas, 'Edible Squirrels', *The Spectator* (5 January 1945), p. 14; 'Absence of Rabbits'.
43 Rice Gaither, 'Squirrels Across the Sea', *New York Times* (28 November 1943); Middleton, *Grey Squirrel*, pp. 93–4.
44 Frances M. Rowe, 'Squirrel Pie' (LTTE), *The Times* (19 February 1941).
45 *Diaries, April 1942–October 1942*, 28 April 1942, Diarist 5365, Mass-Observation Archive, University of Sussex, www.massobservation.amdigital.co.uk.
46 'The Naturalist', *The Field*, XXX/777 (16 November 1867), p. 395.
47 F. Winch to F. E. Charlton, 21 October 1943, MAF 44/45.
48 Kenneth Pipe, 'Ministry thinks SQUIRREL is "Rather Like Rabbit"', *Daily Express* (1 October 1943).
49 'Friends and Enemies of the Farmer: Rat and Rabbit Pests', *The Times* (23 May 1944).
50 Monica Shorten, 'A Survey of the Distribution of the American Grey Squirrel (*Sciurus carolinensis*) and the British Red Squirrel (*S. Vulgaris leucourus*) in England and Wales in 1944–5', *Journal of Animal Ecology*, XV/1 (1946), pp. 82–92.
51 Monica Shorten, *Squirrels* (London, 1954), p. 58.
52 Ibid.
53 Philip Howell and Hilda Kean, 'The Dogs That Didn't Bark in the Blitz: Transpecies and Transpersonal Emotional Geographies on the British Home Front', *Journal of Historical Geography*, LXI (2018), pp. 44–5, 47–8, 50.
54 Stephen Tallents, 'Grey Squirrel Meat' (LTTE), *The Times* (2 December 1946); 'Casserole of Squirrel', in Tallents, *Green Thoughts* (London, 1952), p. 126.
55 Tallents, *Green Thoughts*, pp. 126–8.
56 'Grey Squirrels (Destruction) [HC]', *Hansard*, 415 (12 November 1945), c.1749; 'Grey Squirrel Shooting Clubs [HC]', ibid., 432 (27 January 1947), cc.607–8; MAF, Minute Sheets, January 1948, MAF 131/53.
57 'Great Britain: Tarnished Grandeur', *Time*, http://content.time.com, 4 March 1946; 'Squirrel Pies: Two Ministries Have Recipes', *Daily Telegraph* (7 March 1946).
58 'Squirrel Hot-Pot', *Manchester Guardian* (2 February 1946); Tallents, *Green Thoughts*, p. 126.
59 'London Diary', *Times of India* (6 March 1946).

60 MAF, Minute Sheets, 4 February 1943, MAF 44/45.
61 F. E. Charlton to Robert Hudson, 'Grey Squirrels Act 1943', 18 October 1943, MAF 44/45.
62 Kenneth Pipe, 'Kippers Departing, Squirrels Arriving', *Daily Express* (30 September 1943); Hailsham to Mr. Engholm (MOF), 2 October 1943, MAF 44/45.
63 'Squirrel Pie', *The Times* (4 December 1947).
64 MAF (ICD), to W. McAuley Gracie (MAF, ICD), 5 February 1952, MAF 130/81.
65 'Squirrel Pie', *Daily Telegraph* (11 March 1953).
66 MAF, Minute Note, 17 October 1946, MAF 44/45; Forestry Commission (FC), 'War on the Grey Squirrel', 10 March 1953, MAF 131/53; FC, 'Hints on Controlling Grey Squirrels' (1955); FC, CGS, Minutes of Meeting on 4 May 1955; FC, CGS. Minutes of Meeting on 26 October 1955, MAF 131/53.
67 D. E. Peacock, 'The Grey Squirrel *Sciurus carolinensis* in Adelaide, South Australia: Its Introduction and Eradication', *Victorian Naturalist*, CXXVI/4 (2009), pp. 150–51.
68 'The Squirrel Trouble', *New York Times* (9 March 1952).
69 Ibid.
70 Ibid.
71 Roy L. Robinson, 'Grey Squirrels: Forestry Commission's View' (LTTE), *The Times* (9 July 1937).
72 'Land Reserves for Forestry "Disquietingly Low"', *Manchester Guardian* (22 May 1952); 'Restoring Britain's Thinning Woodlands: Experiments in Hardwood Regeneration', *Manchester Guardian* (27 August 1951). Scotland's greys remained confined to the central lowlands.
73 Forestry Commission, 'The Grey Squirrel: A Woodland Pest', Leaflet No. 31 (1954), MAF 149/116; 'Forestry Commission Joins in War on Grey Squirrel: "Natural Pets" That Destroy Britain's Trees', *Manchester Guardian* (8 January 1954).
74 W. H. Marston, 'Damage by Squirrels', *Journal of the Forestry Commission*, XIX (1948), p. 102.
75 'National Drive Against Grey Squirrels: Forestry Commission to Pay Shilling a Tail', *Manchester Guardian* (11 March 1953).
76 FC, 'War on the Grey Squirrel: New Bonus Scheme for Tails', February 1953, MAF 131/88; John Sheail, 'The Grey Squirrel (*Sciurus carolinensis*) – A UK Historical Perspective on a Vertebrate Pest Species', *Journal of Environmental Management*, LV/3 (1999), p. 149.
77 'National Drive Against Grey Squirrels'.
78 Lord Elton, 'Control of Grey Squirrels [HL]', *Hansard*, 181 (21 April 1953), c.988.
79 Gerald Williams, 'Vermin (Destruction) [HC]', *Hansard*, 517 (2 July 1953), cc.572–3. A fixed term on payments was regarded as a deterrent to the cultivation of a breeding population: CGS, Minutes, Meeting, 26 October 1955, 4, MAF 131/53.
80 MAF 131/88; Joanna Toye, *The Archers Miscellany* (London, 2009), p. 34; Rosie Dillon, *The Archers: An Unofficial Companion* (Chichester, 2011), p. 15; Stephen Harris, 'The Red Has Lost – So Accept the Grey', BBC *Wildlife*, XXIV/9 (September 2006), p. 37.
81 'National Drive Against Grey Squirrels'; 'Grey Squirrel Tails by the

Thousand', *Manchester Guardian* (18 June 1953).
82 Ministry of Agriculture, Fisheries and Food (MAFF), 'The Campaign Against the Grey Squirrel: More Than 750,000 Kills in Two Years' (press release), 21 June 1955, MAF 217/30.
83 'Grey Market', *Manchester Guardian* (5 June 1953).
84 MAFF, 'Grey Squirrel Campaign: Progress Report for Six Months Ended 30th September, 1955', October 1955, MAF 217/30; 'Campaign Against Grey Squirrel: Satisfactory, But . . .', *Manchester Guardian* (5 November 1953).
85 'Forestry Commission Joins in War'.
86 'Miscellany: Squirrels and Blitz Sites', *Manchester Guardian* (2 July 1957).
87 CGS, Minutes, Meeting, 26 October 1955, 3, MAF 131/53.
88 J.K.A., 'A Country Diary', *Manchester Guardian* (1 October 1952); A.W.B., ' A Country Diary', *Manchester Guardian* (30 December 1952).
89 CGS, 'The Campaign Against the Grey Squirrel', 12 January 1956, MAF 131/53.
90 Geoffrey Beven, 'The Distribution of the Grey Squirrel in the London Area (1953–1956)', *London Naturalist*, 36 (1957), pp. 6–7.
91 W. G. Teagle, 'Mammals in the London Area: A Report for 1960', *London Naturalist*, 42 (1963), pp. 50–51.
92 W. G. Teagle, 'Mammals in the London Area 1961', *London Naturalist*, 43 (July 1964), pp. 127–9; W. G. Teagle, 'Mammals in the London Area 1962', *London Naturalist*, 44 (July 1965), pp. 53–4.
93 Teagle, 'Mammals in London Area 1962', p. 53; Teagle, 'Mammals in London Area: A Report for 1960', p. 50.
94 Teagle, 'Mammals in London Area: A Report for 1960', p. 50.
95 Teagle, 'Mammals in London Area 1961', pp. 127–9. See also Richard Fitter, 'Surprising Citizens', *The Observer* (8 September 1963).
96 Teagle, 'Mammals in London Area 1961', p. 128.
97 FC, CGS, Minutes, Meeting, 4 May 1955, MAF 131/53; MAFF, 'The Campaign Against the Grey Squirrel: More Than 750,000 Kills in Two Years'; '750,000 Squirrels Killed: Result of the Bonus Scheme', *Manchester Guardian* (22 June 1955). Another source gave slightly lower figures. From the bounty's inception in March 1953 to the end of September 1955, 563,510 tails were presented for cash and a further 167,459 for cartridges – a grand total of 730,969: MAFF, 'Grey Squirrel Campaign: Progress Report for Six Months Ended 30th September, 1955', October 1955, MAF 217/30. According to a third source, the kill rate between 1 October 1953 and 30 September 1954 ('forest year' 1954) was 406,903, with the bounty claimed on 75 per cent. In 'forest year' 1955 (1 October 1954 to 30 September 1955), the count dropped to 236,783, with the shilling claimed on 85 per cent: CGS, Minutes, Meeting, 26 October 1955, MAF 131/53.
98 CGS, Minutes, Meeting, 26 October 1955; CGS, 'The Campaign Against the Grey Squirrel', 12 January 1956.
99 CGS, 'The Campaign Against the Grey Squirrel', 12 January 1956.
100 CGS, Item 3, 'Grey Squirrel Bounty Scheme – Report by Mrs. Vizoso', Meeting, 5 May 1954; CGS, Minutes, Meeting, 26 October 1955, MAF 131/53.
101 Ibid.

102 As 'the right targets become fewer', champions of native squirrels feared they would become collateral damage in areas of coexistence: 'Grey Squirrels', *Manchester Guardian* (6 December 1955).
103 'Double Rate for Grey Squirrels: Now "Two Bob a Tail"', *Manchester Guardian* (6 December 1955).
104 'British Double Squirrel Bounty', *New York Times* (6 December 1955); 'Gray Squirrel Tails', *New York Herald Tribune* (11 November 1956).
105 'U.S. Squirrels Lose Battle in Britain', *New York Times* (4 March 1956).
106 FC, 'Grey Squirrel Bonus to End' (press release), 31 December 1957, MAF 131/53.
107 Harry V. Thompson (MAFF) and T. R. Peace (FC), 'The Grey Squirrel Problem' (1960), MAF 131/53.
108 FC, 'Grey Squirrel Bonus to End'. See also 'Squirrel Bonus to End', *Manchester Guardian* (1 January 1958).
109 'Control of Grey Squirrels [HL]', *Hansard*, 220 (15 December 1959), c.375.
110 Harold Macmillan, *War Diaries: Politics and War in the Mediterranean, January 1943–May 1945* (London, 1984), p. 43.
111 Minute (M.) 12/60 (3 January 1960), Prime Minister's Papers (PREM) 11/3196, TNA.
112 Hare to Prime Minister (5 January 1960) PREM 11/3196.
113 M. 47/60 (29 February 1960), PREM 11/3196.
114 M. 306/60 (15 August 1960); Soames to Prime Minister (17 August 1960); A. J. Phelps to J. R. Moss (30 September 1960), PREM 11/3196.
115 'Spoilsports', *Manchester Guardian* (29 March 1955); John Steinbeck, 'Report on America', *Punch*, CCXXVIII/5989 (22 June 1955), pp. 754–5.
116 'Squirrels Die of Hunger: Little Animals in White House Grounds Starving', *Washington Post* (31 August 1909); 'White House Squirrels Dead', *Washington Post* (3 August 1916); 'Houses Are Provided White House Squirrels', *Washington Post* (19 January 1933).
117 'Pete, Pet Squirrel at the Executive Mansion, Is Causing Laddie Boy to Look to His Laurels' (10 October 1922), National Photo Company Collection, Prints & Photographs Division, Library of Congress, www.loc.gov/pictures/item/2002712422.
118 'White House Squirrel Digs Up Bulbs for Nuts', *Washington Post* (17 March 1923).
119 Richard L. Lyons, 'Neuberger Put to Save White House Squirrels', *Washington Post* (23 March 1955); 'Putting Green: Play It Where It Lies, Mr President', The White House Museum, www.whitehousemuseum.org, accessed 22 August 2022.
120 Merriman Smith, 'White House Squirrels Being Deported for Taking Divots from Putting Green', *Washington Post* (15 March 1955).
121 'Deportation of Squirrels News to Ike: Great Squirrel Mystery', *Washington Post* (25 March 1955); Steinbeck, 'Report on America'; 'The Presidency: And Then the Squirrels', *Time* (11 April 1955).
122 'Heidi: A Profile', *Sports Illustrated*, https://vault.si.com/vault/1958/07/21/events-discoveries, 21 July 1958.
123 Richard Neuberger, 'Trapping of Squirrels at the White House', *Congressional Record* (Senate) (25 March 1955), pp. 3744–5, www.govinfo.gov; T. S. Palmer and R. W.

Williams, *Farmers' Bulletin No. 265: Game Laws for 1906* (Washington, DC, 1906), p. 7.

124 'Spoilsports'; Lyons, 'Neuberger Put to Save Squirrels'; Douglas Perry, 'How an Oregon Senator Challenged the President Over His Treatment of the White House's Squirrels', *The Oregonian*, www.oregonlive.com, 15 November 2016.

125 Edward T. Folliard, 'White House Halts Eviction of Squirrels: Ike Calls Off Squirrel Hunt', *Washington Post* (26 March 1955).

126 Lyons, 'Neuberger Put to Save Squirrels'.

127 M. 306/60 (15 August 1960); Soames to Prime Minister (17 August 1960); Phelps to J. R. Moss (30 September 1960); Soames to Prime Minister (5 October 1960), D. H. Andrews to Phelps (21 December 1960), B.J.T. to Soames (22 December 1960), PREM 11/3196.

128 In the 1980s, Jessica Holm focused on harsh weather's impact on the survival rates of young squirrels, red and grey. Cold and wet springtime conditions could soak a drey, its entire litter perishing from exposure. In her Isle of Wight study area, 67 per cent of red kits died before they left the drey: Holm, *Squirrels* (London, 1987), pp. 38–9; Holm, 'Son of Nutkin', *BBC Wildlife*, CIV/3 (March 1987), pp. 103–6.

129 Garth Christian, 'The Problem of Britain's Squirrels', *Country Life*, CXXIX/3335 (2 February 1961), pp. 226–7.

130 PREM 11/3196; 'Grey Squirrels [HC]', *Hansard*, 581 (7 February 1958), c.214.

131 Sheail, 'Grey Squirrel', p. 153.

132 Forestry Commission, *Report on Forest Research for the Year Ended March, 1960* (London, 1961), p. 71; FC/MAFF, Minutes, Meeting on Grey Squirrels, 10 August 1961, MAF 131/154.

133 'Field Trials with Warfarin in Crown Woodlands in England 1968–1970', in 'Grey Squirrel Bait: Product Assessment Report of a Biocidal Product for National Authorisation Applications, Case Number BC-VD032513–48, Evaluating Competent Authority: United Kingdom (2 January 2019)', pp. 29–36.

134 W. McAuley Gracie to J. W. Evans, 11 February 1952, MAF 131/154.

135 'Grey Squirrels [HC]', *Hansard*, 854 (12 April 1973), cc.1630–31.

136 'Squirrels: Control by Warfarin [HL]', *Hansard*, 596 (26 January 1999), cc.871–5.

137 'Agriculture (Miscellaneous Provisions) Bill [HL]', *Hansard*, 326 (30 November 1971), cc.169–70.

138 Ibid., cc.170–73.

139 Ibid., c.171.

140 Ibid., c.174.

141 Ibid., cc.175–7, 185.

142 Ibid., cc.175–6.

143 'Grey Squirrels [HC]', *Hansard*, 837 (16 May 1972), cc.226–7; Tam Dalyell, 'Westminster Scene – Ratting on Squirrels', *New Scientist*, LIV/798 (1 June 1972), p. 511.

144 'Control of Squirrels by Shooting Is a Failure', *The Guardian* (21 March 1973).

145 'Grey Squirrels [HC]', *Hansard*, 843 (26 October 1972), c.410.

146 H. W. Pepper, *Grey Squirrel Damage Control with Warfarin*, Research Information Note 180 (Farnham, Surrey, 1996), pp. 1, 3.

147 'Grey Squirrels [HC]', *Hansard*, 854 (12 April 1973), cc.1630–31; 'Poison Approved for Grey Squirrels', *The Guardian* (18 April 1973).

148 MAF, 'The Grey Squirrel' (February 1953), MAF 131/83.
149 'Grey Squirrel Control in Scotland [HL]', *Hansard*, 397 (18 January 1979), cc.1160–62.
150 FC, 'Appraisal of Squirrel Status in the United Kingdom' (February 1981); 'Forestry [HL]', *Hansard*, 417 (23 February 1981), cc.944, 931–2.
151 Monica Shorten, 'Introduced Menace: American Grey Squirrel Poses Threat to British Woodlands', *Natural History*, LXXIII/10 (December 1964), p. 45.
152 Ibid., pp. 47–8.
153 Colin Luckhurst, 'A Country Diary', *The Guardian* (28 June 1969).
154 John C. Moore, *The Waters Under the Earth* (London, 1965), p. 9.
155 Ibid., pp. 10, 12, 13.
156 Ibid., pp. 13–14.
157 Shorten, 'Introduced Menace', pp. 45–6.

6 WANTED: RED AND ALIVE

1 'Biodiversity [HL]', *Hansard*, 764 (16 July 2015), c.91.
2 'British Hope to Save Red Squirrel from Its Aggressive U.S. Cousin', *Phoenix Gazette* (1 August 1996).
3 'Nutty Policy: Brits Declare War on American Squirrels', *Columbus Dispatch* (17 August 1996).
4 Petronella Wyatt, 'The Red and the Grey', *The Spectator* (29 December 2001), p. 49.
5 Veronica Heath, 'A Country Diary', *The Guardian* (19 November 1993).
6 Karin Brulliard, 'In Squirrelly British Politics, Better Red than Dead', *Washington Post* (8 March 2017).
7 Joe Shute, 'Tail of the Unexpected: The Resurgence of the Red Squirrel', *Daily Telegraph*, www.telegraph.co.uk, 21 September 2013; Clive Aslet, 'Future Less Grey for Squirrels', *Daily Telegraph*, www.telegraph.co.uk, 16 November 2011.
8 Janet Wickens, 'RSST Update', *Squirrel*, 24 (April 2012), p. 5; 'A Sad Farewell', *Squirrel*, 25 (October 2021), p. 6. *Squirrel* is the European Squirrel Initiative's journal.
9 Squirrel Appreciation Day was inaugurated in the USA by Christy Hargrove of North Carolina on 21 January 2001: 'Squirrel Appreciation Day' [undated], *HolidayInsights*, www.holidayinsights.com, accessed 5 December 2022.
10 'Charles Lets Red Squirrels Roam Around His Home', *Daily Telegraph* (21 January 2021). Charles first spoke of his indoor hosting and feeding in 2009: 'Prince Charles Entertains Red Squirrels at Balmoral', *Country Life*, www.countrylife.co.uk/news, 6 April 2009.
11 'Charles' Red Squirrel "Infatuation" Revealed in Country Life Magazine', *ITV News*, www.itv.com/news, 13 November 2018; 'Prince Charles Is "Completely Infatuated" with Squirrels', *House and Garden*, www.houseandgarden.co.uk, 15 November 2018.
12 G.A.F., 'A Country Diary', *The Guardian* (4 November 1960).
13 John Betjeman, *Summoned by Bells* (London, 1960), p. 3.
14 Betjeman, 'Called to Account by 31 Highgate West Hill', *The Guardian* (28 November 1960).
15 'Five Old Friends on the New 9p Stamps', *Daily Mail* (29 September 1977); 'Picture Gallery', *The Times* (24 August 1977); Samuel A. Tower, 'Stamps: British Flowers, Dogs and Other Prized Animals', *New York Times* (25 March 1979).

16 *Daily Mail* (7 September 2004).
17 'Red Squirrels [HL]', *Hansard*, 587 (25 March 1998), c.1316.
18 Patrick Barkham, '"Kill Them, Kill Them, Kill Them": The Volunteer Army Plotting to Wipe Out Britain's Grey Squirrels', *The Guardian*, www.theguardian.com, 2 June 2017.
19 Quoted in Brulliard, 'To Save Their Beloved Red Squirrels, Brits Will Monitor and Kill Invasive American Ones', *Washington Post*, www.washingtonpost.com, 6 March 2017. Crowley et al. identify three (not always mutually exclusive) 'modes of killing' greys. First, reparative (the moral and ecological duty to kill for conservation/biodiversity and to atone for the error of introduction). Second, killing without qualm or gratification; and third, categorical killing (as vermin/pests/invasives). Under the first mode – which fits Bailey – they identify nostalgic affection for reds as embodiments of national identity and feelings of protectiveness for 'our' squirrel as victim: Sarah L. Crowley, Steve Hinchliffe and Robbie A. McDonald, 'Killing Squirrels: Exploring Motivations and Practices of Lethal Wildlife Management', *Environment and Planning E: Nature and Space*, 1/1–2 (2018), pp. 120–43.
20 Marlena Spieler, 'Saving a Squirrel by Eating One', *New York Times*, www.nytimes.com, 6 January 2009.
21 'Squirrel Nutkin to the Life: An Engaging Animal Study in a Current London Photographic Exhibition', *Illustrated London News*, 6184 (14 December 1957), p. 1034.
22 'Wildlife and Countryside Bill [HL]', *Hansard*, 416 (3 February 1981), c.1075.
23 ITV London's *Survival* series, 'The Fall of Squirrel Nutkin', aired 19 December 1989.
24 James Wentworth Day, 'Hunting the Grey Devil in the Woods', *Daily Mail* (13 September 1980).
25 William Beach Thomas, 'Island Antipodes', *The Observer* (29 November 1953).
26 'Tufty' [1986?]; 'Happy Birthday Tufty' [1978?], ROSPA Archive, Birmingham, UK (ROSPA).
27 British Pathé, 'Tufty Club' (Manor House Nursery School, Merton, Surrey) (1962), www.britishpathe.com/video/tufty-club.
28 'ROSPA to Launch New National "Tufty" Club', *Safety News* (November 1961), p. 4; 'Drive to Cut Toddler Road Toll', *Safety News* (January 1962), p. 1; 'Tufty Is Warmly Welcome', *Safety News* (July 1962), p. 10, ROSPA.
29 'Tufty' [1986?].
30 'Tufty Fluffytail Arrives to Save Children', *Daily Mirror* (5 December 1961). Circa 10,000 under-eights were killed or injured on British roads annually. See also 'Tufty Turns 60: Happy Birthday to a Red Squirrel that Helped to Improve Road Safety for Children', *Daily Telegraph* (10 April 2013).
31 'Tufty Club Activity', *Safety News* (March 1962), p. 2; Elsie Mills, *The Tufty Club: Stories of Tufty Fluffytail and His Furryfolk Friends* (London, 1961); *The Second Tufty Club Book: More Stories of Tufty and His Furryfolk Friends* (London, 1963), ROSPA.
32 'All Our Yesterdays: Look Back in D-anger: The Tufty Years', *Safety Education* (Summer 1998), p. 18; 'Foreword, The Tufty Club Introduction Notes: Suggestions

for Leaders of Tufty Clubs' (1973), p. 1; 'Foreword, The Tufty Club Introduction Notes: Suggested Guide Lines for Tufty Club Leaders' (1978), p. 1, ROSPA; ROSPA, 'Tufty', www.rospa.com, accessed 20 August 2022.

33 'Foreword, The Tufty Club Introduction Notes' (1978), p. 1; Elsie Mills, *The Tufty Annual* (London, 1971), p. 25, ROSPA.

34 Carey Blyton, 'The Furryfolk on Holiday', https://careyblyton.com, accessed 20 August 2022; 'Red Squirrels: Potential Extinction [HC]', *Hansard*, 662 (3 July 2019), c.540.

35 'All Our Yesterdays: Look Back in D-anger'.

36 'Tufty's Fame is Spreading Rapidly as the Guardian of the Under-Fives', *Safety News* (July 1962), p. 1. On Beatrix Potter collectibles ('merch') – ceramics, calendars, stuffed toys, hand puppets, figurines and music boxes – see Debby DuBay and Kara Sewall, *Beatrix Potter Collectibles: The Peter Rabbit Story Characters* (Atglen, PA, 2005).

37 'Tufty Exchange – Fur Flies Over Tufty', *Tufty Club Newsletter* (Autumn 1978), p. 5, ROSPA.

38 'Tufty – Past, Present and Future' [presentation to Northern Ireland Tufty Club Association], 21 November 1981, p. 5, ROSPA.

39 Graham Jones, 'Road-Safety Tufty is "a Sexist, Racial Squirrel"', *Daily Telegraph* (20 September 1984). Another paper quoted the Conservative leader as follows: 'I suppose Tufty would be approved of if he was gay and black.' See Peter Hardy, 'Nuts to Tufty!: Storm Over "Racist and Snobbish" Road Safety Squirrel', *Daily Express* (20 September 1984). See also Rupert Morris, 'Rotten Luck, Tufty's in the Dock', *The Times* (20 September 1984).

40 'Editorial' and 'Tufty Relaunch', *Tufty* (Spring/Summer 1993), pp. 2–3; 'All Change for the 90's!', *Tufty Tales*, 1 (April 1993), p. 3, ROSPA.

41 All Our Yesterdays: Look Back in D-anger'.

42 Margaret Collins, 'Kids Eye-View', *Safety Education* (Spring 1995), pp. 18–19, ROSPA.

43 'Tufty' [1986?].

44 Collins, 'Kids Eye-View: Tufty', *Safety Education* (Autumn 1995), pp. 17–20, ROSPA.

45 Andrew Wilson, 'Tufty's Last Stand', *Daily Mail* (5 December 1998).

46 Chris Webber, 'Now Tufty Gets a Helping Hand Crossing the Road', *Northern Echo* (26 February 2001). In parts of northern England, over half of adult red mortality is attributable to traffic: Anna L. Meredith and Claudia Romeo, 'Disease and Causes of Mortality in Red Squirrel Populations', in *Red Squirrels: Ecology, Conservation and Management in Europe*, ed. Craig M. Shuttleworth et al. (Stoneleigh Park, Warks, 2015), p. 122.

47 P.W.W. Lurz, 'Changing "Red to Grey": Alien Species Introductions to Britain and the Displacement and Loss of Native Wildlife from Our Landscapes', in *Displaced Heritage: Responses to Disaster, Trauma and Loss*, ed. Ian Convery et al. (Woodbridge, Suffolk, 2014), p. 269.

48 David Thomas, 'Worra Story!', *Daily Mail* (10 February 2006).

49 'Anti-Litter Campaign Poster [HC]', *Hansard*, 569 (30 April 1957), cc.12–13; A.P.G. Brown, 'Minister's Visit to the Anti-Litter Conference, 19 March', memo, 11 March 1958, Keep Britain Tidy Campaign: Departmental

Involvement, Ministry of Housing and Local Government, HLG 51/1132, TNA.
50 'Empty Beer Cans and Litter [HL]', *Hansard*, 205 (25 June 1957), cc.104–6.
51 L. Harrison Matthews, 'Teddy Boy of the Trees', *The Observer* (14 April 1957).
52 'Anti-Litter Campaign Poster [HC]', *Hansard*; 'Pot Urging the Kettle to Be Clean: Red Unsuitable for Litter Campaign', *Manchester Guardian* (1 May 1957); Christopher Hollis, 'Essence of Parliament', *Punch*, CCXXXII/6089 (8 May 1957), p. 596.
53 'The Course of Nature: Tidy and Untidy Animals', *The Times* (13 August 1957).
54 'Jewellery and Silverware [HC]', *Hansard*, 513 (20 March 1953), c.379.
55 'Wildlife and Countryside Bill [HL]', *Hansard*, 416 (3 February 1981), cc.1074–5.
56 'Red Squirrels [HL]', *Hansard*, 587 (25 March 1998), c.1322.
57 Humphrey Drummond, 'The Unequal Struggle of the Red Squirrel', *The Field*, CCLXV/6880 (1 December 1984), pp. 48–9.
58 'Nature Conservancy Council [HL]', *Hansard*, 493 (17 February 1988), c.693.
59 'Red Squirrels [HC]', *Hansard*, 272 (21 February 1996), cc.331–3.
60 Ibid., c.331.
61 Ibid., c.334.
62 'Survival of the Red Squirrel in the UK', EDM 278, https://edm.parliament.uk/early-day-motion/13053, accessed 13 March 2022.
63 'Red Squirrels [HL]', *Hansard* (25 March 1998), c.1316.
64 Norman Toulson, *The Squirrel and the Clock: National Provident Institution, 1835–1985* (London, 1985), pp. 82, 32, 35, 76–7.
65 'Red Squirrels [HL]', *Hansard* (25 March 1998), c.1316.
66 Ibid., cc.1317–18.
67 Ibid., c.1323.
68 Ibid., c.1328; 'Red Squirrels [HC]', *Hansard* (21 February 1996), c.338.
69 'Squirrels [HL]', *Hansard*, 680 (23 March 2006), c.365.
70 Ibid., c.355.
71 Ibid., cc.362–3.
72 'The Nature Conservancy [HL]', *Hansard*, 237 (28 February 1962), c.1005. For a critique of the automatic assumption that a native species is inherently superior in value, as well as more desirable and beautiful, not to mention more 'correct' in a particular place, see Matthew K. Chew and Andrew L. Hamilton, 'The Rise and Fall of Biotic Nativeness: A Historical Perspective', in *Fifty Years of Invasion Biology: The Legacy of Charles Elton*, ed. David M. Richardson (Chichester, 2011), p. 42. Environmental journalist Fred Pearce also queries the conviction that 'natives [are] obviously more at home' than non-native species: Pearce, *The New Wild: Why Invasive Species Will Be Nature's Salvation* (Boston, MA, 2015), p. 78.
73 'Squirrels [HL]', *Hansard* (23 March 2006), c.368.
74 Ibid., cc.373–4.
75 Ibid., c.369.
76 Ibid., c.360.
77 'Squirrels: Control by Warfarin [HL]', *Hansard*, 596 (26 January 1999), c.874.
78 Tim Adams, 'They Shoot Squirrels, Don't They?', *The Observer*, www.theguardian.com, 19 October 2008.
79 'Squirrels [HL]', *Hansard* (23 March 2006), cc.359–60.
80 Scott, 'Help Is At Hand for Nutkin's Last Stand', *Daily Mail* (19 July 1990).

81 John Ezard, 'Grey Day for Red Squirrels', *The Guardian* (16 November 1995).
82 John Bonner, 'Virus Blamed on Invading Squirrels', *New Scientist*, CXLVIII/1999 (14 October 1995), p. 10; John Kelly, 'Brits Rally to Save Red Squirrels from Onslaught of Invasive Grays', *Washington Post* (11 April 2012).
83 Alex Strauss, Andy White and Mike Boots, 'Invading with Biological Weapons: The Importance of Disease-Mediated Invasions', *Functional Ecology*, XXVI/6 (2012), p. 1257. Paul Duff and Anna Meredith, 'Disease and Mortality of the Grey Squirrel', in *The Grey Squirrel: Ecology and Management of an Invasive Species in Europe*, ed. Craig M. Shuttleworth et al. (Stoneleigh Park, Warks, 2016), p. 158.
84 Michael B. Usher, Terence J. Crawford and Jean L. Banwell, 'An American Invasion of Great Britain: The Case of the Native and Alien Squirrel (*Sciurus*) Species', *Conservation Biology*, VI/1 (March 1992), pp. 108–15; Timothy Daniel Dale, 'Transmission Dynamics and Pathogenesis of Squirrelpox in UK Red (*Sciurus vulgaris*) and Grey Squirrels (*Sciurus carolinensis*)', unpublished doctoral thesis, University of Liverpool, 2013, p. 8.
85 S. P. Rushton et al., 'Disease Threats Posed by Alien Species: The Role of a Poxvirus in the Decline of the Native Red Squirrel in Britain', *Epidemiology and Infection*, 134 (2006), p. 522; Strauss et al., 'Invading with Biological Weapons', pp. 1249, 1251, 1256–9.
86 Rushton, 'Disease Threats Posed by Alien Species', pp. 521, 530; Matthew Beard, 'Red Squirrels Face Deadly Virus Spread by Rival Greys', *The Independent* (20 June 2005).
87 Roger Highfield, 'Drug Therapy to Make Grey Squirrels Docile', *Daily Telegraph* (3 November 1995).
88 R. Kenward and J. L. Holm, 'On the Replacement of the Red Squirrel in Britain: A Phytotoxic Explanation', *Proceedings of the Royal Society of London, Series B, Biological Sciences*, 251 (1993), pp. 192, 187, 188.
89 John Lucas, 'Come in, You Reds', *Daily Telegraph* (25 July 1992); 'Bright Eyed, Bushy Tailed and in Danger', *Daily Mail* (1 August 1996).
90 'Red Squirrels [HC]', *Hansard*, 272 (21 February 1996), c.337.
91 'Squirrels from U.S. Are Driving British Nuts', *Orlando Sentinel* (16 August 1996).
92 Tony Samstag, 'Return of the Native Red Squirrel', *The Times* (26 October 1984).
93 'Red Squirrels [HC]', *Hansard* (21 February 1996), c.332; Alexander MacLeod, 'Rescuing the Red Squirrel: British Launch Campaign to Save Their Native, Tufted-Eared Species', *Christian Science Monitor* (8 April 1997); 'Poison Tables Turned', *Daily Mail* (22 February 1996).
94 A. J. McIlroy, 'Red Alert in Forest Helps Save Squirrels', *Daily Telegraph* (23 May 1994); John Gurnell and Janie Steele, *Grey Squirrel Control for Red Squirrel Conservation: A Study in Thetford Forest*, English Nature Research Report No. 453 (February 2002), p. 19; 'Squirrels [HL]', *Hansard*, 665 (11 October 2004), c.WA42.
95 Richenda Bland, *Country Life*, LXXVI/1968 (6 October 1934), p. 374; 'Establishing the Red Squirrel', *Country Life*, LXXVI/1978 (15 December 1934), p. 654.

96 William Beach Thomas, 'Flourishing Squirrels', *The Spectator* (27 November 1936), p. 947.

97 Angus Watson, 'Red Squirrels', *The Spectator* (8 April 1938), p. 631.

98 L.G.M., 'Saving the Squirrel', *Daily Mail* (6 November 1928).

99 'How the Red Squirrels Came Back', *Daily Mail* (19 March 1936).

100 'Red Squirrels at Whipsnade', *The Observer* (13 November 1932); 'Red Squirrels: Four Pairs Liberated at Whipsnade', *The Times* (22 October 1932).

101 William Beach Thomas, 'More Brown Squirrels', *The Spectator* (19 April 1935), p. 656.

102 H.M.S., 'Zoo's Pet Squirrel', *Daily Telegraph* (11 June 1940).

103 B.C.R. Bertram and D.-P. Moltu, 'Re-Introducing Red Squirrels into Regent's Park', *Mammal Review*, XVI/2 (June 1986), pp. 83–4.

104 Ibid., p. 82; 'Radio Call of the Wild', *Daily Mail* (27 October 1984).

105 *Reports of the Council and Auditors of the Zoological Society of London for the Year 1987* (London, 1988), p. 2.

106 Tony Samstag, 'Return of the Native Red Squirrel', *The Times* (26 October 1984). These cities include Berlin, Budapest, Hamburg, Munich, Warsaw and Zurich.

107 Brian C. R. Bertram and David-Petter Moltu, *The Reintroduction of Red Squirrels into Regent's Park* (London, 1987), pp. 3–4.

108 Bertram and Moltu, 'Re-Introducing Red Squirrels', p. 87; James Thompson, 'Red Alert', *Daily Mail* (23 May 1987); Tim Radford, 'Cars Crush Hope for Red Squirrels', *The Guardian* (20 April 1989); 'Red Squirrels Fail to Recolonise Regent's Park', *New Scientist*, CXIV/1559 (7 May 1989), p. 23.

109 Bertram and Moltu, 'Re-Introducing Red Squirrels', pp. 82–3.

110 The only form of chase conceivably interfering with the reds' welfare was sexual pursuit. Though the two species are reproductively incompatible, vigorous pursuit of red females during the customary, prolonged mating chase may have prevented red males from breeding: Bertram and Moltu, *Reintroduction*, pp. 4–5, 6.

111 Hannah E. Jones et al., 'Mathematical Model of Grey Squirrel Invasion: A Case Study on Anglesey', in *Grey Squirrel*, ed. Shuttleworth et al., p. 237.

112 Craig Shuttleworth, 'Nutkin Ventured, Nutkin Gained', 'National and International Perspectives on Red Squirrel Conservation', University of Exeter, 19 April 2013.

113 'Red Squirrels Recolonise Grey Area', *The Independent* (21 September 2004).

114 Richard Fletcher, 'Father and Son Champion Grey Squirrels', *The Journal* [Newcastle] (9 December 2009).

115 'Victimising Grey Squirrels', Version 2 (December 2009), available at www.grey-squirrel.org.uk.

116 Olivia Blair, 'Grey Squirrels Now Extinct in Anglesey', *The Independent*, www.independent.co.uk, 26 October 2015.

117 Stephen Harris, Carl D. Soulsbury and Graziella Iossa, 'Is the Culling of Grey Squirrels a Viable Tactic to Conserve Red Squirrel Populations?' (University of Bristol, 2006), pp. 12, 18.

118 Anna L. Signorile and Craig M. Shuttleworth, 'Genetic Evidence of the Effectiveness of Grey Squirrel Control Operations: Lessons from the Isle of Anglesey', in *Grey Squirrel*, ed. Shuttleworth et al., p. 440.

119 Ibid., pp. 479–80.

120 Craig Shuttleworth, 'Anglesey: Where Did it Go Wrong?', *Squirrel*, 30 (April 2015), p. 10; Pia Schuchert et al., 'Landscape Scale Impacts of Culling upon a European Grey Squirrel Population: Can Trapping Reduce Population Size and Decrease the Threat of Squirrelpox Virus Infection for the Native Red Squirrel?', *Biological Invasions*, 16 (2014), p. 2390.

121 F. E. Williams et al., *The Economic Cost of Invasive Non-Native Species on Great Britain* (Wallingford, Oxon, 2010), pp. 181, 184.

122 'UK Squirrels Endangered by Brawny American Cousins', *Jerusalem Post* (10 November 2005); Red Squirrels Northern England (RSNE), 'Conservation and Strongholds', www.rsne.org.uk/conservation-and-strongholds, accessed 20 August 2022. RSNE (not to be confused with RANE), a group established in 2011, and partnered with bodies such as the Forestry Commission, Northumberland Wildife Trust and Cumbria Wildlife Trust, became mainly responsible for managing red squirrel conservation strategies: 'Red Squirrels Northern England (RSNE)', in *Saving the Red Squirrel: Landscape Scale Recovery*, ed. Craig M. Shuttleworth et al. for Red Squirrel Survival Trust (London, 2020), pp. 175–85.

123 'Red Squirrels: Potential Extinction [HC]', *Hansard*, 662 (3 July 2019), c.524.

124 Charles Clover, 'No More Grey Areas, the Squirrels Have to Die', *Daily Telegraph* (9 November 2009).

125 'Britain Trying to Save the Red Squirrel', *San Jose Mercury* (9 November 2005).

126 D. T. Max, 'The Squirrel Wars', *New York Times* (7 October 2007).

127 Kelly, 'Brits Rally'.

128 Harry Pepper and Gordon Patterson, *Forestry Commission Practice Note 5 – Red Squirrel Conservation* (Edinburgh, September 1998), pp. 2, 4, 5.

129 R.S.R. Fitter, 'The Distribution of the Grey Squirrel in the London Area', *The London Naturalist for the Year 1938* (London, 1939), p. 7.

130 Herbert L. Edlin, *The Changing Wild Life of Britain* (London, 1952), pp. 42, 44–5.

131 'Red Squirrels [HL]', *Hansard* (25 March 1998), c.1327.

132 Joyce Quin, 'Red Squirrels (North-East England) [HC]', *Hansard*, 424 (8 September 2004), c.317.

133 Chris Tighe, 'Curb on Oak Trees in Red Squirrel Habitat', *Financial Times* (20 September 2003); 'Red Alert as Foresters Plant Obstacle in Way of Grey Tide', *Northern Echo* (22 September 2003).

134 The Wildlife Trusts, *Red or Dead* (1 January 1999); 'Assembly Secretary Goes Wild in the Woods', *Forestry and British Timber* (1 January 1999), p. 7.

135 J. Bryce, S. Cartmel and C. P. Quine, 'Habitat Use by Red and Grey Squirrels: Results of Two Recent Studies and Implications for Management' (Edinburgh, October 2005); 'Peace in the Forest', *New Scientist*, 2113 (20 December 1997), p. 5.

136 Mark Townsend, 'Red Squirrel Saved from Extinction: Forest Barriers Will Help Repel Invading American Greys', *The Observer* (22 September 2002).

137 'Doomed in 12 Years, the Red Squirrel', *Daily Mail* (23 September 1998).

138 James Gorman, 'Putting Nature on the Pill: Wildlife Managers Look to

Contraception as a Way to Control Nuisance Animals', *New York Times* (31 August 2004).

139 Harvey Elliott, 'Farmers Will Put Squirrels on the Pill', *Daily Mail* (2 October 1972); Anthony Michaelis, 'The Squirrel Pill', *Daily Telegraph* (3 September 1973).

140 Elizabeth Johnson and A. J. Tait, 'Prospects for the Chemical Control of Reproduction in the Grey Squirrel', *Mammal Review*, 13 (1983), pp. 170–72.

141 Malcolm Smith, 'Squirreled Away', *The Guardian* (20 October 1994); 'Swallow a Bitter Pill and Survive, Squirrels', *Daily Mail* (20 October 1994).

142 'Grey Squirrels: Population Control [HL]', *Hansard*, 555 (15 June 1994), c.1691.

143 H.D.M. Moore et al., 'Immunocontraception in Rodents: A Review of the Development of a Sperm-Based Immunocontraceptive Vaccine for the Grey Squirrel (*Sciurus carolinensis*)', *Reproduction, Fertility and Development*, IX/1 (1997), pp. 125–9; Alastair Clay, "Rampaging Squirrels to Go on the Pill', *Independent on Sunday* (14 June 1998).

144 Harry Pepper and Harry Moore, 'Squirrel Management 1: Development of a Contraceptive Vaccine for the Grey Squirrel', in *Forest Research Annual Report and Accounts, 1999–2000* (Edinburgh, 2001), p. 25.

145 Stephen P. Rushton et al., 'Modeling Impacts and Costs of Gray Squirrel Control Regimes on the Viability of Red Squirrel Populations', *Journal of Wildlife Management*, LXVI/3 (2002), pp. 684, 694.

146 Kay Haw, 'A New Hope for Grey Squirrel Management – Fertility Control Research Update', *Pest Control News* (2019), www.pestcontrolnews.com. The National Wildlife Management Centre of the UK government's Animal and Plant Health Agency (APHA) recently completed a Defra/UKSA funded research programme (2018–22) on oral immunocontraception and species-specific delivery methods: Flavie Vial, 'Plight of the Red Squirrel', APHA science blog (9 October 2018), https://aphascience.blog.gov.uk. On 11 July 2022 a breakthrough was widely reported: the successful trial of a grey-specific feeder dispensing contraceptive-laced hazelnut paste, ready for deployment in the wild in a couple of years: Helena Horton, 'Oral Contraceptives Could Help Reduce Grey Squirrel Numbers, Research Finds', *The Guardian*, www.theguardian.com/environment, 11 July 2022; 'Grey Squirrels Go on the Pill to Save Their Red Cousins', *The Telegraph*, www.telegraph.co.uk/news, 11 July 2022.

147 Gus McFarlane, Bruce Whitelaw and S. Lillico, 'CRISPA-Based Gene Drives for Pest Control', *Trends Biotechnology*, XXXVI/2 (2018), pp. 130–33; Jonathan Leake, 'First They Cloned Dolly the Sheep. Now They're Targeting Grey Squirrels', *Sunday Times* (5 January 2020); 'ESI Update' and 'DIGB: Directed Inheritance of Gender Bias', *Squirrel*, 38 (December 2020), pp. 2, 5–7.

148 Jessica Holm, 'The Real Squirrel Nutkin', *BBC Wildlife*, III/7 (July 1985), pp. 337, 340.

149 Paul Lewis, 'Putting the "Tough" Back into Red Tufties', *The Observer* (19 September 1993).

150 'And in the Red Corner, Super Squirrel', *Daily Mail* (8 May 1999);

Jaya Narain, 'Super Squirrel on Red Alert', *Daily Mail* (28 June 2000).

151 James Fair, 'Formby Reds in Crisis', BBC *Wildlife*, XXVI/9 (Summer 2008), p. 39; 'Liverpool Research to Save Red Squirrels', https://news.liverpool.ac.uk, 1 October 2009.

152 Dale, 'Transmission Dynamics and Pathogenesis', pp. iii, 86, 220.

153 James Gillespie, 'Red "Super Squirrel" Clark Kent Beats the Virus', *The Times* (3 November 2013); Steve Graves, 'Watch: Red Squirrels on Merseyside Inspiring Hope Species Can Beat Deadly Virus', *Liverpool Echo* (19 June 2014).

154 Craig Shuttleworth, 'Grey Squirrels Are Bad for the British Countryside, Full Stop', *Squirrel*, 34 (April 2017), p. 7.

155 Charlotte Craw, 'The Ecology of Emblem Eating: Environmentalism, Nationalism and the Framing of Kangaroo Consumption', *Media International Australia*, CXXVII/1 (2008), pp. 82–95.

156 'Food Prices Are Up 24pc Since General Election', *The Guardian* (16 February 1973); 'Food Prices [HC]', *Hansard*, 850 (15 February 1973), c.1420.

157 Jillian Robertson, 'Squirrels Baroque', *The Observer* (25 November 1979); Joanna Coles and Helen Nowicka, 'Lord in a Stew Over Squirrels', *The Guardian* (2 June 1995).

158 Alphonso Van Marsh, 'Squirrel Wars in the UK', CNN *London*, www.youtube.com/watch, 3 September 2008. The sale of grey squirrel meat attracted criticism from groups such as Vegetarians' International Voice for Animals (Viva!): 'Anger over Squirrel Meat on Sale in North London', BBC *News*, www.bbc.co.uk, 29 July 2010.

159 *Nutkin's Last Stand* (PBS) was directed and produced by Nicholas Berger, a graduate student at Stanford University, http://archive.pov.org/nutkinslaststand. See also 'Help Save Tufty . . . Eat a Grey Squirrel', *Not Delia*, www.notdelia.co.uk, 26 July 2008.

160 'Squirrels [HL]', *Hansard* (23 March 2006), cc.369, 371.

161 Ibid., c. 371; Max, 'Squirrel Wars'.

162 'Squirrels [HL]', *Hansard*, cc.379, 384.

163 'Red Squirrel Population [HC]', *Hansard*, 448 (28 June 2006), c.116; Spieler, 'Saving a Squirrel'. For squirrel recipes, see Vester Presley and Nancy Rooks, *The Presley Family Cookbook* (Memphis, TN, 1980) and Donna Presley et al., *The Presley Family and Friends Cookbook* (Nashville, TN, 1998). See also BBC (*Arena*), 'The Burger and the King', www.bbc.co.uk, 1 January 1996.

164 H. N. Southern, 'The Importance of Birds and Beasts of Prey', *Country Life*, CXII/2912 (7 November 1952), p. 1469.

165 'Polecats and Pine Martens', *Country Life*, CXII/2915 (28 November 1952), p. 1718.

166 'Forestry Commission Joins in War on Grey Squirrel', *Manchester Guardian* (8 January 1954). The pine marten's 'special taste for squirrels' is demonstrated in Felix Salten's *Perri: The Youth of a Squirrel* (London, 1938). The story begins with Perri's mother crash-landing in the lap of a three-year-old girl (Annerle) as she evades a marten that has already taken two of her kits. The threat of marten predation continues to lurk. 'Squirrels taste wonderful', pronounces the marten that Perri considers her 'worst enemy': *Perri*, pp. 13–16, 182, 202.

167 MAF, 'Report on Introducing the Pine Martens into New Areas to Control the Grey Squirrel' (1952), MAF 131/82.
168 MAF/Forestry Commission, 'Squirrel Film' (1959), transcript, p. 1, MAF 131/174.
169 Emma Sheehy and Colin Lawton, 'Population Crash in an Invasive Species Following the Recovery of a Native Predator: The Case of the American Grey Squirrel and the European Pine Marten in Ireland', *Biodiversity and Conservation*, XXIII/4 (2014), pp. 753–74; George Monbiot, 'How to Eradicate Grey Squirrels Without Firing a Shot', *The Guardian*, www.theguardian.com, 30 January 2015; Emma Sheehy et al., 'The Enemy of My Enemy Is My Friend: Native Pine Marten Recovery Reverses the Decline of the Red Squirrel by Supressing Grey Squirrel Populations', *Proceedings of the Royal Society B*, CCLXXXV/1874 (March 2018), pp. 1–9.
170 Sheehy, 'Enemy of My Enemy', pp. 6–8. British Red Squirrel, 'Pine Martens Confirmed as Key to Reversing Grey Squirrel Invasion' (2022), www.britishredsquirrel.org/grey-squirrels/pine-martin.
171 Joshua P. Twining et al., 'Native and Invasive Squirrels Show Different Behavioural Responses to Scent of a Shared Native Predator', *Royal Society Open Science*, 7 (2020), pp. 1–9; Joshua P. Twining, 'Grey Squirrels Are Oblivious to Threat from Pine Martens – Giving Native Reds the Advantage', *The Conversation*, https://theconversation.com, 26 February 2020.
172 'A Plea For the Squirrel', *Country Life*, XXXI/798 (20 April 1912), p. 562.
173 Mike Dunn et al., 'Public Attitudes towards "Pest" Management: Perceptions on Squirrel Management Strategies in the UK', *Biological Conservation*, 222 (2018), p. 57.
174 Joshua Twining et al., 'Declining Invasive Grey Squirrel Populations May Persist in Refugia as Native Predator Recovery Reverses Squirrel Species Replacement', *Journal of Applied Ecology*, LVIII (2021), pp. 250, 254, 256–7; Patrick Barkham, 'Pine Martens Dash Hopes of Curbing Grey Squirrels by Avoiding City', *The Guardian*, www.theguardian.com, 15 June 2020.
175 'Foreword', in *Red Squirrels*, ed. Shuttleworth et al., p. 5.
176 William D. Montalbano, 'British Wage War on U.S. "Invaders"; Imported Gray Squirrels, Reviled as "Tree Rats", Are Pushing the Native Red Variety Toward Extinction', *Los Angeles Times* (2 January 1997).
177 On endling, see Robert M. Webster and Bruce Erickson, 'The Last Word?', *Nature*, 380 (4 April 1996), p. 386; Dolly Jørgensen, 'Endling, the Power of the Last in an Extinction-Prone World', *Environmental Philosophy*, XIV/1 (2017), pp. 119–38.
178 Megan Lane, 'The Moment Britain Became an Island', *BBC News*, www.bbc.co.uk, 15 February 2011.
179 Montalbano, 'British Wage War'.
180 Adams, 'They Shoot Squirrels'.
181 Ibid.
182 Ibid.
183 Max, 'Squirrel Wars'.
184 Adams, 'They Shoot Squirrels'.
185 Max, 'Squirrel Wars'.
186 Adams, 'They Shoot Squirrels'.
187 Ibid.
188 Harry Mount, 'Tufty Terminator', *Daily Mail* (13 June 2009).

189 Adams, 'They Shoot Squirrels'.
190 Fred Barbash, 'Redcoats Turn Tail at Nutty American Invasion', *Washington Post* (5 August 1996).
191 http://archive.pov.org/nutkinslaststand, accessed 20 August 2022.
192 Anna Richardson, 'Doreen Sees Red Over Bank's Squirrel Poster' and 'Angry Doreen Tears Down Bank's Grey Squirrel Advert', *Carlisle News and Star*, www.newsandstar.co.uk, 5 April 2006; 'Bank's Sorry to Woman Who Saw Red Over Grey Squirrel Advert', ibid. (2 May 2006); 'Doreen Gets Bank to Change Squirrel Ads', ibid. (12 August 2006).
193 Adams, 'They Shoot Squirrels'.
194 'Nutkin's Last Stand: Behind the Lens', PBS/Points of View video interview with Nicholas Berger, www.youtube.com, uploaded 29 October 2015.
195 Nancy deWolf Smith, *Wall Street Journal*, 14 August 2009.
196 D. W. Yalden, *The History of British Mammals* (London, 1999), p. 204.
197 'Squirrel Wars in the UK' (CNN).
198 John L. Koprowski et al., 'Gray Not Grey: Ecology of *Sciurus carolinensis* in their Native Range in North America', in *Grey Squirrel*, ed. Shuttleworth et al, p. 4; Michael A. Steele and Lucas A. Wauters, 'Diet and Food Hoarding in Eastern Grey Squirrels (*Sciurus carolinensis*): Implications for an Invasive Advantage', in *Grey Squirrel*, ed. Shuttleworth et al, pp. 97–8.
199 Ernie Gordon, *The Adventures of Rusty Redcoat*, vol. 1 (Ilfracombe, Devon, 2007); Matthew Moore, 'Red Squirrel Attacks Children's Writer', *Daily Telegraph* (10 November 2008).
200 'Rusty Redcoat Volume II in Bookstores Now', *Squirrel*, 30 (April 2015), p. 11.
201 Adam Luke, 'Rusty Redcoat Book Released to Highlight Plight of Red Squirrels in the UK', *Chronicle Live*, www.chroniclelive.co.uk, 25 November 2014.
202 Ibid.
203 Royal Central, 'Prince Charles and David Cameron Thank School Pupils for Their Fight for Red Squirrels', https://royalcentral.co.uk, 3 February 2015.
204 Ernie Gordon, 'Red Squirrel Appeal', www.rustyredcoat.co.uk/newsandvideo.html, October 2014.
205 Tim and Pat Cook, *Stumpy: Hero of the Lakes* (Kendal, Cumbria, 2008), pp. 19, 1–2.
206 Ibid., pp. 22, 46.
207 Ibid., pp. 1–2, 6, 20, 13, 53–4, 50.
208 Ibid., p. 19.
209 On 'the importance of being cute', see Hal Herzog, *Some We Love, Some We Hate, Some We Eat: Why It's So Hard to Thing Straight About Animals* (London, 2010), pp. 37–66. American podcaster Josh Clark – based in Atlanta, Georgia, where eastern greys are near-ubiquitous – on becoming visually acquainted with Britain's native squirrel (via Saving Scotland's Red Squirrels's website, https://scottishsquirrels.org.uk/about) pronounced them just about the cutest squirrels you'll ever see: Josh Clark and Charles W. 'Chuck' Bryant, 'Squirrels, Ahoy!', *'Stuff You Should Know' Podcast* (iHeartPodcasts, 17 May 2022).
210 Cook, *Stumpy*, pp. 46, 20, 19, 23, 40.
211 David Macdonald and Dawn Burnham, *The State of Britain's Mammals 2011* (Oxford, 2011), pp. 7, 9; Chris Parsons, 'Red Squirrel Could Be Extinct in Two Decades as Numbers Fall by 50% in Last 50

Years', *Daily Mail* online, www.dailymail.co.uk, 26 September 2011.
212 Toni Bunnell, *Eden Species Report: Top Ten Threatened Animal Species in the UK – Their Distribution and What You Can Do to Conserve Them* (August 2011), pp. 3, 9.
213 Robin Page, 'We Must Save the Red Squirrel: Exclusive. With Our Native Squirrel Under Threat as Never Before, HRH The Prince of Wales Tells Robin Page Why He Refuses to Stand Idly By', *Daily Telegraph* (19 March 2011).
214 Robin Page, *Why the Squirrel Hides Its Nuts* (Barton, Cambridge, 2011).
215 Robin Page, 'Squirrels on the Pill? You Must Be Joking. Put Them in a Casserole!', *Daily Mail* (4 October 2007).
216 Robin Page, 'Red Squirrels Fight for Survival', *The Telegraph* (21 September 2011).
217 Robin Page, 'Squirrel Nutkin and the Scandal of Britain's Vanishing Red Squirrels', https://skylarkwarrior.wordpress.com, 18 July 2019.
218 Ibid.; 'Grey Control Continues in Bid to Release Reds in Cornwall', *Squirrel*, 29 (October 2014), p. 8; 'Prince Orders Grey Squirrel Cull to Save Reds', *Daily Telegraph* (20 October 2014).
219 Robin Page, 'Bring Back the Lead Aspirin!', *Mail on Sunday* (7 February 2010); Page, 'Red Squirrels Fight for Survival'; Louise Gray, 'Eco-Xenophobia on the Rise Warns Conservationist', *The Telegraph* (9 October 2009).
220 'Prince's Talks with Man Who Says Britain's Overcrowded', *Daily Mail* (15 December 2016); Stephen Hurrell, 'Robin Page Sacked by *The Telegraph* Says Country People Are "Britain's Most Endangered Minority"',

CambridgeshireLive, www.cambridge-news.co.uk, 7 August 2016.
221 Janice Atkinson-Small, 'Why It's Time We Went to War with the Grey Squirrel', *Mail Online*, http://atkinsonsmallblog.dailymail.co.uk, 5 October 2012.
222 Ibid.
223 http://britishdemocraticparty.org/grey-squirrels-and-red-squirrels, accessed 10 October 2014.
224 'Why Squirrels Fear Brexit', www.youtube.com, 1 March 2019.
225 Bryony Jewell, '"Prophecy Prepper" Spends Thousands on Food, Army Rations and an AXE as He Fears Riots and Disorder Will Break Out After Brexit', *Daily Mail* online, www.dailymail.co.uk, 31 December 2018.
226 Alison Rowat, 'All Show, No Punch: Time to Pull the Plug on TV Debates?', *Glasgow Herald* (21 November 2019).
227 Joel Golby, 'This Week's Biggest Twitter Controversy? Jo Swinson's Squirrel Problem', *The Guardian* (21 November 2019); Andy Gregory, 'Lib Dem Leader Jo Swinson Forced to Deny Shooting Stones at Squirrels After Spoof Story Goes Viral', *The Independent* (19 November 2019); Helena Horton, 'Jo Swinson Warns of Dangers of Fake News After Story About Her Attacking Squirrels Went Viral', *The Telegraph* (19 November 2019).
228 Hannah Boland and Laurence Dodds, 'Are Twitter and Facebook Really Taking Responsibility for their Role in Influencing Elections?', *The Telegraph* (14 December 2019).
229 Chris Smyth, 'Swinson Forced to Deny Torturing Squirrels', *The Times* (20 November 2019).

230 Michael and Anne Heseltine, *Thenford: The Creation of an English Garden* (London, 2016), p. 317. See also 'We Shot 350 Squirrels – Absolutely Awful Things', *Squirrel*, 33 (November 2016), p. 10.
231 Heseltine, *Thenford*, p. 317.
232 Sean MacConnell, 'New Group Aims to Wipe Out Grey Squirrels', *Irish Times* (27 December 2004).
233 Sandro Bertolino et al., *Eurasian Red Squirrel (Sciurus vulgaris): The IUCN Red List of Threatened Species 2007*, available at www.iucnredlist.org.
234 Cook, *Stumpy*, p. 19.
235 Sandro Bertolino and Piero Genovese, 'Spread and Attempted Eradication of the Grey Squirrel (*Sciurus carolinensis*) in Italy, and Consequences for the Red Squirrel (*Sciurus vulgaris*) in Eurasia', *Biological Conservation*, CIX/3 (2003), p. 352; Anna Lisa Signorile and Julian Evans, 'Damage Caused by the American Grey Squirrel (*Sciurus carolinensis*) to Agricultural Crops, Poplar Plantations and Semi-Natural Woodland in Piedmont, Italy', *Forestry*, LXXX/1 (2007), p. 89.
236 Fabrizio Brignone [trans. Filippo Gautier], 'An American Rodent – the Grey Squirrel Is Now Frightening Hazelnut Growers', *Squirrel*, 25 (October 2012), p. 10.
237 'Predicting Grey Spread', *Forestry and British Timber* (1 March 2005), p. 11.
238 Jerry Langton, 'Alien Invader from North America: Eastern Grey Squirrel at Home in Europe', *Toronto Star* (1 April 2006).
239 Sandro Bertolino et al., 'Native and Alien Squirrels in Italy', *Hystrix: The Italian Journal of Mammalogy*, XI/2 (2000), p. 66.
240 Marcus Hall, 'Creatures that Do Not Belong: Confronting the Exotic Species Problem in the Mediterranean', in *Views from the South: Environmental Stories from the Mediterranean World (19th – 20th Centuries)*, ed. Marco Armiero (Naples, 2006), pp. 177–8; Alice Andreoli, 'La Dura Vita dello Scoiattolo Italiano', *Il Venerdi di Repubblica* (18 June 2004).
241 Sandro Bertolino, 'Introduction of the American Grey Squirrel (*Sciurus carolinensis*) in Europe: A Case Study in Biological Invasion', *Current Science*, XCV/7 (2008), p. 905.
242 'Clearing Up a Grey Area', *Forestry and British Timber* (1 March 2004), p. 10.
243 S. Bertolino et al., 'Predicting the Spread of the American Grey Squirrel (*Sciurus carolinensis*) in Europe: A Call for a Co-Ordinated European Approach', *Biological Conservation*, 141 (2008), pp. 2564–75.
244 Lisa Signorile et al., 'How Population Genetics Can Contribute to Management of Grey Squirrel Invasions', in *Red Squirrels*, ed. Shuttleworth et al., p. 110; Fiona Harvey, 'Red Squirrel Populations Wiped Out in Northern Italy', *The Guardian*, www.theguardian.com, 27 September 2012.
245 Bertolino et al., *Eurasian Red Squirrel. IUCN Red List 2007*; S. Shar et al., *Eurasian Red Squirrel (Sciurus vulgaris): The IUCN Red List of Threatened Species 2016*, www.iucnredlist.org.
246 Montalbano, 'British Wage War'.
247 Peter Lurz, *Red Squirrels: Naturally Scottish* (Battleby, Perthshire, 2010), p. 1.
248 Kelly, 'Brits Rally'.
249 Roseanna Cunningham, 'Foreword', in Lurz, *Red Squirrels*; 'Deer Most

Recognised Scottish Animal, Survey Finds', BBC News, www.bbc.co.uk/news/uk-scotland-highlands-islands, 23 November 2011.
50 'Golden Eagle Voted "Scotland's Favourite" Wild Animal', BBC News, 1 November 2013, www.bbc.co.uk/news/uk-scotland.
51 'Highland Spring Ad', www.youtube.com, 20 March 2012.
52 'Highland Spring in Hot Water Over Grey Squirrel TV Advert', *Deadline*, www.deadlinenews.co.uk, 12 April 2012; Stephen Wilkie, 'Green Group Sees Red Over Advert's Grey Squirrel Star', *Daily Star* (13 April 2012); 'Highland Spring Defends Use of Grey Squirrel in New TV Advert', *Daily Record*, www.dailyrecord.co.uk, 20 April 2012.
53 Scottish Wildlife Trust, 'Give Scotland's Red Squirrels a Helping Hand', https://scottishwildlifetrust.org.uk, 25 September 2018.

7 LEARNING TO LIVE WITH (AND TO LOVE) THE GREY

1 C. Denton Hornby, 'The Grey Squirrel', *Journal of the Ministry of Agriculture*, 49–50 (1942–4), p. 234.
2 'Squirrel Pie', *The Times* (4 December 1947).
3 Cecily M. Rutley, *Greykin, A Squirrel* (London, 1949), pp. 3, 10.
4 Garth Christian, 'Flowers for Jane', *Manchester Guardian* (1 June 1950); J. M. Wilson, 'Suggestions for an Approach to (a) Parents and (b) Schoolmasters with the Object of Enlisting Their Help in a Campaign Against the Robbing of Birds Nests by Children', RSPB, Minutes of Council Meeting, 4 October 1950, Richard Fitter Papers (1937–1976),

Special Collections, Bodleian Library, Oxford University.
5 G. Rowbottom, 'Black Out Grey Squirrels' (LTTE), *Daily Mail* (6 September 1974).
6 A.W.B., 'A Country Diary', *Manchester Guardian* (20 March 1945).
7 'Red Squirrel Population [HC]', *Hansard*, 448 (28 June 2006), cc.104, 108.
8 'Red Squirrels [HL]', *Hansard*, 587 (25 March 1998), c.1323.
9 Ibid., c.1328.
10 'RSPB's Big Garden Birdwatch Results', *Squirrel*, 28 (April 2014), p. 1.
11 'Red Squirrel Population [HC]', *Hansard* (28 June 2006), c.108.
12 'Foreword', in *The Grey Squirrel: Ecology and Management of an Invasive Species in Europe*, ed. Craig M. Shuttleworth et al. (Stoneleigh Park, Warks, 2016), p. viii.
13 Melissa J. Merrick et al., 'Urban Grey Squirrel Ecology, Associated Impacts, and Management Challenges', in *Grey Squirrel*, ed. Shuttleworth et al., p. 66.
14 'Red Squirrel Conservation [HL]', *Hansard*, 591 (23 June 1998), c.114.
15 'House Sparrows [HL]', *Hansard*, 596 (3 February 1999), c.1493.
16 Helena Stroud, 'Red Squirrels Northern England Launched by HRH The Prince of Wales', *Squirrel*, 22 (April 2011), p. 7.
17 R. E. Kenward and J. L. Holm, 'What Future for British Red Squirrels?', *Biological Journal of the Linnean Society*, XXXVIII/1 (1989), p. 88.
18 John Bryant, 'Grey Power' (LTTE), *The Guardian* (17 October 1994).
19 Cole Moreton, 'Red Alert: News Analysis: Saving the Squirrel', *Independent on Sunday* (13 November 2005).

20 Chris Packham, 'Ecological Cleansing', *Wildlife Extra*.com (2008), accessed 10 October 2018. As quoted in Animal Aid, 'Grey Squirrels' (factfile) [undated], p. 2, https://issuu.com/vivaweb5/docs/squirrel.
21 Oscar Rickett, 'Prince Charles's Attack on Grey Squirrels is Nostalgia Posing as Environmentalism', *The Guardian*, www.theguardian.com, 9 May 2014.
22 Richard Fitter, 'Britain's Beasts', *The Observer* (1 March 1964).
23 'Victimising Grey Squirrels', Version 2, December 2009, pp. 1–2, www.grey-squirrel.org.uk. See also 'Nazi Conservationists', email from Angus Macmillan to Sally Hardy (Ponteland [Northumberland] Red Squirrels), March 2008, at 'Professor's Acorn's "We're As Native As You!"', www.grey-squirrel.org.uk/sallyhardy.php.
24 Andrew Tyler, 'Scapegoating the Aliens' [c. 2000], www.animalaid.org.uk, accessed 2 December 2022.
25 'Red Squirrels: Potential Extinction [HC]', *Hansard*, 662 (3 July 2019), cc.524, 541; 'Cumbrian MP Denies Being a "Squirrel Racist" in Commons Debate', *News and Star*, www.newsandstar.co.uk, 4 July 2019; Andrew Gimson and Paul Goodman, 'Interview: Thérèse Coffey', *conservativehome*, 19 March 2021, https://conservativehome.com.
26 Kat Lay, 'Red Squirrel Group: We're Called Racist for Shooting Greys', *The Times* (10 July 2017). For Charles Moore, former editor of *The Spectator*, *Sunday Telegraph* and *Daily Telegraph*, calling those who kill greys racist is another instance of 'the trend to import the discourse of racism into animal affairs'; 'soon it will be "hate speech" to stigmatise animals of foreign descent and oppose their free movement': Moore, 'The Spectator's Notes', *The Spectator* (3 February 2018).
27 Keith MacInnes, 'What Happened to the Bounty on the Tail of the Pesky Grey Squirrel?', *Daily Telegraph* (22 July 2010).
28 A.W.B., 'A Country Diary', *Manchester Guardian* (25 September 1945); William Beach Thomas, 'Squirrels in the Garden', *The Observer* (24 November 1946).
29 Kenneth Grenville Myer, *The Story of Bushy Squirrel and Me* (London, c. 1940), pp. 3, 16.
30 John Allan May, 'Come What May: Mutiny of the (Squirrel Bounty)', *Christian Science Monitor* (17 March 1953).
31 'Grey Squirrel Tails [HC]', *Hansard*, 514 (23 April 1953), c.1383; 'Forestry Commission Joins in War on Grey Squirrel', *Manchester Guardian* (8 January 1954).
32 Richard Mabey, *The Unofficial Countryside* (London, 1973), pp. 76–7.
33 'Hare Coursing Bill [HL]', *Hansard*, 366 (16 December 1975), cc.1392–3.
34 Arthur Middleton to Lord Woolton (Minister for Food), 11 January 1943, MAF 44/45.
35 Mr Lower (Secretary to the Committee, Agricultural Executive Committee, County of Surrey) to MAF (Lytham St Annes, Lancs.), 'Destruction of Grey Squirrels', 4 March 1943, MAF, 44/45.
36 MAF 131/53.
37 Margaret P. Williams, to MAFF Minister, 15 May 1961, MAF 131/90.
38 Dorothy McCardle, 'The Subject Was Squirrels', *Washington Post* (13 December 1970). See also Eugene

Kinkead, *Squirrel Book* (New York, 1980), p. 88.
39 'Wily Americans Win – Annenberg Suffers', *Atlanta Constitution* (13 December 1970).
40 Alexander F. Miller, 'Remedy for Squirrels' (LTTE), *New York Times* (12 December 1970).
41 'Squirrels Win', *Washington Post* (3 December 1970); McCardle, 'Subject Was Squirrels'.
42 'Wily Americans Win'.
43 R.S.R. Fitter, 'The Distribution of the Grey Squirrel in the London Area', in *The London Naturalist for the Year 1938* (London, 1939), pp. 7, 14; Fitter, *London's Natural History* (London, 1945), p. 216.
44 Kenward and Holm, 'What Future?', pp. 84–6.
45 The Mersey Forest, 'The Sefton Coast Woodlands: A 20 Year Woodland Working Plan, 2003–2023, Volume 1 (Overview), Final Version', January 2003, p. 3.
46 Fitter, 'Distribution of Grey', p. 14.
47 Jessica Holm, 'The Real Squirrel Nutkin', *BBC Wildlife*, III/7 (July 1985), pp. 334, 338.
48 Garth Christian, 'The Problem of Britain's Squirrels', *Country Life*, CXXIX/3335 (2 February 1961), pp. 226–7.
49 L. P. Samuels, 'A Country Diary', *The Guardian* (3 November 1970).
50 Kenward and Holm, 'What Future?', p. 86; Holm, 'Real Squirrel Nutkin', p. 339.
51 Kathleen MacKinnon, 'Competition Between Red and Grey Squirrels', *Mammal Review*, VIII/4 (1978), pp. 185–90; Jessica Holm, *Squirrels* (London, 1987), p. 115.
52 M. L. Hale et al., 'Patterns of Genetic Diversity in the Red Squirrel (*Sciurus vulgaris* L.): Footprints of Biogeographic History and Artificial Introductions', *Conservation Genetics*, 5 (2004), pp. 167–79.
53 MacKinnon, 'Competition Between Red and Grey', p. 190.
54 Jan Taylor, *Sciurus: The Story of a Grey Squirrel* (London, 1978), pp. 72, 77, 53, 83; June Southworth, 'The Beauty in Love with Red Squirrels', *Daily Mail* (19 March 1987).
55 Holm, *Squirrels*, p. 102.
56 'Don't Be Beastly to the Squirrel' was a section heading in Auberon Waugh, *Country Topics: Some Essays* (London, 1971), p. 154.
57 Auberon Waugh, 'A Terrible Risk', *Daily Telegraph* (18 January 1992); William Cook, 'Introduction', *Kiss Me, Chudleigh: The World According to Auberon Waugh*, ed. Cook (London, 2010), p. 6.
58 Waugh, *Country Topics*, p. 103.
59 R. E. Tyrrell, 'An Anti-American Poseur', *American Spectator*, XXIII/6 (June 1990), p. 8.
60 Waugh, *Country Topics*, p. 155.
61 Ibid., p. 51.
62 Bradford, 'The Grey Squirrel as a Timber Pest', *The Times* (11 October 1971); Waugh, *Country Topics*, p. 51.
63 P. N. Thomson, 'The Grey Squirrel' (LTTE), *The Times* (21 October 1971).
64 Waugh, 'Conservationist Menace', *The Spectator* (13 April 1985), p. 6; *Kiss Me, Chudleigh*, pp. 158–9; Waugh, 'Saving the Countryside', *Daily Telegraph* (8 May 1994).
65 Charles Moore, 'Diary', *The Spectator* (20 May 1989), p. 7.
66 Waugh, *Country Topics*, p. 51.
67 Bradford, 'The English Landscape', *The Times* (18 November 1972); Waugh, *Country Topics*, pp. 154–5.

68 Waugh, *Country Topics*, pp. 155–6.
69 'Grey Squirrels: Population Control [HL]', *Hansard*, 555 (15 June 1994), c.1691.
70 Woodrow Wyatt, *The Further Exploits of Mr Saucy Squirrel* (London, 1977), pp. 10–12, 25, 92, 94, 41, 70.
71 Ibid., pp. 78–80.
72 'Grey squirrels: Control Studies [HL]', *Hansard*, 569 (22 February 1996), c.1140.
73 Matthew Norman, 'Diary', *The Guardian* (29 April 1997).
74 Petronella Wyatt, 'The Red and the Grey', *The Spectator* (29 December 2001), p. 49.
75 Michael Tod, *The Second Wave* (London, 1995), pp. 56, 59, 60.
76 Jonathan Guthrie, 'Nuts to Red Squirrels', *Financial Times* (24 March 2006).
77 William Montalbano, 'British Wage War on U.S. "Invaders"', *Los Angeles Times* (2 January 1997).
78 William Greaves, 'The Battle of Britain's Beer!', *Daily Mail* (24 November 1979).
79 Gareth Huw Davies, 'Bright Eyed, Bushy Tailed and Brilliant!', *Radio Times*, CCLVIII/3378 (25 August 1988), pp. 12–13. On the problem-solving abilities of a range of squirrel species, see *The Super Squirrels*, a BBC TV (Natural World series) documentary, broadcast on 19 June 2018.
80 Jack Bell, 'Cashing In: The Beeb Makes a Packet Out of TV Ads', *Daily Mirror* (17 February 1990). 'Stars Go Nuts Over Squirrel', *Daily Mirror* (22 February 1990); Barry Baker, 'Cyril the Squirrel Drives Them Nuts at TV Award Show', *Daily Mail* (22 February 1990); 'Star Squirrel', *The Observer* (25 February 1990).
81 Muriel Laskey, *Cyril the Squirrel* (New York, 1946); Ivy O. Eastwick, 'A Squirrel Called Cyril', *Christian Science Monitor* (13 January 1973); 'Cyril the Squirrel Is Nuts About Diana's Flowers', *Northern Echo* (12 December 2001).
82 'Grey Future in Forester's Fight', *Food and Beverage Today* (1 May 2003), p. 17.
83 Polly Ghazi, 'Red Squirrel May Be Gone Within 10 Years', *The Observer* (23 August 1992).
84 Nancy Banks-Smith, *The Guardian* (14 February 1992). See also Lynne Truss, 'Hellraisers in Among the Nuts', *The Times* (6 April 1991). To mark the 2022 World Cup in Qatar, a Hertfordshire man, Steve Barley, built a football-themed obstacle course in his back garden (he had rigged one up previously for the 2020 Tokyo Olympics). A 1-minute-28-second video can be viewed at 'World Cup: Man Builds Squirrel Course in Hitchin Garden', BBC News, 17 November 2022, www.bbc.co.uk/news/av/uk-england-beds-bucks-herts. For a longer video Barley made ('Squirrel Football – SIFA [Squirrel International Football Association] World Cup'), see www.youtube.com.
85 C. M. Hewson and R. J. Fuller, *Impacts of Grey Squirrels on Woodland Birds: An Important Predator of Eggs and Young?* (BTO research report no. 328) (Thetford, Norfolk, June 2003), pp. 3, 17.
86 Anne Wareham, 'Squirrels the Ultimate Pest', *The Spectator* (21 March 2015); Wareham, *Outwitting Squirrels and Other Garden Pests and Nuisances* (London, 2015), pp. 11–20.
87 Claire Ellicott, 'Hauled to Court, Forced to Pay £1,500 and Branded a

Criminal ... For Drowning a Grey Squirrel', *Daily Mail* online, www.dailymail.co.uk/news, 20 July 2010; Quentin Letts, 'Why I Hate Squirrels!', *Daily Mail*, www.dailymail.co.uk, 21 July 2010; Jamie Foster, 'Update on the Lawful Control of Grey Squirrels', *Squirrel*, 24 (April 2012), p. 6. The following month, the chairperson of Morpeth [Northumberland] Red Squirrels, retired military policeman Norris Atthey, drowned a grey to demonstrate that drowning is a humane method that takes just thirty seconds: 'Northumberland Man Held Over "Grey Squirrel Drowning"', *BBC News*, https://www.bbc.co.uk/news/uk-england-tyne, 19 August 2010; Jamie Foster, 'RSPCA Continues to Bring Prosecutions', *Squirrel*, 26 (April 2013), p. 4.

88 RSPB, 'Deterrents for Squirrels', www.rspb.org.uk, accessed 12 August 2022; 'How Chilli and Plastic Keep Squirrels Away', *Daily Telegraph* (22 July 2010).

89 For a rigorous dissection of the dichotomy between native and non-native biota, pivoting on whether 'nativeness' is 'conceptually defensible' – the authors conclude that 'its categorical meaning and significance both dissolve under scrutiny' and that 'biotic nativeness is theoretically weak and internally inconsistent' (p. 36) – see Matthew K. Chew and Andrew L. Hamilton, 'The Rise and Fall of Biotic Nativeness: A Historical Perspective', in *Fifty Years of Invasion Biology: The Legacy of Charles Elton*, ed. David M. Richardson (Chichester, 2011), pp. 35–47.

90 'Victimising Grey Squirrels', pp. 1–2; email from Angus Macmillan to Sally Hardy, 'Not so "Native"', January 2008, www.grey-squirrel.org.uk/sallyhardy.php.

91 Michael Leapman, 'It's Just Nuts', *Daily Mail* (10 November 2005).

92 'A Kiss from the Last Red Squirrel', https://ualresearchonline.arts.ac.uk/id/eprint/15221; 'A Kiss from the Last Grey [sic] Squirrel', *the ashden directory*, www.ashdendirectory.org.uk, accessed 19 August 2022. For discussions of animal citizenship and 'denizenship', viz. permanent resident non-citizenship, especially for 'liminal' (urban wildlife) species such as squirrels and pigeons that occupy a position between the categories of wild animals and domesticated animals, see William A. Edmundson, 'Do Animals Need Citizenship?', *I.CON (International Journal of Constitutional Law)*, 13 (2015), pp. 751, 755; Will Kymlicka and Sue Donaldson, 'Animals and the Frontiers of Citizenship', *Oxford Journal of Legal Studies*, XXXIV/2 (2014), p. 203.

93 Thomas Katheder, 'The Squirrel Wars' (LTTE), *New York Times* (21 October 2007). Katheder exaggerated the imported fox's long-term impact on its native counterpart, whose population is currently stable and its range expanding: Canid Specialist Group/WildCRU, 'Gray Fox, *Urocyon cinereoargenteus*' (2016), www.canids.org.

94 Norine Harris and Bill Ortel, 'The Squirrel Wars', *New York Times* (21 October 2007). Americans were not as tolerant of sparrows and starlings as they implied. See Peter Coates, 'Eastenders Go West: English Sparrows, Immigrants, and

the Nature of Fear', *Journal of American Studies*, XXXIX/3 (2005), pp. 431–62; Coates, *American Perceptions of Immigrant and Invasive Species: Strangers on the Land* (Berkeley, CA, 2006), pp. 28–70.
95 Mark Blazis, 'At Play or in a Stew, Squirrels Are Highly Enjoyable', *Telegram and Gazette* (15 September 2009).
96 On how nineteenth-century ships' rats literally left their mark on maritime documents such as logbooks, see Kaori Nagai, 'Vermin Writing', *Journal for Maritime Research*, XXII/1–2 (2020), pp. 59–74.
97 Piran C. L. White and Stephen Harris, 'Economic and Environmental Costs of Alien Vertebrate Species in Britain', in *Biological Invasions: Economic and Environmental Costs of Alien Plant, Animal, and Microbe Species*, ed. David Pimentel (Boca Raton, FL, 2002), p. 140.
98 A. L. Signorile et al., 'Do Founder Size, Genetic Diversity and Structure Influence Rates of Expansion of North American Grey Squirrels in Europe?', *Diversity and Distributions*, 20 (2014), pp. 918–30.
99 Gail Wilson, 'Protecting Mainland Europe from an Invasion of Grey Squirrels', *Imperial College London News*, 5 June 2014, www.imperial.ac.uk/news; 'Fears Over Grey Squirrels in Europe', *Belfast Telegraph* (5 June 2014). For 'extraordinary friendliness', see Sandro Bertolino et al., 'The Grey Squirrel in Italy: Impacts and Management', in *Red Squirrels: Ecology, Conservation and Management in Europe*, ed. Craig M. Shuttleworth et al. (Stoneleigh Park, Warks, 2015), pp. 163–4.
100 A. L. Signorile et al., 'Mixture or Mosaic? Genetic Patterns in UK Grey Squirrels Support a Human-Mediated "Long-Jump" Invasion Mechanism', *Diversity and Distributions*, 22 (2016), pp. 566–7.
101 Ibid., pp. 570, 573.
102 Hayley Dunning, 'Don't Blame Grey Squirrels', *Imperial College London News*, www.imperial.ac.uk, 26 January 2016.
103 Jason Gilchrist, 'In Defence of the Grey Squirrel, Britain's Most Unpopular Invader', *The Conversation*, 8 March 2017, https://theconversation.com.
104 Esther Rantzen, 'Damn These Tree Rats', *Daily Mail* (15 June 2006); Angus Macmillan, 'A Time to Kill' and G. E. Fitzgerald, 'How to Take on the "Tree Rat" and Win' (letters to the editor), *Daily Mail* (21 June 2006).
105 'Exceptions to Presumption in Section 34(2) [Wildlife and Countryside Act] [HL]', *Hansard*, 619 (23 November 2000), c.1058.
106 Colin Fernandez, 'Aristocrats Aided Grey Squirrel Takeover', *Daily Mail* (27 January 2016).
107 Sarah Knapton, 'March of the Grey Squirrels Is Down to Meddling Victorians', *Daily Telegraph* (27 January 2016).
108 A. L. Signorile et al., 'Using DNA Profiling to Investigate Human-Mediated Translocations of an Invasive Species', *Biological Conservation*, 195 (2016), pp. 98, 100; 'Skye High Squirrel Fear', *Sunday Mail* (Scotland) (9 January 2011).
109 John M. Mercer and V. Louise Roth, 'The Effects of Cenozoic Global Change on Squirrel Phylogeny', *Science*, CCXCIX/5612 (7 March 2003), pp. 1568–9.

110 Monte Basgall, 'Squirrels' Evolutionary Family Tree Reveals Influence of Climate, Geology', *Campus*, 20 February 2003, https://today.duke.edu; Elizabeth Pennisi, 'How Global Change Shaped the Squirrel Family', *Science*, CCXCIX/5610 (21 February 2003), pp. 1165, 1167.

111 Mercer and Roth, 'Effects of Cenozoic Global Change', p. 1571; Basgall, 'Squirrels' Evolutionary Tree'.

112 Tim Radford, 'Red Squirrel Has U.S. Family Tree', *The Guardian* (21 February 2003).

113 Marie Woolf, 'Squirrel "Apartheid" Plan Boosting Red Against Grey', *The Independent* (17 August 2005).

114 'Squirrel Island', www.mockduck.co.uk/work/squirrel-island, accessed 20 July 2022.

115 'Wild Creatures and Wild Plants Protection Bill [HC]', *Hansard*, 884 (24 January 1975), c.2140.

116 Annexe I, 'The Red Squirrel Species Action Plan' (31 July 2003), pp. 9–10.

117 'Red squirrel Population [HC]', *Hansard* (28 June 2006), c.109.

118 Neil Whitman, 'Squirrel Island: An Interview with Astrid Goldsmith', *Skwigly: Online Animation Magazine*, www.skwigly.co.uk, 9 October 2015.

119 Letts, 'Why I Hate Squirrels!'.

120 Keith Laidler, *Squirrels in Britain* (Newton Abbot, Devon, 1980), pp. 155–6.

121 Iktis, 'Rats in Nutkin's Clothing: The Grey Squirrel's Charm and How It May Be Resisted', *The Field*, CCLXI/6757 (21 July 1982), p. 138.

122 Robin Lane Fox, 'Enemy Forces in Your Flowerbeds', *Financial Times* (27/28 January 2001).

123 'Squirrel Pie "A Real Delicacy"', *Daily Telegraph* (26 September 1984).

124 Geoffrey Hollis, 'Talking Point', *Farmers Weekly* (26 August–1 September 2005), p. 76.

125 F. E. Williams et al., *The Economic Cost of Invasive Non-Native Species on Great Britain: Headline Figures* (Wallingford, Oxon, November 2010), p. 2.

126 Pat Morris, 'How To . . . Tell the Difference: Squirrels', *BBC Wildlife*, XXVI/10 (September 2008), p. 72.

127 Earl of Northesk, 'The Nature Conservancy [HL]', *Hansard*, 237 (28 February 1962), c.1003.

128 H. W. Shepheard-Walwyn, *The Spirit of the Wild* (London, 1924), pp. 4, 5–6.

129 W. Beach Thomas, 'Blackbird Versus Squirrel', *The Spectator* (18 July 1931), p. 82.

130 Fitter, 'Distribution of Grey Squirrel in London Area', p. 6.

131 B. T., 'The Open Air: Winter Nests', *The Observer* (20 December 1931); Peter W. W. Lurz, John Gurnell and Louise Magris, 'Sciurus Vulgaris', *Mammalian Species*, 769 (15 July 2005), p. 2.

132 C. Day Lewis, 'Grey Squirrel: Greenwich Park', *Punch*, CCXLIII/6377 (28 November 1962), p. 781.

133 Monica Shorten, *Squirrels* (London, 1954), p. 40; *The Handbook of British Mammals*, ed. H. N. Southern (Oxford, 1964), p. 272.

134 B. T., 'Open Air'; A.W.B., 'A Country Diary', *Manchester Guardian* (4 July 1933); E. M. Fordham, 'Cross-Bred Squirrels' (LTTE), *The Field*, CCLXV/6880 (1 December 1984), p. 72; D. A. Orton, 'Cross-Bred Squirrels' (LTTE), *The Field*, CCLXV/6884 (29 December 1984), p. 50. These letters specified dark grey squirrels with red heads and greys with a 'chestnut-red line down the back'.

135 Stefan Bosch and Peter W. W. Lurz, *The Eurasian Red Squirrel, Sciurus vulgaris* (Hohenswarleben, Germany, 2012), p. 137.
136 Dagny Krauze-Gryz and Jakub Gryz, 'A Review of the Diet of the Red Squirrel (*Sciurus vulgaris*) in Different Types of Habitats', in *Red Squirrels*, ed. Shuttleworth et al., p. 46; Shuttleworth, 'The Foraging Behaviour and Diet of Red Squirrels *Sciurus vulgaris* Receiving Supplemental Feeding', *Wildlife Biology*, 6 (2000), pp. 149–56.
137 National Trust, 'Help Us Protect the Red Squirrels at Formby', www.nationaltrust.org.uk, accessed 21 August 2022. That Formby's reds (a highly isolated inbred population) are less efficient than other red populations in Britain at incisor gnawing might have something to do with a long period of supplementary feeding with peanuts, which are relatively soft compared to hazelnuts and pine cones. See Philip G. Cox, Philip J. R. Morris and Andrew C. Kitchener, 'Population Fragmentation Leads to Morpho-Functional Variation in British Red Squirrels (*Sciurus vulgaris*)', 30 March 2019, nwww.biorxiv.org/content/10.1101/593319v1.full.pdf
138 Bosch and Lurz, *Eurasian Red*, p. 98; *Handbook of British Mammals*, ed. Southern, p. 273.
139 *Handbook of British Mammals*, ed. Southern, p. 273; Bosch and Lurz, *Eurasian Red*, pp. 45, 98.
140 Bosch and Lurz, *Eurasian Red*, p. 99; W. J. Stokoe, *The Observer's Book of Wild Animals of the British Isles* (London, 1958 [1938]), p. 155; 'Squirrel's Epic Swim Across Lake [Ullswater]', *BBC News*, http://news.bbc.co.uk, 15 November 2007.
141 Ted Hughes and R. J. Lloyd, *The Cat and the Cuckoo* (Bideford, Devon, 1987), p. 5.

Select Bibliography

Barratt, E. M., and J. Gurnell, G. Malarky, R. Deaville and M. W. Bruford, 'Genetic Structure of Fragmented Population of Red Squirrel (*Sciurus vulgaris*) in the UK', *Molecular Ecology*, 8 (1999), pp. 55–65
Benson, Etienne, 'The Urbanization of the Eastern Gray Squirrel in the United States', *Journal of American History*, C/3 (December 2013), pp. 691–710
Bosch, Stefan, and Peter W. W. Lurz, *The Eurasian Red Squirrel, Sciurus vulgaris* (Hohenwarsleben, 2012)
Coates, Peter, *Red and Grey: Toward a Natural History of British Anti-Americanism* (London, 2013)
——, 'A Tale of Two Squirrels: A British Case Study of the Sociocultural Dimensions of Debates over Invasive Species', in *Invasive Species in a Globalized World: Ecological, Social, and Legal Perspectives on Policy*, ed. Reuben P. Keller, Marc W. Cadotte and Glenn Sandiford (Chicago, IL, 2015), pp. 44–71
Crowley, Sarah L., Steve Hinchliffe and Robbie A. McDonald, 'Killing Squirrels: Exploring Motivations and Practices of Lethal Wildlife Management', *Environment and Planning E: Nature and Space*, I/1–2 (2018), pp. 120–43
Dunn, Mike, Mariella Marzano, Jack Foster and Robin M. A. Gill, 'Public Attitudes towards "Pest" Management: Perceptions on Squirrel Management Strategies in the UK', *Biological Conservation*, 222 (2018), pp. 52–63
Gurnell, John, *The Natural History of Squirrels* (London, 1987)
Hale, Marie L., and Peter W. W. Lurz, 'Morphological Changes in a British Mammal as a Result of Introductions and Changes in Landscape Management: The Red Squirrel (*Sciurus vulgaris*)', *Journal of Zoology*, CCVX/2 (2003), pp. 159–67
——, —— and K. Wolff, 'Patterns of Genetic Diversity in the Red Squirrel (*Sciurus vulgaris* L.): Footprints of Biogeographic History and Artificial Introductions', *Conservation Genetics*, 5 (2004), pp. 167–79

Harvie-Brown, J. A., 'The History of the Squirrel in Great Britain', *Proceedings of the Royal Physical Society of Edinburgh*, 5 (1880–81), pp. 343–48; ibid., 6 (1880–81), pp. 31–63; 115–83
—, *The History of the Squirrel in Great Britain* (Edinburgh, 1881)
Holm, Jessica, *Squirrels* (London, 1987)
Holmes, Matthew, 'The Perfect Pest: Natural History and the Red Squirrel in Nineteenth-Century Scotland', *Archives of Natural History*, XLII/1 (2015), pp. 113–25
Kean, Hilda, 'Save "Our" Red Squirrel: Kill the American Tree Rat: An Exploration of the Role of the Red and Grey Squirrel in Constructing Ideas of Englishness', in *Seeing History: Public History in Britain Now*, ed. Hilda Kean, Paul Martin and Sally Morgan (London, 2000), pp. 51–64
Kenward, R. E., and J. L. Holm, 'What Future for British Red Squirrels?', *Biological Journal of the Linnean Society*, XXXVIII/1 (1989), pp. 83–9
Laidler, Keith, *Squirrels in Britain* (Newton Abbot, Devon, 1980)
Lloyd, H. G., 'Past and Present Distribution of Red and Grey Squirrels', *Mammal Review*, 13 (1983), pp. 69–80
Lowe, V.P.W., and A. S. Gardiner, 'Is the British Squirrel (*Sciurus vulgaris leucourus* Kerr) British?', *Mammal Review*, 13 (1983), pp. 57–67; 183–95
Lurz, P.W.W., 'Changing "Red to Grey": Alien Species Introductions to Britain and the Displacement and Loss of Native Wildlife from our Landscapes', in *Displaced Heritage: Responses to Disaster, Trauma and Loss*, ed. Ian Convery, Gerard Corsane and Peter Davis (Woodbridge, Suffolk, 2014), pp. 265–72
MacKinnon, Kathleen, 'Competition Between Red and Grey Squirrels', *Mammal Review*, VIII/4 (1978), pp. 185–90
Middleton, A. D., 'The Ecology of the American Grey Squirrel (*Sciurus carolinensis* Gmelin) in the British Isles', *Proceedings of the Zoological Society of London*, C/3 (1930), pp. 809–43
—, *The Grey Squirrel: The Introduction and Spread of the American Grey Squirrel in the British Isles* (London, 1931)
Sheail, John, 'The Grey Squirrel (*Sciurus carolinensis*) – A UK Historical Perspective on a Vertebrate Pest Species', *Journal of Environmental Management*, 55 (1999), pp. 145–56
Shorten, Monica, 'A Survey of the Distribution of the American Grey Squirrel (*Sciurus carolinensis*) and the British Red Squirrel (*S. Vulgaris leucourus*) in England and Wales in 1944–5', *Journal of Animal Ecology*, XV/1 (1946), pp. 82–92
—, *Squirrels* (London, 1954)
—, 'Introduced Menace: American Grey Squirrel Poses Threat to British Woodlands', *Natural History*, LXXIII/10 (December 1964), pp. 42–9
Shuttleworth, Craig M., Peter Lurz and Matthew W. Hayward, eds, *Red Squirrels: Ecology, Conservation and Management in Europe* (Stoneleigh Park, Warks, 2015)

Select Bibliography

Shuttleworth, Craig M., Peter W. W. Lurz and John Gurnell, eds, *The Grey Squirrel: Ecology and Management of an Invasive Species in Europe* (Stoneleigh Park, Warks, 2016)

Sidorowicz, Jerzy, 'Problems of Subspecific Taxonomy of Squirrel (*Sciurus vulgaris L.*) in Palaearctic', *Zoologischer Anzeiger*, CLXXXVII/3–4 (1971), pp. 123–42

Signorile, A. L., et al., 'Do Founder Size, Genetic Diversity and Structure Influence Rates of Expansion of North American Grey Squirrels in Europe?', *Diversity and Distributions*, 20 (2014), pp. 918–30

Signorile, A. L., et al., 'Mixture or Mosaic? Genetic Patterns in UK Grey Squirrels Support a Human-Mediated "Long-Jump" Invasion Mechanism', *Diversity and Distributions*, 22 (2016), pp. 566–77

Swart, Sandra, 'The Other Citizens: Nationalism and Animals', in *The Routledge Companion to Animal–Human History,* ed. Hilda Kean and Philip Howell (London, 2018), pp. 31–52

Watt, Hugh Boyd, 'On the American Grey Squirrel (*Sciurus carolinensis*) in the British Isles', *Essex Naturalist*, 20 (1923), pp. 189–204

Acknowledgements

Over the past few months, I have been able to check references in *Country Life* and *The Field* from the comfort of home, by consulting the magazines' digitized versions. This underscores just how long it has taken me to produce *Squirrel Nation*. Ten or more years ago, when the exploratory reading began, I was taking the train to Reading regularly, walking across town from the station and then pulling tremendously heavy bound copies off high shelves in frigid storage rooms at the library of the Museum of English Rural Life (MERL). So my first debt is to the University of Bristol's Arts Faculty's research fund, for reimbursing all those trips on GWR. At least one period of study leave from Bristol's School of Humanities helped advance the project (though I think I was supposed to be writing a book that included squirrels, but was not just about them). I also wish to thank the following: David Gerrard, for supplying a photo of a red squirrel crossing sign from west Cumbria, for a reference and for reminding me that the football team we supported when we were boys, Formby FC, was nicknamed 'The Squirrels' (red ones, obviously); Marianna Dudley, for commissioning Sarah March to make an exquisite pointillist drawing of a squirrel as a retirement gift from my department; Joanne Lello, erstwhile squirrel researcher, for clarifying her research on drug therapy for greys; Craig Shuttleworth for responding to questions; David Bickerton of the Lancashire & Cheshire Fauna Society for helping to sort out an ambiguous reference; Caroline Walter of Special Collections at Exeter University's Old Library for contacting the R. J. Lloyd Estate on my behalf; and Louise Lloyd for waiving a reproduction fee. For squirrel talk and snippets of information, I thank Richard Harrad and Harry Marshall (and the gift of a first edition of *The Tale of Timmy Tiptoes* was much appreciated, Harry); also Harriet Evans and Harriet's friend, Erik Mueggler, who, in Michigan, sees American reds 'chasing grey squirrels all around my yard and trees all the time', 'just had to pay someone to remove the red squirrels from inside the walls of my house' and has a 'permanent feud going with a gray squirrel named Matt . . . who manages to find a way to raid my bird feeder no matter how many anti-squirrel protections I put on it.' Jemima

Acknowledgements

Elliott kept me abreast of grey squirrel movements in her neck of the woods in the Scottish borders. I would also like to mention Ian Cullimore, not for squirrel chats, though we had plenty of those. Thanks for being my swim comrade on those monthly dips. I always returned to my desk from Clevedon with a clearer head and renewed appetite for squirrels. A special word of thanks goes to the badgering Dave Watkins (formerly of Reaktion Books), who convinced me that it was my destiny to write a book about squirrels. At Reaktion, I thank Michael Leaman for his encouragement and comments; Alex Ciobanu, picture editor, not least for coming up with additional suggestions for suitable images; and Amy Salter for her astute and incisive editing. I've discussed squirrels, on and off, with Marcus Hall, since 2003, and I thank him for the pdf version of an essay of his in a publication that was hard to locate. My younger daughter, Ivana, not only consistently showed a well above average level of interest in squirrels, but specifically alerted me to an American podcast on squirrels ('Squirrels, Ahoy!') in the *Stuff You Should Know* series. She and her big sister kept me supplied with squirrel-themed cards and gifts, among them a Squirrel Whisperer sweatshirt and a wood carving incorporating a bowl for nuts. Graziella deserves my gratitude for not violating the sanctity of what became The Squirrel Room too many times a day on days she was also working from home. (My family thinks I am obsessed with squirrels, but things could be much worse. I have not yet joined The Squirrel Lover's Club, an international organization founded in 1995 and based in Lower Burrell, Pennsylvania; annual membership dues for those outside the United States, by the way, are $31.) Alice Would let me know that hard copies of BBC *Wildlife* magazine were available in Bristol University's Medical Library (of all places). Julia Adeney Thomas brought an American squirrel recipe to my attention. Eric Mendelsohn confirmed that he was indeed the Eric Mendelsohn that wrote a letter to the *New York Times*. Anita Plumb, the Royal Society for the Prevention of Accidents' long-serving information enquiries officer (recently retired), helped enormously, not least by arranging the visit of a grey squirrel to a bird feeder outside the window of the lower ground floor Infocentre at ROSPA headquarters in Edgbaston, Birmingham, while I was sorting through materials about Tufty Fluffytail. And I thank ROSPA's Gavin McNab for granting permission to reproduce images from the Tufty archive, as well as Joe Tinkler, for help with the scanner. I also thank staff at The National Archives, MERL, Cambridge University Library, Oxford's Bodleian and the Zoological Society of London, as well as the inter-library loans staff at Bristol University's Arts and Social Sciences Library.

I was going to dedicate this book to Otto, who still lives in hope of one day catching one. The squirrels of Telegraph Hill Park, Brockley, emboldened by the lower park's dog-free status, drove Ivana's and Harry's young German shorthaired pointer crazy one memorable autumn Sunday afternoon. But then Giuliana's and Doug's Mila arrived at the end of June 2022, when I was applying the finishing touches. (Sorry, Otto.) As she was slung sleeping over my shoulder, making peculiar noises, I checked for typos and managed to type a few words, even sentences, one-fingered. Mila the Squirrel Girl, I'm trying not to call her – Squirrel Girl being the superhuman, crime-fighting protagonist of *The Unbeatable Squirrel Girl*, a Marvel Comics series since January 2015. Squirrel Girl (aka Doreen Green) has been around as a character since 1991 and enjoys various squirrel-like powers, among them extraordinary agility and gripping and climbing abilities. She can communicate perfectly with actual squirrels, which she mobilizes as allies. Squirrel Girl is recognizably red. But her squirrel-sized, acorn smoothie-loving sidekick (since 2005), Tippy-Toe, is recognizably grey (shades of Potter's *Tale of Timmy Tiptoes*, though Tippy-Toe is female). Giuliana, Ivana and their teenage friends used to call me 'Salmon Man'. I suspect that Mila, to whom I dedicate this book, will soon be calling me 'Squirrel Man'.

Photo Acknowledgements

The author and publishers wish to express their thanks to the sources listed below for illustrative material and/or permission to reproduce it:

From C. E. Bowen, *Frisky the Squirrel* (New York, 1889), photo Library of Congress, Washington, DC: p. 74; photos Peter Coates: pp. 12, 13, 135 (*bottom*), 169, 250; collection of the author: p. 129; Pixabay: pp. 131 (*bottom*; Elliott Day), 239 (Sara J. Price); Flickr: pp. 6 (photo Peter Allen, CC BY-SA 2.0), 131 (*top*; photo Peter Trimming, CC BY 2.0), 133 (photo Peter Trimming, CC BY 2.0); Geograph Britain and Ireland: p. 253 (photo Richard Sutcliffe, CC BY-SA 2.0); photo David Gerrard: p. 185; from Oliver Goldsmith, *A History of the Earth and Animated Nature* (London, 1862): p. 72; iStock.com: p. 52 (Yarkovoy); The J. Paul Getty Museum, Los Angeles: p. 43; Library of Congress, Prints and Photographs Division, Washington, DC: pp. 94, 96, 97, 167; © R. J. Lloyd, reproduced courtesy of the Lloyd Estate, photo courtesy Special Collections, University of Exeter (EUL MS 450): p. 136; The National Archives, Kew (MAF 217/30): p. 134; National Gallery of Art, Washington, DC: p. 132; The New York Public Library: p. 130; Royal Society for the Prevention of Accidents (ROSPA), Birmingham (Tufty Club Records): pp. 135 (*top*), 180, 181, 182, 183; from A. Thorburn, *British Mammals*, vol. 1 (London and New York, 1920), photo Field Museum of Natural History Library, Chicago: p. 107; from Adam White, *The Instructive Picture Book: Or, a Few Attractive Lessons from the Natural History of Animals*, 3rd edn (Edinburgh, 1859), photo Archives of American Art, Smithsonian Institution, Washington, DC: p. 48.

Index

Page numbers in *italics* refer to illustrations

Abbey (National Savings Accounts) 211–12
acorns 41, 64, 65, 75, 100, 143, 162, 163, 165, 194, 214, 248, 253
Adams, David 126
Adams, Tim 208, 209, 212
Addison, Christopher 122, 123
advertisements 190, 211–12, 222, 238–40
Advocates for Animals 44, 198
agency, non-human 243–4
Agnew, Stuart 217
Agriculture (Miscellaneous Provisions) Bill (1971) 171
Ahmed, Sara 30
Alexander, Danny 22–3
Aliperti, Jaclyn 10
Alverstone Mead Nature Reserve, Isle of Wight *131*
American red squirrel 68, 73, 75, 81–2, 118, 147, 159, 206, 247, 320
analogies involving squirrels
  greys 34, 37–8, 87, 103–4, 127, 138–9, 144–5, 186–7, 192, 210, 217, 225, 228, 235, 237, 238
  reds 60, 237, 238
Anderson, Mark 67
Andrews, D. H. 168
Anglesey 161, 198–9, 216, 228
Animal Aid 45, 63, 227, 228
animal history 29–30

Animal Welfare Act (2006) 241
Anne of Brittany 57–8
Annenberg, Leonore and Walter 231
Apsley, Lord (Cirencester Park) 204
Archers, The 160
Athol, Duke of (Blair Castle) 42, 50
Atkins, Peter 91
Atkins, Robert 189
Atkinson, Peter 25, 188–9
Atkinson-Small, Janice 35, 217
Attenborough, David 105–6
attitudes (British), to Americans and America 13, 14, 31–7, 120, 143–4, 145–6, 234, 242, 243, 246
Audubon, John James, Eastern grey squirrel' (lithograph) *130*
Australia 28, 67, 83, 158, 204, 213, 247
Aylestone, Lord 230

Bach, Lord 205
Bailey, Julie 178–9
Bailey, Vernon 105
bark stripping (barking) 30, 64, 65, 66, 68, 119, 225
  damage or threat to trees 11, 18, 41, 65, 139, 237
Barkham, Patrick 18–19, 178
Barne, Miles 33, 221

Index

Battle of Britain (documentary film) 213
Beaumont, Lord 225
Bedford, Duke of 88–9, 96, 99, 101, 121, 197, 246–7
Beebee the squirrel 85
beer (ale) *12*, 29, 238
Beith, Alan 225, 226
Benson, Etienne 92
Berger paint can *13*
Berger, Nicholas, *Nutkin's Last Stand* 211–14
Bern Convention (1982) 220
Bertram, Brian 197
Bertram, Julius 106
Betjeman, John 178
Bevan, Emlyn 190
Bevins, Reginald 186–7
Biodiversity Action Plan (BAP) 176
Birch Grove, Sussex 164–5, 170
birds
    bird feeder (table) 9, 36, 209, 238, *239*, 241, 320
    squirrels as threat to 32, 49, 63, 99, 101, 104, 115–16, 119, 140, 225, 235, 237, 240
    bird sanctuaries 111, 122–3
Birkhall (Balmoral) 177, 216
Bland, Richenda 195–6
Blazis, Mark 243
Board of Agriculture and Fisheries 109, 207
'bob a brush' *see* bounty payments
bounty payments
    Australia 158
    'bob a brush' *134* (poster), 158–65, 180, 206, 230
    greys 65, 67, 150, 166, 169, 170, 186, 231, 235
    North American colonies 82–3, 160
    reds 15, 67–8, 125, 164
Bowen, Charlotte Elizabeth, *Frisky the Squirrel* 73, 74

Boyd, Arnold W. 139, 147–8, 225
Bradby, Tom 218
Bradford, Lord (Weston Park) 234–5
Brexit 29, 31, 218–19
Bridgen, Andrew 24
Brightwell, Leonard Robert 123, 145–6
British Democratic Party (British Democrats) 217
British Museum of Natural History 14, 54, 56, 71, 73, 105, 115, 245
British National Party (BNP) 217
British Trust for Ornithology (BTO) 240
British Wildlife Centre 216
Britishness 28, 29, 40, 53–6, 204, 212, 241–2, 243
    see also Englishness
Brocklehurst, Thomas Unett 86, 87, 88, 153, 233, 241
Brocklehurst, William Walter 89
Brownsea Island, Dorset 34, 185, 248
Bryant, John 227
Buccleuch, Duchess of (Dalkeith Park) 50
Buckmaster, Lord 110
Bundy, Pete (Longleat) 204
Bunnell, Toni, *Eden Species Report* 215–16
'Bunny' the squirrel 96
Bureau of Animal Population 69, 118
Burton, Lord 51
Burton, Maurice 56, 106

cages, for squirrels 56, 58, 59–60, 61, 74, 84, 87, 88, 145, 146, 208
Calvert Trust 21
Cameron, David 214
Campbell, W. D. 34
Carling Black Label 238–40
Carnell, Veronica 211

carnivorousness of squirrels 34, 49, 77, 104, 105, 139, 140, 179, 236
castration 14, 118
Central Park, New York City 31, 87, 92, *94*, 96, 98, 100
Champneys, Adelaide Mary 196
Chapman, Sydney 172
Charles, Prince 16, 39, 176, 177, 200, 214, 216–17, 222, 226
Chatterbox, 'Squirrels at Play' *59*
children's books 22, 49, 73, 82, 89, 117, 128, 137–8, 151, 214–15, 216, 224–5, 229–30, 235–6, 240
  D. B., 'The Squirrel' (poem) *62*
  see also Potter, Beatrix
Chorley, Lord 192
Christian, Garth 170, 225
Church, Richard, *A Squirrel Called Rufus* 151
Churchill, Winston 16, 158, 165, 192, 210
Clappison, James 191, 194–5
Clark Kent the squirrel 204
Clark, Lord 54
Clegg, Don 210
Clegg, Nick 20, 22–3, 127
coccidiosis 70
Cohen, Stanley 16, 186, 210
Coffey, Thérèse 182, 228
Collins, Amanda 249
Collins, Margaret 184
colour of squirrels 70–73, 91, 118, 184–5, 228, 229, 239, 247, 250–51, 252, 254
Colvin, Sidney 71
'common' (red) squirrel 41, 47, *48*, 56, 57, 67, 68, 70–71, 73, 86, 91, 100, 101, 105, 251
comparability of red and grey 251–4
competition between red and grey 31, 44, 91, 99, 100–106, 117, 118–23, 128, 141–8, 163, 197, 210, 233, 237, 248
Condry, William 32, 224

conservation, of red squirrels 18, 21, 44, 176, 199, 201, 207, 215–16, 227
  'killing for conservation' 46, 198
  Red List (IUCN) 220–21
contraception *see* reproductive control
control of greys 7, 15, 18, 31, 83, 121–8, 150, 152, 158–65, 172–3, 177, 207, 222, 231
  see also reproductive control
Cook, Tim and Pat, *Stumpy: Hero of the Lakes* 214–15, 216, 220
Cory, Harper 140
Cotgreave, Peter 17, 221–2
Cotterell, Richard 158, 162
Council for the Preservation of Rural England 113, 192
County War Agricultural Executive Committee (WAEC) 152
Courthope, George 128
Coward, Thomas A. 67, 120, 144, 147
Cranbrook, Earl of 173
Crawford, Earl of 41
criminalization of squirrels 15, 69, 93
Cross's Menagerie, Liverpool 88
culling
  greys 8, 18, 24, 25, 46, 111, 112, 123, 194, 198–9, 208–12, 220, 237
  reds 42, 46, 64, 67–8, 125, 139
Cumbria 17, 54–5, 177, 179, 185, 187, 190, 195, 205, 211, 228
Cunningham, Roseanna 222
'Cuthbert' the squirrel 97
Cyril the Squirrel 239–40

Dale, Ian 219
David, Baroness 179
Day-Lewis, Cecil 252
Daylight Robbery! Portrait of the Grey Squirrel as Thief (documentary film) 238–40
Daylight Robbery 2 (documentary film) 240

Index

de-coniferization 191, 200
Deep England 28, 29
Deer, Patrick 35
deforestation 42, 49, 50, 69, 88, 236–7
Defra (Department for Environment, Food and Rural Affairs) 24, 80, 182, 204, 205, 211, 228
demonization, of squirrels 14, 15, 249
Denby, Edwin, feeding 'Pete' *167*
Destructive Imported Animals Act (DIAA, 1932) 22, 124
disease 24, 69–70, 119–20, 124, 143, 193, 197, 214, 215, 217
disease-resistant reds 194, 203–4
DNA 42, 55–6, 244, 245, 247
Dorchester, Lord 158
drey (squirrel nest) 74, 117, 124, 125, 146, 197, 236, 253
Drummond, (Capt.) Humphrey 188
Dugdale, Thomas 230
Dulverton, Lord (Batsford Park) 171, 172
Dutch elm disease 172, 235

East Anglian Red Squirrel Group 216
Eastern grey squirrel in North America 14, 31, 35, 37, 63, 73, 80–84, 86–7, 88, 92, 100, 105, *130*, 160, 166, 247
eating squirrel
  'eat a grey to save a red' 194, 204–5
  greys 9, 15, 81, 150, 152–8, 230
  North America 83
  recipes 153, 154, 155, 157
  reds 56–7, 153–4
  squirrel pie 57, 153, 154, 155, 156, 158, 250
  U.S. military personnel 155
Edlin, Herbert 200

Eisenhower, Dwight D., president 166–8, 231
Elliot, Raymond 241
Elliot, Walter 124
Elton, Charles 118
emotions, and squirrels 7, 12, 13, 17, 23–4, 30, 35, 38, 39–40, 106, 123, 141, 251
endling 208
enemy release hypothesis 85, 147
English Nature 38, 195
Englishness 28, 29, 40, 186, 236
Epping Forest, Essex 53–4, 158, 186
estates (rural) 34, 49, 50, 51, 63, 65, 66, 71, 75, 86, 87, 88, 99, 101, 102, 113, 115, 125, 126, 128, 139, 146, 164, 165, 171, 177, 188, 191, 204, 206, 217, 219, 220, 221, 234, 235
Eurasian (red) squirrel 7, 51, *52*, 54, 57, 68, 70, 81, *132*, 200, 211, 220, 221, 233, 241, 244, 247
  British subspecies 42, 43, 44, 52–6, 73, 222
European Squirrel Initiative (ESI) 176, 203, 220, 242

Farrington, Baroness 191, 225–6
Fauna Preservation Society 41
Feversham, Earl of 126, 127, 128
Filbert the squirrel 60
film 53, 179, 182, 183, 184, 190, 206, 209–14, 249
Finlay, G. B. 186
Finn, Frank 14, 41, 91–2
Finnart (Loch Long) 89
First World War 96, 99, 100, 109, 142, 152, 156
Fitter, Richard (R.S.R.) 64, 123, 128, 200, 232, 252
fluctuations in squirrel populations 69, 120, 162, 173
folk devils 16, 210
food dispensers (hoppers) 194–5, 197

327

foresters 14, 17, 41, 42, 47, 49, 50, 63, 64, 65, 83, 102, 109, 119, 124, 164, 171, 213, 220, 228
forestry 15, 26, 63, 64, 67, 114, 116, 119, 128, 164, 171, 192, 195, 199, 200, 217, 232, 240, 250
Forestry Commission 27, 41, 54, 64, 106, 109, 116, 124, 127, 128, 150, 158–64, 168, 170, 173, 176, 199, 200, 201, 202, 248
Formby, Merseyside 20, 21, 42, 53, 139, 182, 185, 228, 320
  National Trust reserve 6, 8, 10, 20, 21, 203, 232, 252–3
fossils 54, 247–8
Foster, Marcia Lane 180
Franklin, Benjamin and Deborah 84–5
Fraser, Murdo 23
free cartridges 152, 156, 159, 160
frontier of grey advance 141, 145, 147, 175, 190, 192, 193, 218
'furryfolk' 180, *182*
  see also Tufty Fluffytail and Tufty Club

gamekeepers
  and greys 14, 125, 159, 165, 174, 219, 225
  and reds 41, 50, 63, 70, 79, 213, 228
gardeners 32, 86, 95, 113, 231, 240
gardens (private) 9, 11, 17, 22, 26, 27, 31, 32, 36, 63, 86, 98, 100, 101, 102, 104, 105, 113, 117, 127, 138, 161, 172, 191, 216, 219, 226, 229, 230, 235, 237, 238, 241, 247, 249
Gardeners' Question Time 45
Gardiner, Barry 205
Gardiner, Linda 115, 116
Gerrard, Steven 8
Gibson-Watt, David 187
Gibson-Watt, Lord 173
Gilchrist, Jason 246
Godber, Joseph 204
Goebbels, Joseph 157

Golden Break Advertising Awards 239
Goldsmith, Astrid, *Squirrel Island* (animated film) 248–9
Gordon, Ernie, *The Adventures of Rusty Redcoat* 214
Gordon-Cumming, Alexander (Baronet of Altyre) 65
Gordon-Cumming, Constance Frederica 65
Gordon-Stables, Lovat Lionel and 'Master' Brian 98
Gosling, Arthur 159
Green, C. E. 54
Grey, George 51
Grey Squirrels (Prohibition of Importation and Keeping) Order (1937) 22–4, 124–8, 140, 150
Grey Squirrels Order (1943) 152, 157
Grey Squirrels (Warfarin) Order (1973) 150, 172
Gurnell, John 226
Guthrie, Jonathan 237
Guthrie, Thomas Anstey 95

habitat redesign 194, 200–201
Hailsham, Viscount 157
Haldane, Viscount 110
Hale, Lord 26
Hallsworth, Doreen 211–12
Haraway, Donna 46
Hare, John 164–5
Harcourt, Lewis Vernon 106–7
Harman, Harriet 23
Harris, Norine 243
Harris, Stephen 44, 198, 244
Harrison, Trudy 199, 228
Harvie-Brown, John Alexander 49, 50, 51, 57, 58, 63, 66
Hayward, Eva 46
hazelnuts 11, 31, 35, 75, 94, *131*, 140, 143, 190, 197, 208, 220, 238, 253
Heald, Oliver 24, 25
Heidi, the dog 167

Index

Henbury Park, Cheshire 86, 88, 89, 109
Henry VIII 68, 236–7
Herbert, William 56
Heritage Lottery Fund (Scotland) 223
Herzog, Hal, *Some We Love, Some We Hate, Some We Eat* 204
Heseltine, Michael and Anne (Thenford) 219–20
Highland Spring 222–3
Highland Squirrel Club 67–8, 116, 119, 139
Hill, Mark, *Audit of Non-Native Species in England* 38
Hill, Octavia 70–71
Hinton, M.A.C. 73, 90
Hodgkinson, Andrew 228–9
Hoffmann, Hans, 'Red Squirrel' *132*
Hogarth, A. Moore 116
Holbein the Younger, Hans 58
Holden, Margaret, *Near Neighbours* 117
Holm, Jessica 233, 238, 240
home 13–15, 23, 25–6, 28, 29, 30, 33, 63, 72, 84, 113, 119, 128, 177–8, 184, 213–14, 225–6, 231–2, 233
'home beauty' 114
Home Counties 28, 42, 112, 123, 188, 232
'home' squirrel 29, 144
Horace 108
Hornaday, William T. 81–2, 100, 153
Howard, Russell, *The Russell Howard Hour* 218
Howard-Vyse, H. (Stoke Place) 111
Hughes, Ted, 'Squirrel' (artwork) *136*, 254
Humane Education Society 230
Huxley, Julian 149

Inglewood, Lord (Hutton-in-the-Forest) 171, 178, 190, 191, 192, 194, 205, 208, 226

introductions
 greys 16, 19, 33, 34, 83, 84, 86, 87, 88, 98, 100, 116, 121, 126, 227, 244–5
 internal translocations of greys 14, 83, 116, 127, 245, 247
 (re)introductions of reds 27, 42, 47, 49–51, 53–6, 65, 66–7, 90, 194, 195–7, 198–9, 217
invasion 27–8, 33–4, 35, 113, 120, 125, 146, 193, 196, 199, 203, 212, 218, 221, 243, 244
invasive species 9, 24, 28, 30, 38, 46, 118, 146–7, 152, 221, 243, 245
Ireland 42, 43, 44, 49, 83, 89, 188, 206, 208, 220
Italy 44, 52, 55, 68, 83, 220–21, 244–5

Jeffries, Peter 219
Jersey, Lord 234–5
'Jimmy' (squirrel) 97
Johnson, Paul 36, 37
Joint Nature Conservation Committee 210

Katheder, Thomas 242, 243
Kean, Hilda 29
Kearton, Richard, *Squirrel* 43
Keep Britain Tidy (anti-litter campaign) 186–7
Kelway, Phyllis 126–7, 143
Kerr, Robert 53
Kew Gardens, London 68–9, 71, 72, 95–6, 97, 102, 105, 123, 146
Kidd, William 60–61
Kielder Forest, Northumberland 21, 55, 192, 200, 204, 210
Kilbracken, Lord (Killegar estate) 188
Kinkead, Eugene 31–2
Kinnoull, Earl of 176
Knowles, R. E. 87

Laddie Boy (dog) 166
lager 17, 238–40

Lake District 9, 17, 18, 28, 37, 54, 74, 75, 79, 104, 114, 120, *133*, 171, 176, 178, 191, 205, 206, 211, 214
Lambeth Council (safety subcommittee) 184
Lancum, F. Howard 140
Lane Fox, Robin 249–50
Langland, Joseph 17
Lascelles, Gerald 65
Lawton, (Capt.) E. G. 102, 105
Leadenhall Market, London 54, 56, 157, 195
League Against Cruel Sports (LACS) 227
Leapman, Michael 242
Leatherland, Lord 172
Lello, Joanne 194
Letts, Quentin 249
levelling up policies, for reds 193–204
Lindsay, Earl of 237
Livergant, Elyssa, *A Kiss from the Last Red Squirrel* 242
Liverpool, Earl 192
Lloyd, R. J. *136*, 254
Lodge, George Edward 64
London, grey presence 60, 63, 87, 88, 91–9, 103, 107, 111, 115, 117, 122, 128, 138, 139–40, 150, 178, 188, 189, 212
London Natural History Society (Ecological Section) 161–2
London Zoo 58, 72–3, 79, 80, 89, 92, 93, 94, 99, 121, 142, 145, 197
Lonsdale, Earl of 172
Lovat, Lady (Beaufort Castle) 49, 65–6
Lurz, Peter 17, 56, 185, 200
Lynd, Robert Wilson 121
Lyly, John, *Endymion* 58
Lyons, John 35
Lyster, Simon 201

Mabey, Richard 230
McCartney, Paul 20, 21, 22
McCluskey, Lord 173
MacKinnon, Kathleen 233
McLean, Mark 218
Macmillan, Harold 143, 164–6, 168–70
Mackie, Lord 226
Mainland, Leslie G. 94
Maisie (animation) 240
Matthews, Leonard Harrison 186
Mammal Society 7, 53, 226, 227
Martin, Harrison 212
Massingham, Harold John 111–12
May, John Allan 230
Mendelsohn, Eric 87
merchandise, squirrel-related *12*, *13*, *129*, 178–9, 183
Mercer, John 247
Mersey, Viscount 164
Middleton, Arthur Douglas (A. D.) 31–2, 34, 47, 69, 82, 86, 113, 118–20, 125, 142, 152, 155, 232, 245
Middleton, Lord 30
migration of squirrels 51, 98, 106, 112, 123, 145, 161, 162, 208, 221, 241, 245
Miller, Rachel 204
Mills, Elsie 180–181, 182, 186
Mills, Peter 172
Ministry of Agriculture and Fisheries (MAF) 15, 18, 22, 24, 49, 57, 109, 121, 122, 124, 125, 126, 127, 128, 144, 145, 147, 150, 152, 153, 155, 156, 157, 158, 159, 160, 173, 224, 229, 230
Ministry of Agriculture, Fisheries and Food (MAFF) 24, 26, 70, 150, 162, 164, 166, 168, 170, 172, 202, 204, 206, 231
Mitchell, Louise 21
Mitchell, Philippa 211
Mitchell, (Sir) Peter Chalmers 116
modernity, and greys 113–14, 174, 178
Montalbano, William 17, 36–7, 208

Montgomery, (Sir) David 150
Moore, Harry 202
Moore, John Cecil, *The Waters Under the Earth* 173–5
moral panic 16, 210
Moreton, Cole 227
Mowbray, Lord 187–8
Mungo the squirrel 84–5
mural, by Zase and Dekor *135*
Myer, Kenneth Grenville, *The Story of Bushy Squirrel and Me* 229–30

Nance, Susan 30
National Anti-Grey Squirrel Campaign (NAGS) 118, 121–2, 144
National Farmers' Union (NFU) 112, 122
National Provident Institution (NPI) 190, 201, 211
National Provident Life 211
National Red Squirrel Captive Breeding Programme 177
National Trust 10, *12*, 20, 70, 122, *129*, 176, 203
nationalism and nationality 10, 13–14, 18–19, 24, 27, 28–9, 30, 31, 32, 35, 38–9, 40, 46, 73, 91, 103, 141, 151, 172, 179, 186, 219, 222, 231, 242, 243, 251
patriotism 27, 114, 158, 212
Neil, Ken 223
Nelson, Edward W. 105
Neuberger, Richard 167–8
New Forest 28, *59*, 64, 153, 159, 196, 227, 247, 248
Newbould, Frank, *Your Britain: Fight for it Now* (poster) 27, 148
Nicholson, Carri 211
Nicholson, Edward Max 112, 120–21
non-native species 9, 21, 24, 28, 30, 38, 39, 40, 42, 46, 55, 61, 66, 83, 85, 89, 92, 108, 114, 148, 192, 193, 198, 200, 201, 208, 226, 227, 232, 241, 242, 246, 251
Northern Ireland 26, 46, 207, 244
Northumberland 21, 25, 28, 51, 55, 161, 176, 177, 187, 188, 192, 193, 196, 205, 208–9, 211, 214, 225, 244

Office of Works 111–12, 115, 122, 150
Onwurah, Chi 25
Orritt, Sally 203
Ortel, Bill 243

Packham, Chris 227
Page, George Shepherd 87–8
Page, Robin 216–17
Why the Squirrel Hides Its Nuts 216
Paley, John 216
Palmer, Alfred Herbert 71
Panton, Jane Ellen 49
parapoxvirus 70
Parker, Paul 208–13
parks (public) 83, 91, 101, 104, 107, 250, *253*
in London 36, 60, 71, 72, 83, 88, 91–9, 100, 101, 102, 103, 105–7, 110, 111, 112, 115, 121, 122, 123, 125, 128, 138, 139, 140, 143, 145, 150, 159, 161–2, 167, 196–7, 212, 219, 231, 252, 253
see also Regent's Park
Patterdale and District Red Squirrel Group 214
peanuts 93–4, 95, 162, 210, 252
Pearson, Chris 40
Peel, Earl 191
Peers, Debbie 20–21
Penrith and District Red Squirrel Group 178, 228–9
persecution of squirrels 15, 41, 49, 51, 56, 68, 70, 90, 104, 105, 109, 173, 179, 217, 233
Perth and Kinross Red Squirrel Group 222

pests, squirrels as 15, 22, 24, 26, 41, 42, 46, 47, 63, 66, 82, 83, 89, 105, 106, 110, 116, 119, 122, 150, 152, 155, 160, 164, 170, 171, 172, 173, 208, 211, 220, 230, 243–4
'Pete' the squirrel 166, *167*
Peterson, Roger Tory 243
pets, squirrels as 25, 45, 57–61, *62*, 63, 68, 70, 72, 74, 77, 85, 88, 90, 91, 96, *97*, 106, 116, 123, 161, 172, 177, 211, 219, 230
Phelps, A. J. 166, 168
Phillimore, Lord 27
pine marten 8, 51, 67, 90, 104, 119, 125, 152, 194, 205–7
Pitt, Frances 145, 147, 206
Pizzi, Romain 15
Plant a Tree in '73 (National Tree Planting Year) 172
Plumb, Lord 191
Pocock, Constance Innes 99
Pocock, Reginald Innes 89, 92, 104
poems 17, *62*, 72, 117, *136*, 254
Poots, Simon 222–3
Portal, Major M.E.B. 115, 116, 158
postage stamps 178
Potter, Beatrix 18, 28, 33, 39, 54, 61, 63, 74, 76, 78, 79, 80, 114, 120, 179, 182, 188, 192, 199, 207, 211, 214, 249–50
  The Tale of Squirrel Nutkin 15, 21, 33, 67, 75–8, *77*, 80, 177, 180, 192, 214, 239, 249
  The Tale of Timmy Tiptoes 33, *77*, 78, 79–80, 89, 320
Pow, Richard 199
Powell, Enoch 234, 237
Pratten, Albra, *Winkie: The Grey Squirrel* 229
Presley, Elvis 205
Price, William 204
Pringle, Peter 35
'Professor Acorn' 198, 216, 227, 241–2

Protection of Animals Act (1911) 170
Pullar, Polly 15, 16, 75

Quin, Joyce 189

rabbit 9, 18, 45, 57, 61, 67, 78, 147, 152, 153, 154, 155, 156, 163, 180, 201, 222, 227, 251
Radford, Edmund 126
Ramsbotham, Herwald 126
Randy (Ran) the squirrel 36
Rantzen, Esther 246
rat 25, 32, *48*, 99, 142, 146, 147, 152, 153, 157, 172, 204, 251
rat and sparrow clubs 152
Rau, William H., 'Bunny' (photo) *96*
Readman, Paul, *Storied Ground* 28
Red Alert Northern England (RANE) 179, 199
Red Alert North East 21, 176, 177, 188
Red Alert North West 190
Red Squirrel Appreciation Day 177
Red Squirrel Protection Partnership (RSPP) 176, 204, 205, 208, 209–10, 211, 212, 216
Red Squirrel Survival Trust (RSST) 24, 39, 176, 177, 198, 214, 216, 222
Red Squirrel Week 21, 176
Red Squirrels Northern England (RSNE) 176, 177
Red Squirrels United 177
Redesdale, Lord (Rede Valley) 192–3, 204–5, 208–13
red-squirrelians 177, 187, 189, 237
Regent's Park, London 71, 92–5, 96, 97, 98, 99, 100, 102, 104, 106, 110, 112, 121, 123, 145, 161, 162, 196–7, 231, 252–3
reproductive control 16, 194, 201–3
Reynolds, Jason 201
Rickett, Oscar 227

Index

Ridley, Viscount (Blagdon Hall) 188
Ritchie, James 51, 67
road accidents, involving squirrels 8, *185*, 193, 197
Roberts, Harry 138–9
Rolleston, Lady (Watnall Hall) 139–40
Rollit, (Sir) Albert Kaye 100, 101–2
Rooney, Coleen and Wayne 20–21, 42, 186
Rorimer, James 57–8
Ross, Jonathan 239
Ross, Stephen 248
Roth, Louise 247
Rotherham, Ian 217
Rotherwick, Lord 226
Rowallan, Lord (Rowallan Castle) 191
Royal Society for the Prevention of Accidents (ROSPA) 180–84, 186
Royal Society for the Protection of Birds (RSPB) 101, 115, 122, 225, 226, 241, 246
RSPCA 8–9, 139, 241, 248
Ruskin, John 60–61
Rutley, Cecily 224–5

Sachs, Andrew 179
St Katharine's Lodge Hospital, London 96–7, *97*
St Quintin, William Herbert (Scampston Hall) 101
Saltoun, Lady 191–2, 205, 209
sanctification 15, 41, 177–87
Sanders, Marc (The Famous Wild Boar Hotel) 205
Sanderson, Dorothy 209
Save Our Squirrels (SOS) 179, 205, 211
Saving Scotland's Red Squirrels (SSRS) 222, 223
Scaramouch the squirrel 60–61

Scott, Walter (Abbotsford) 50
Second World War 39, 41, 49, 75, 81, 149–52, 164, 174, 230
Shipley, Georgiana 84–5
Scottish Highlands 7, 23, 41–2, 67, 206–7, 218, 232
Scottish Natural Heritage (SNH) 201, 222
Scottish Woodland Owners' Association 173
Scud: The Life Story of a Squirrel (T. C. Bridges) 49, 61, 63, 73, 89–90
'Scribble a Squirrel' 21
Selborne, Earl of 246
Seton, Ernest Thompson 82, 99
  Bannertail: The Story of a Gray Squirrel 82
Sewell, Luke, *Squirrel Wars: Red versus Grey* (documentary film) 209, 210–12
Shepheard-Walwyn, H. W., *The Spirit of the Wild* 143, 251–2
Shorten, Monica 27, 32, 69, 77, 78, 87–8, 155–6, 161, 165, 173, 175, 245
Shrapnel, Norman 7
Shuttleworth, Craig 45, 198, 199, 216
Sid the squirrel 9, 10
Sidorowicz, Jerzy 52
Signorile, Lisa 244–5, 246–7
Simonis, Giuseppe Casimiro 244–5
Skuggy the squirrel 61
Snowdon (Yr Wyddfa) 9, 17
Soames, Christopher 165, 168, 169–70
sociozoologic scale 46
sparrow (house) 67, 116, 120, 138, 142, 146, 152, 243
'Squexit' 218
'squirrel apartheid' 248
Squirrel Girl 322
'squirrel racism' 18, 128, 210, 217, 226–9, 233, 234

squirrel shooting clubs 67–8, 116, 119, 139, 152, 156, 159, 160, 182, 186
Squirrel Nutkin 10, 32, 44, 61, 73, 75–80, 120, 178, 179, 181, 182–3, 184, 188, 191, 192, 193, 195, 198, 199, 201, 205, 207, 208, 222, 239, 248, 249–50, 251
see also Potter, Beatrix
squirrelpox virus (SQPV) 24, 192, 193, 194, 203, 204, 214, 237, 246, 249
squirrelscapes 28, 30, 39–40, 45, 161
Stapleford, David 27, 177
Stewart, James, in Oliver Goldsmith, *A History of the Earth and Animated Nature* 72
Stillman, William James, *Billy and Hans* 68, 91
Stoddart, Lord 235
Stokoe, William John 63
stronghold strategy 182, 191, 194, 199–200, 201, 248
Stubbs, F. J. 54
'Stripe' the squirrel 8, 9
superiority of native squirrel 31, 148, 192, 200, 213
super-squirrel 237, 244, 246
'Survival of the Red Squirrel in the UK' (early day motion) 189
Swainson, Laurance (L. W.) 121–2
Swanston, John (Abbotsford) 50
Swart, Sandra 38, 46
Swinson, Jo 218–19

Tague, Ingrid 30
Tallents, Stephen George 156–7, 192
Taylor, Jan, *Sciurus: The Story of a Grey Squirrel* 11, 233
Taylor, William Ling 41
teddy boys 16, 186, 187

Tew, Tom 210
Thetford Forest 44, 195
Thomas, Oldfield 53, 105, 115
Thomas, William Beach 82, 99, 110–11, 113–14, 115, 116–17, 121, 142, 146–7, 154, 174, 229, 252
Thompson, Ronald 179
Thorburn, Archibald, *British Mammals* 105–6, *107*
Timber Growers' Association 172
Timber Growers UK 240
'Tiny' the squirrel 97
Tod, Michael, *The Silver Tide* 33–4, 214, 237, 248
Topsell, Edward 58, 59
'tree rat' 14, 32, 120, 121, 140, 143, 147, 154, 157, 159, 172, 196, 215, 216, 240, 251
Treves, (Sir) Frederick 102, 103
Truman, Harry, president 166
Tufty Club 15–16, *135*, 180–85, *180*, *181*, *182*, *183*, 208
Tufty Fluffytail (or Tufty) 181, 182, 184, 185, 186, 203, 227, 249–50
Turner, Andrew 249
Twinkleberry the squirrel 61, 79, 192, 249–50
Tyler, Andrew 45, 228

UK Red Squirrel Group (UKRSG) 176, 208
UK Strategy for Red Squirrel Conservation 176
Unicorn Tapestries (*The Hunt of the Unicorn*) 31, 57–8
United Kingdom Independence Party (UKIP) 217, 218
United Kingdom Squirrel Accord (UKSA) 176
Usborne, Henry 187

Van Dooren, Thom 46
Van Marsh, Alphonso, *Squirrel Wars in the UK* (CNN) 209, 213

Index

vermin 41, 46, 67, 68, 91, 104, 109, 110, 125, 152, 170, 171, 204, 215
Viney, (Revd) Basil 139–40
Vizoso, Monica *see* Shorten, Monica

Walker-Meikle, Kathleen 58
Warburg, J. C. 93, 94, 95
Wareham, Anne, *Outwitting Squirrels and Other Garden Pests and Nuisances* 240–41
'war'
  greys against reds 23, 112, 121, 128, 137–8, 145, 149, 151, 159, 163, 196, 214–15, 242
  on greys, by humans 23, 109, 112, 121, 127, 128, 144, 148, 149–75, 193–204, 212, 213, 217, 230, 246
  on reds, by humans 63
warfarin 25, 150, 170–72, 173, 176, 191, 195, 202, 214
Watt, Hugh Boyd 47, 90, 101, 104, 105
Watney, Vernon James 115, 116
Waugh, Auberon 234–5
Wentworth Day, James 229
Whipsnade Zoo 196
White, Gilbert, *The Natural History and Antiquities of Selborne* 56
White, Piran 244
White House squirrels 166–8
Wickens, Janet 24–5, 32
Wiggin, Bill 80, 205
Wight, Isle of 54, 110, *131*, 199, 219, 233, 248–9
Wildlife and Countryside Act (1981) 24, 46
Wildlife Conservation Research Unit (WildCRU) 215
'Wildlife to Work' survey 45
Wildlife Trusts 21, 176, 188, 191, 199, 201, 204, 222
Wilhelmina the squirrel 22
William, Prince 177
Williams, Gerald 160
Williams, Helen Vaughan, *Squirrel War; or, The Fight for the Doll's House* 128, 137–8
Williams, Iolo 207
Wilson, Peter 240
Wirral the Squirrel 22, 251
Wise, Lord 141–2
Woburn Park, Bedfordshire 34, 79, 83, 88–9, 92, 101, 109, 196, 245
Wolfe, Humbert, 'The Grey Squirrel' (poem) 117
Wood, J. G., and Theodore 71
Woodcock, John 228
Wordsworth, William 114, 120, 148, 233
Wright, Jo, *Cyril the Squirrel* 240
Wyatt, Petronella 177, 237
Wyatt, Woodrow (Baron) 235–7
  The Exploits of Mr Saucy Squirrel 235, 237
  The Further Exploits of Mr Saucy Squirrel 235–6, 237
saucy, as squirrel attribute 74–5, 76, 235, 253

Xia Mingfang 26

Yalden, Derek 213
Younghusband, Francis 114

Zoological Society of London (ZSL) 16, 17, 44, 92, 99, 103–4, 116, 121, 142, 145, 186, 196, 221, 244